AF750257

BIRMINGH

Penguin Education
Penguin Library of Physical Sciences

Free-Electron Physics

P. S. Farago

Advisory Editor
V. S. Griffiths

General Editors
Physics: N. Feather, F.R.S.
Physical Chemistry: W. H. Lee
Inorganic Chemistry: A. K. Holliday
Organic Chemistry: G. H. Williams

Free-Electron Physics

P. S. Farago

Penguin Books

Penguin Books Ltd, Harmondsworth,
Middlesex, England
Penguin Books Inc., 7110 Ambassador Road,
Baltimore, Md 21207, U.S.A.
Penguin Books Australia Ltd, Ringwood,
Victoria, Australia

First published 1970
Copyright © P. S. Farago, 1970

Filmset in Monophoto Times by Oliver Burridge
Filmsetting Ltd, Crawley, England, and made and
printed by Compton Printing Ltd,
London and Aylesbury

This book is sold subject to the condition that
it shall not, by way of trade or otherwise, be lent,
re-sold, hired out, or otherwise circulated without
the publisher's prior consent in any form of
binding or cover other than that in which it is
published and without a similar condition
including this condition being imposed on the
subsequent purchaser

QC775.3
F
0785,5761

BIRMINGHAM UNIVERSITY LIBRARY

0140800999

To Elizabeth

Contents

Editorial Foreword 9

Preface 11

1 Motion of Charged Particles in Static Electromagnetic Fields 13

1.1 Introduction 13
1.2 Conservation laws 17
Further reading 34

2 Geometrical Electron Optics 35

2.1 Introduction 35
2.2 Electrons in magnetic fields 36
2.3 Imaging properties of static fields of rotational symmetry 49
2.4 Examples of rotationally symmetric electron lenses 58
2.5 Quadrupole lenses 65
2.6 Fields with one plane of mirror symmetry 70
2.7 Dispersion and resolving power 84
2.8 Beam focusing and beam confinement – general conditions 90
2.9 Motion of neutral dipole particles in static fields 99
Further reading 108

3 Motion of Charged Particles in Time-Dependent Electromagnetic Fields 109

3.1 Introduction 109
3.2 Charged particles in time-dependent transverse fields 117
3.3 Velocity modulation and phase focusing 131
Further reading 137

4 Particle Accelerators 138

4.1 Introduction 138
4.2 Linear accelerators 139
4.3 Cyclic accelerators 153
4.4 Limits of the energy attainable by particle accelerators 175
Further reading 178

5 Electronic Amplifiers and Oscillators 179

5.1 Introduction 179
5.2 Some features of the operation of classical thermionic valves 185
5.3 Klystrons and reflex klystrons 195
5.4 Space-charge waves and cyclotron waves 206
5.5 Amplification by coupling of modes of wave propagation (travelling-wave tubes) 216
5.6 The magneton 222
5.7 Parametric amplifiers 231
5.8 Concluding remarks on electronic amplifiers and oscillators 241
Further reading 246

Appendix 247

A.1 Simple field distributions 247
A.2 Some properties of the matrix relating object and image space in geometrical electron optics 250
A.3 The Lagrange invariant of geometrical electron optics 251
A.4 The effect of finite angular momentum on image formation 252
A.5 Hamiltonian equations of trajectories 254
A.6 Floquet's theorem 257
A.7 Useful relativistic relationships 258
A.8 The Hamiltonian of a charged particle in a travelling electromagnetic wave 262

Index 265

Editorial Foreword

For many years, now, the teaching of physics at the first-degree level has posed a problem of organization and selection of material of ever-increasing difficulty. From the teacher's point of view, to pay scant attention to the groundwork is patently to court disaster; from the student's, to be denied the excitement of a journey to the frontiers of knowledge is to be denied his birthright. The remedy is not easy to come by. Certainly, the physics section of the Penguin Library of Physical Sciences does not claim to provide any ready-made solution of the problem. What it is designed to do, instead, is to bring together a collection of compact texts, written by teachers of wide experience, around which undergraduate courses of a 'modern', even of an adventurous, character may be built.

The texts are organized generally at three levels of treatment, corresponding to the three years of an honours curriculum, but there is nothing sacrosanct in this classification. Very probably, most teachers will regard all the first-year topics as obligatory in any course, but, in respect of the others, many patterns of interweaving may commend themselves, and prove equally valid in practice. The list of projected third-year titles is necessarily the longest of the three, and the invitation to discriminating choice is wider, but even here care has been taken to avoid, as far as possible, the post-graduate monograph. The series as a whole (some five first-year, six second-year and fourteen third-year titles) is directed primarily to the undergraduate; it is designed to help the teacher to resist the temptation to overload his course, either with the outmoded legacies of the nineteenth century, or with the more speculative digressions of the twentieth. It is expository, only: it does not attempt to provide either student or teacher with exercises for his tutorial classes, or with mass-produced questions for examinations. Important as this provision may be, responsibility for it must surely lie ultimately with the teacher: he alone knows the precise needs of his students – as they change from year to year.

Within the broad framework of the series, individual authors have rightly regarded themselves as free to adopt a personal approach to the choice and presentation of subject matter. To impose a rigid conformity on a writer is to dull the impact of the written word. This general licence has been extended even to the matter of units. There is much to be said, in theory, in favour of a single system of units of measurement – and it has not been overlooked that national policy in advanced countries is moving rapidly towards uniformity under the *Système International* (S.I. units) – but fluency in the use of many systems is not to be despised: indeed, its acquisition may further, rather than retard, the physicist's education.

A general editor's foreword, almost by definition, is first written when the series for which he is responsible is more nearly complete in his imagination (or the publisher's) than as a row of books on his bookshelf. As these words are penned, that is the nature of the relevant situation: hope has inspired the present tense, in what has just been written, when the future would have been the more realistic. Optimism is the one attitude that a general editor must never disown!

January 1968 N. FEATHER

Preface

To see what is general in what is particular and what is permanent in what is transitory is the aim of scientific thought.

A. N. Whitehead

Electron physics describes the behaviour of free electrons under a great variety of circumstances. In all cases this behaviour is determined by the interaction between the electrons and electromagnetic fields. The understanding of these interactions and their application led to the development of a large number of electronic devices. Some are the sophisticated tools of research in science and technology, some belong to the tool kit of engineers and others are household appliances in common use. In view of the great significance of the subject it is only natural that there are numerous monographs which deal with different aspects of electron physics at an advanced level. The purpose of this book is to review the fundamental principles which form a common basis for the development of the subject, and to show how a relatively small number of basic ideas relate widely different devices to one another.

The first two chapters are devoted to *static electron optics*, where the entire treatment is restricted to first-order approximation. An attempt is made to deal with a large variety of problems from a unified point of view and considerations leading to detailed design are not presented. The extension of the treatment not only to charged particles other than electrons but also to neutral spin particles was a temptation the author could not resist.

The remaining three chapters deal with problems involving interaction between charged particles and time-dependent electromagnetic fields. A consideration of the energy balance divides the subject in a natural way, leading to separate treatments of particle accelerators and beam-type amplifiers. Beyond a qualitative description of the mechanism of *particle accelerators*, some quantitative discussion of the trapping of charged particles in a travelling wave is included, and an attempt is made to investigate the problem of beam confinement and stability, using those concepts of electron optics previously introduced. The treatment of *amplification* by beam-type devices leads to a discussion of the interaction between coupled modes of wave propagation. Two classes of amplifiers ('conventional' and 'parametric') are distinguished by the radiation process responsible for their gain.

Problems which require quantum-physical considerations (at least for the justification of a classical description), such as scattering problems, the motion of spin particles in high-frequency fields, or the fundamentals of quantum electronics, have been avoided. Similarly, no discussion of random fluctuations and noise problems is included. Even so, in order to discuss such a diverse field in a coherent manner, the basic principles must be expressed in a very general but, at the same time, manageable mathematical form. From this point of view the task may appear too ambitious for an undergraduate text. Honours physics students, however, are taught *classical mechanics*, and the *Hamiltonian formalism* is not unfamiliar to them. Nevertheless the Hamiltonian formalism is recapitulated here and some fruitful classical ideas are revived; some matters of detail are summarized in the Appendix. Students, at an early stage of their studies, become acquainted with the fundamentals of Maxwell's *theory of electromagnetic fields* and the rudiments of the *special theory of relativity*. Again, what appeared as a necessary extension in this field is treated in the Appendix. The author believes that on the whole the required mathematical tools do not go beyond those which honours physics students can be expected to handle competently. Extensive use is made of the methods of matrix algebra, and some of the problems of wave propagation in periodic structures are likely to be familiar from solid-state physics.

If one seriously believes that students – at any level – should not store their knowledge of different 'subjects' in tightly sealed compartments, then one should not hesitate to challenge them to apply their knowledge in a purposeful way to problems which arise in a somewhat unconventional context. This is all that this volume sets out to do.

The author wishes to express his gratitude to Professor N. Feather, F.R.S., for his encouragement and helpful remarks; to Dr P. J. Kennedy and Mr R. M. Sillitto for a number of illuminating discussions on matters of content and presentation. Thanks are due to Dr H. M. Brash and Mr D. M. Campbell for their help in proof reading and indexing.

Chapter 1
Motion of Charged Particles in Static Electromagnetic Fields

It is time we should now return to the laws of motion, and examine more closely, in the light of what we have learned, one or two of the more prominent ideas which they embody.

P. G. Tait

1.1 Introduction

The purpose of this volume is to discuss certain problems concerning the interaction between free electrons and electromagnetic fields. The electron is usually looked upon as a *particle* of charge $e = -1{\cdot}602 \times 10^{-19}$ coulombs and rest mass $m = 9{\cdot}108 \times 10^{-31}$ kilograms. It is well known, however, that under certain circumstances the electron exhibits *wave* properties, and its wavelength (de Broglie wavelength) is given by $\lambda = h/p$, where $h = 6{\cdot}625 \times 10^{-34}$ W s^2 is Planck's constant and p is the momentum of the electron. In addition the electron has an intrinsic angular momentum and a magnetic moment. The approach followed in describing the interaction between electrons and electromagnetic fields depends on whether the wave nature or the corpuscular nature of the electron provides the dominant features of the interaction process. General quantum-mechanical arguments teach us that in an interaction with an electromagnetic field, the electron's wave nature plays a part only if the field intensity varies over a region whose linear dimensions are comparable with the de Broglie wavelength. Such inhomogeneous fields occur in the vicinity of atomic particles, and to take into account the wave nature of the electrons, one must use the methods of wave mechanics (quantum mechanics) in the study of the scattering of electrons by atoms, molecules, or matter in general.

We shall confine our discussion to problems where the electromagnetic fields produced by macroscopic devices can be considered constant over regions whose linear dimensions are large compared with the de Broglie wavelength. Under such circumstances the interaction between the electrons and fields are adequately described by the methods of classical physics. To avoid any misunderstanding it must be emphasized that the term 'classical' does not preclude the necessity of satisfying the requirements of the special theory of relativity.

If an electron is in an electromagnetic field, forces other than of electromagnetic origin are negligible because of the very high charge-to-mass ratio of the electron ($-e/m = 1{\cdot}759 \times 10^{11}$ C kg^{-1}). On the other hand electromagnetic fields of moderate intensities produce large acceleration, and therefore electrons easily reach velocities approaching that of light. The energy equivalent to the rest mass of the electron is

$$m_0 c^2 = 0{\cdot}511 \text{ MeV}$$

and if electrons move in an electric field between points whose potential difference is about ten thousand electron volts or more, relativistic effects become noticeable. Thus the equation which correctly describes the motion of the electron in an electric field **E** and magnetic field of flux density **B** is

$$\frac{d\mathbf{p}}{dt} = e(\mathbf{E}+\mathbf{v}\times\mathbf{B}), \qquad \mathbf{1.1}$$

$$\text{where} \quad \mathbf{p} = m_0\mathbf{v}\left(1-\frac{v^2}{c^2}\right)^{-\frac{1}{2}}$$

is the momentum of the electron and, in general, $d\mathbf{p}/dt$ is not equal to mass times acceleration. However in many situations of practical importance one can use the equation of motion in its non-relativistic form. The conditions for this can be obtained as follows.

Forming the scalar product of both sides of equation **1.1** with **v**, the left-hand side gives

$$\mathbf{v}.\frac{d}{dt}\left\{m_0\mathbf{v}\left(1-\frac{v^2}{c^2}\right)^{-\frac{1}{2}}\right\} = \frac{d}{dt}\left\{m_0 c^2\left(1-\frac{v^2}{c^2}\right)^{-\frac{1}{2}}\right\} = \frac{d}{dt}(mc^2),$$

$$\text{where} \quad mc^2 = m_0 c^2\left(1-\frac{v^2}{c^2}\right)^{-\frac{1}{2}}$$

is the total energy of the electron. From the right-hand side of equation **1.1**:

$$e\mathbf{v}.(\mathbf{E}+\mathbf{v}\times\mathbf{B}) = e\mathbf{v}.\mathbf{E}.$$

$$\text{Hence} \quad \frac{d}{dt}(mc^2) = e\mathbf{v}.\mathbf{E}. \qquad \mathbf{1.2}$$

Thus we see that if $\mathbf{v}.\mathbf{E} = 0$, the mass of the electron stays constant, and as a consequence the relation $d\mathbf{p}/dt = m(d\mathbf{v}/dt)$ is valid. This is the case in a pure magnetic field ($\mathbf{E} = 0$), or if the electron velocity is perpendicular to the electric field ($\mathbf{v}.\mathbf{E} = 0$). In both cases it is permissible to use non-relativistic equations of motion, putting for the mass its value appropriate at the velocity. $m = m_0(1-v^2/c^2)^{-\frac{1}{2}}$.

The motion of the electron is completely described by equation **1.1** and the initial conditions, that is, the initial position and velocity of the electrons. If the field distributions are simple, the solution of the equation of motion is straightforward. One recognizes, for example, that in a constant, uniform electric field the motion of the electron is the same as that of a projectile in a uniform gravitational field. In a pure magnetic field the characteristic feature of the motion arises from the fact that the force acting on the electron is velocity dependent, and perpendicular to both the magnetic field and the instantaneous velocity of the electron. In the absence of an electric field, that is, when $E = 0$, equation **1.1** can be written

$$\frac{d\mathbf{p}}{dt} = \frac{e}{m}\mathbf{p}\times\mathbf{B}. \qquad \mathbf{1.3}$$

Although it is easy enough to solve this differential equation directly, a somewhat different approach, which will be found useful in tackling more complicated problems later on, is more instructive. When one describes the distribution of a field, the motion of a particle etc., as a function of coordinates and time, this is done – unless otherwise stated – with reference to a coordinate system which is at rest with respect to the laboratory and the observer. Furthermore, one assumes that this 'laboratory frame' constitutes an inertial frame of reference. Now there are many situations in which it is advantageous to describe the phenomena from the point of view of a moving observer, that is, with reference to a coordinate system which is moving with respect to the 'laboratory frame'. Taking this attitude, one recognizes that equation **1.3** resembles the relation between a vector fixed in a rotating frame of reference and its time rate of change as observed in a frame at rest (see Figure 1).

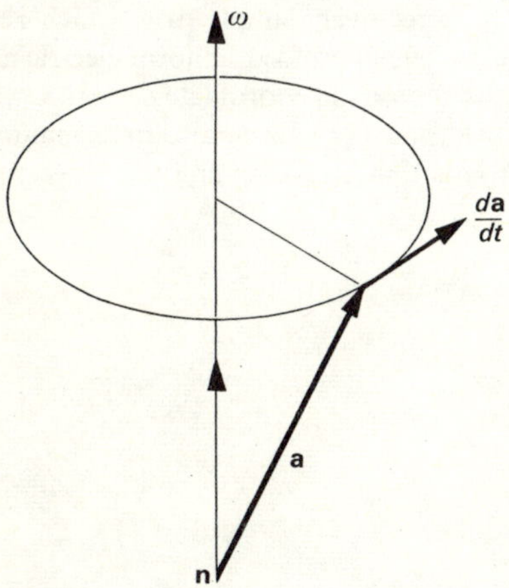

Figure 1

Consider a coordinate system rotating with constant angular velocity ω about the direction of the unit vector $\hat{\mathbf{n}}$ defined in an inertial reference frame, such that $\omega\hat{\mathbf{n}} = \boldsymbol{\omega}$. If $\mathbf{a}$ is a vector fixed in the rotating system its time rate of change, as measured by an observer at rest in the origin of the inertial frame, is

$$\frac{d\mathbf{a}}{dt} = \boldsymbol{\omega}\times\mathbf{a}. \qquad \mathbf{1.4}$$

Equation **1.3** reduces to this form if **B** is independent of the coordinates, that is, if the field is uniform. Thus, in the presence of a uniform magnetic field, an observer fixed in an inertial coordinate system observes the momentum vector of the electron rotating with an angular velocity

$$\omega = -\frac{e}{m}\mathbf{B}.$$

In other words, the momentum **p** has a constant component parallel to the magnetic field and a component which is rotating in a plane perpendicular to the magnetic field. The result of this is that in a constant magnetic field the electron describes a helix of uniform pitch, the axis of which is parallel to the field direction. If the initial momentum of the electron is perpendicular to the magnetic field, the electron orbit is a closed circle of radius

$$R = \frac{v}{\omega} = \frac{p}{eB}.$$

The quantity $RB = p/e$ is often called the magnetic rigidity of the electron trajectories; $\omega = eB/m$ is the 'cyclotron frequency'.

If both electric and magnetic fields are present the motion can be quite complicated even if both fields are uniform. A relatively simple case, which is of practical importance, is the motion of electrons in crossed uniform electric and magnetic fields. To be more specific, let us choose the coordinate system S, with its y-axis parallel to the uniform electric field, and its z-axis parallel to the uniform magnetic field. Taking components for equation **1.1**, assuming that mass m is constant, we obtain

$$\frac{d^2x}{dt^2} = \frac{e}{m}B\frac{dy}{dt},$$

$$\frac{d^2y}{dt^2} = \frac{e}{m}\left(E - B\frac{dx}{dt}\right), \qquad \mathbf{1.5}$$

$$\frac{d^2z}{dt^2} = 0.$$

The third of these equations shows that in crossed uniform fields, there is no acceleration in the direction of the magnetic field, and the second shows that the acceleration due to the electric field is compensated by the magnetic field if $dx/dt = E/B$. This suggests that the motion of the electron will appear greatly simplified if viewed from a coordinate system S', which is moving relative to the frame S with a uniform velocity $u = E/B$ parallel to the x-axis, that is, in a direction perpendicular to the plane defined by the two fields.

If $E/B \ll c$ (the velocity of light), the coordinates in S and S' are related by a Galilean transformation:

$$x = x' + (E/B)t \qquad y = y' \qquad z = z'.$$

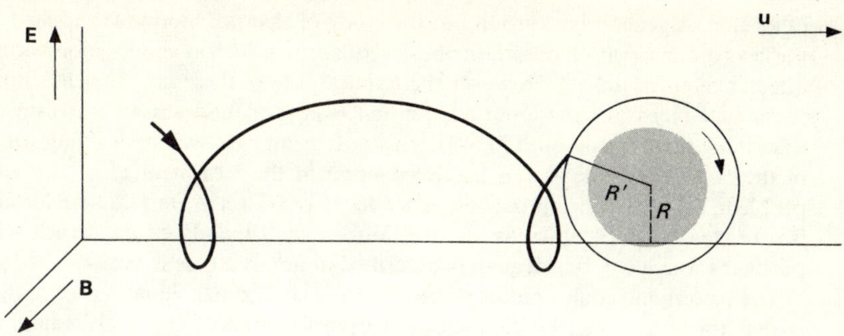

Figure 2

Applying this transformation to equations **1.5**, the resulting equations can be summarized in a single vector relation

$$\frac{d^2\mathbf{r}'}{dt^2} = \frac{e}{m}\dot{\mathbf{r}}' \times \mathbf{B}, \qquad \mathbf{1.6}$$

where $\mathbf{r}'$ is the position vector of the electron in the moving frame.

Equation **1.6** indicates that the motion of the electron in the frame S' can be described as taking place in a pure uniform magnetic field.* The electron motion in the laboratory frame can be determined from simple geometrical considerations by combining the motion in S' and the displacement of S' relative to S. If the initial velocity v_0 of the electron in the laboratory frame has no component parallel to the magnetic field, then the electron will move in a plane. Its initial velocity in the moving frame is $\mathbf{v}_0 - \mathbf{u} = \mathbf{w}$ and in this frame it describes a circle of radius $R' = w/\omega$, $\omega = eB/m$. In the laboratory frame the trajectory is generated by a point on a circle of radius R', rotating with angular velocity ω, whose centre is drifting with velocity E/B. These motions are combined if one considers two concentric circles (Figure 2), one with radius R' and the other with radius $R = E/\omega B$. If the two circles are rigidly coupled and the circle of radius R is rolling along a straight line parallel to the x-axis with velocity E/B, then a point of the circle of radius R' will generate the electron trajectory.

1.2 Conservation laws

In many cases of practical interest the field distributions are rather complicated, and the solution of the equations of motion may present considerable difficulties. These difficulties can be alleviated if the general features of the electron motion are known. Indeed, it is instructive to obtain information about the general pattern of the electron motion before embarking on the solution of the equations

*This statement is more generally valid than it would seem from the argument above. It can be shown that if an electric and a magnetic field in an inertial frame are orthogonal to one another, that is, satisfy the condition $\mathbf{E}.\mathbf{B} = 0$ everywhere in space, even though they are not necessarily uniform, then $(\mathbf{E} \times \mathbf{B})/\mathbf{B}.\mathbf{B}) = \mathbf{u}$, and an observer fixed in a reference frame moving with velocity $\mathbf{u}$ relative to S observes only a magnetic field $\mathbf{B}' = \mathbf{B}(1-u^2/c^2)^{\frac{1}{2}}$.

of motion. A general background to the study of electron motion is supplied by the laws of conservation of certain physical quantities, such as energy, momentum, angular momentum, etc. However the applicability of these laws depends upon the actual physical situation, and the first task is to find some formalism for describing the electron motion, which also offers an easy way of deciding which of the conservation laws can be called upon in the treatment of a particular problem. The tool which is the best suited for this task is the Hamiltonian formalism of classical mechanics. In this section we shall mainly be concerned with problems which can be adequately described in non-relativistic terms.

The underlying condition for setting up the Hamiltonian equations of motion is that the system can be characterized by a function of the coordinates, the velocity components and time, $L = L(x_1, x_2, x_3; \dot{x}_1, \dot{x}_2, \dot{x}_3; t)$, such that the set of equations

$$\frac{d}{dt}\frac{\partial L}{\partial \dot{x}_i} - \frac{\partial L}{\partial x_i} = 0 \qquad (i = 1, 2, 3) \qquad \mathbf{1.7}$$

yields Newton's equations of motion in terms of the coordinates and their time derivatives.

In the case of a *free* particle the Lagrangian is given in a relativistically correct way by

$$L_0 = -m_0 c^2 \left(1 - \frac{v^2}{c^2}\right)^{\frac{1}{2}}. \qquad \mathbf{1.8}$$

Since $v^2 = \dot{x}_1^2 + \dot{x}_2^2 + \dot{x}_3^2$ we readily see that

$$\frac{\partial L}{\partial \dot{x}_i} = m_0 \dot{x}_i \left(1 - \frac{v^2}{c^2}\right)^{-\frac{1}{2}} \equiv p_i \quad (i = 1, 2, 3)$$

that is, the partial derivatives of the Lagrangian of a free particle with respect to the velocity components give the momentum components.

The classical form of the Lagrangian for a free particle is obtained from equation **1.8** by taking the first-order approximation in v^2/c^2 and disregarding the additive constant term $-m_0 c^2$:

$$L_0 = \tfrac{1}{2} m v^2, \qquad \mathbf{1.8a}$$

that is, the classical Lagrangian for a free particle is given by its kinetic energy. In this case one again finds

$$\frac{\partial L_0}{\partial \dot{x}_i} = m\dot{x}_i \equiv p_i.$$

Since L_0 does not depend explicitly on the coordinates, equation **1.7** shows that the momentum of free particles is constant.

Let us now consider a particle moving in an electromagnetic field defined by a scalar potential Φ and a vector potential $\mathbf{A}$. It will be shown in section 3.1, that if the fields are time dependent, both Φ and $\mathbf{A}$ are explicit functions of time and the fields are given by

$$\mathbf{E} = -\nabla\Phi - \frac{\partial \mathbf{A}}{\partial t} \quad \text{and} \quad \mathbf{B} = \nabla \times \mathbf{A}.$$

In the static case neither of the potentials depends explicitly on time and one obtains the familiar result that the electric field is given by the negative gradient of a scalar potential, and the magnetic flux density by the curl of the vector potential; the two fields being constant in time and independent of each other.

For a particle moving in an electromagnetic field – irrespective of its time dependence – the Lagrangian takes the form

$$L = L_0 + L_f, \tag{1.9a}$$

with $L_f = e\mathbf{v}.\mathbf{A} - e\Phi.$ **1.9b**

To show the validity of this assumption one forms the terms $\frac{d}{dt}(\partial L/\partial \dot{x}_i)$ and $\partial L/\partial x_i$ which appear in equation **1.7**.

$$\frac{\partial L_f}{\partial \dot{x}_i} = eA_i$$

and therefore $\dfrac{d}{dt}\dfrac{\partial L_f}{\partial \dot{x}_i} = e\dfrac{\partial A_i}{\partial t} + e\left(\dot{x}_i\dfrac{\partial}{\partial x_i} + \dot{x}_2\dfrac{\partial}{\partial x_2} + \dot{x}_3\dfrac{\partial}{\partial x_3}\right)A_i.$

Further $\dfrac{\partial L_f}{\partial x_i} = -e\dfrac{\partial \Phi}{\partial x_i} + e\left(\dot{x}_i\dfrac{\partial A_1}{\partial x_i} + \dot{x}_2\dfrac{\partial A_2}{\partial x_i} + \dot{x}_3\dfrac{\partial A_3}{\partial x_i}\right).$

Writing out both sets of equations for each component $i = 1, 2, 3$, and forming the differences as required by equation **1.7**, one finds that they can be put into the concise form

$$e\left[(\nabla\Phi)_i + \frac{\partial A_i}{\partial t} - \{\mathbf{v}\times(\nabla\times\mathbf{A})\}_i\right] \equiv -e\{\mathbf{E}_i + (\mathbf{v}\times\mathbf{B})_i\}.$$

Finally, since $\dfrac{\partial L_0}{\partial x_i} = 0$ and $\dfrac{d}{dt}\dfrac{\partial L_0}{\partial \dot{x}_i} = \dfrac{dp_i}{dt}$, one finds that equation **1.7** gives the required equation of motion

$$\frac{d\mathbf{p}}{dt} = e(\mathbf{E} + \mathbf{v}\times\mathbf{B})$$

written out explicitly for each component.

We now wish to introduce the Hamiltonian function. 'Canonical' momentum components are defined with the aid of the Lagrangian $L = L_0 + L_f$ as

$$\pi_i = \frac{\partial L}{\partial \dot{x}_i} = m\dot{x}_i + eA_i \tag{1.10}$$

and hence $\mathbf{v} = \dfrac{1}{m}(\boldsymbol{\pi} - e\mathbf{A}).$ **1.10a**

Note that for free particles, $\pi_i = m\dot{x}_i \equiv p_i$, that is, the canonical momentum becomes identical with the conventional momentum. On the other hand for a particle in an electromagnetic field the quantity $\boldsymbol{\pi}-e\mathbf{A}$ bears the same relationship to the velocity as does the conventional momentum $\mathbf{p}$ in the case of a free particle.

Whereas the Lagrangian is a function of coordinates, velocity components and, in general, time, the Hamiltonian does not depend explicitly on velocity components, but is a function of coordinates, components of the canonical momentum and, in general, time. The Hamiltonian is defined as follows

$$H = \boldsymbol{\pi}.\mathbf{v}-L, \tag{1.11}$$

where $\mathbf{v}$ must be expressed in terms of $\boldsymbol{\pi}$ using equation **1.10a**.

Instead of carrying out this substitution and subsequent calculations, let us rewrite the equation defining H.

$$H = (\boldsymbol{\pi}-e\mathbf{A}).\mathbf{v}+e\mathbf{A}.\mathbf{v}-L_0-L_f.$$

Substituting L_f from equation **1.9b**

$$H = (\boldsymbol{\pi}-e\mathbf{A}).\mathbf{v}-L_0+e\Phi. \tag{1.12}$$

For a free particle $\Phi = 0$, $\mathbf{A} = 0$ and $\boldsymbol{\pi} = \mathbf{p}$. Therefore the Hamiltonian H_0 of a free particle is

$$H_0 = \mathbf{p}.\mathbf{v}-L_0. \tag{1.13}$$

Comparing equations **1.12** and **1.13** shows that if the Hamiltonian of a free particle is known, that of the particle in an electromagnetic field can be obtained by replacing $\mathbf{p}$ by $\boldsymbol{\pi}-e\mathbf{A}$ and adding $e\Phi$, that is

$$H = H_0(\mathbf{p} \rightarrow \boldsymbol{\pi}-e\mathbf{A})+e\Phi. \tag{1.14}$$

From the Lagrangian L_0 given by equation **1.8** and the identity

$$\mathbf{p}.\mathbf{v} = m_0 c^2 \frac{v^2}{c^2}\left(1-\frac{v^2}{c^2}\right)^{-\frac{1}{2}}$$

it follows that

$$H_0 = m_0 c^2\left(1-\frac{v^2}{c^2}\right)^{-\frac{1}{2}}.$$

The right-hand side of this equation is the total relativistic energy of a free particle of rest mass m_0. Therefore, in terms of momentum, the above expression becomes

$$H_0 = c(p^2+m_0^2c^2)^{\frac{1}{2}}. \tag{1.15}$$

In the non-relativistic case $p \ll m_0 c$; therefore, disregarding the additive constant $m_0 c^2$, a first-order expansion in $p/m_0 c$ gives

$$H_0 = \frac{p^2}{2m}. \tag{1.15a}$$

The Hamiltonian for a particle in an electromagnetic field, following equation **1.14**, is

$$H = c[(\boldsymbol{\pi}-e\mathbf{A})^2+m_0^2c^2]+e\Phi. \qquad \mathbf{1.16}$$

In a non-relativistic approximation this reduces to

$$H = \frac{1}{2m}(\boldsymbol{\pi}-e\mathbf{A})^2+e\Phi. \qquad \mathbf{1.16a}$$

The 'Hamiltonian equations of motion' are most easily derived from the general definition of the Hamiltonian in equation **1.11**. Forming the partial derivatives $\partial H/\partial x_i$ and taking into account equation **1.7**, one finds

$$\frac{d\pi_i}{dt} = -\frac{\partial H}{\partial x_i} \qquad (i = 1, 2, 3). \qquad \mathbf{1.17a}$$

Similarly forming the partial derivatives $\partial H/\partial \pi_i$ and taking into account the definition $\pi_i = \partial L/\partial \dot{x}_i$, one obtains

$$\frac{dx_i}{dt} = \frac{\partial H}{\partial \pi_i} \qquad (i = 1, 2, 3). \qquad \mathbf{1.17b}$$

The set of equations **1.17a** and **1.17b** are often called canonical equations, the quantities x_i and π_i which satisfy these equations being called canonically conjugate pairs.

Under certain conditions the Hamiltonian is independent of one coordinate or another, say x_k. In this case $\partial H/\partial x_k = 0$ and equation **1.17a** shows that π_k is constant. This is the most striking way in which a constant of motion, or a conservation law, emerges from the Hamiltonian description of motion.

First it can be readily shown that the total rate of change of the Hamiltonian function is determined by its explicit time dependence alone, namely

$$\frac{dH}{dt} = \frac{\partial H}{\partial t}. \qquad \mathbf{1.18}$$

This follows from the expression

$$\frac{dH}{dt} = \frac{\partial H}{\partial t}+\sum_{i=1}^{3}\left(\frac{\partial H}{\partial x_i}\frac{dx_i}{dt}+\frac{\partial H}{\partial \pi_i}\frac{d\pi_i}{dt}\right),$$

where the sum vanishes by virtue of equations **1.17a** and **1.17b**.

The meaning of equation **1.18** can be expressed by saying that *if the Hamiltonian does not depend explicitly on time*, $\partial H/\partial t = 0$, *then* H *is a constant of motion*, that is, its initial value is conserved throughout the motion.

This is the situation if the particle is *in a static field* because in this case $\Phi(x_i)$ and $\mathbf{A}(x_i)$ are independent of time. The second term of the Hamiltonian, written in the form of equation **1.16**, is now the potential energy of the particle. The first term, often called the kinetic term, gives the energy contribution arising from the motion of the particle. In the non-relativistic case this term is simply the kinetic energy; in the relativistic case it also takes account of the change of mass with velocity. The relation $H = constant$ is then *a statement of energy conservation.*

The applicability of other familiar conservation laws, such as momentum or angular momentum conservation, depends on the field distributions. At this point it must be emphasized that the coordinates x_i can be any set of three quantities suitable for the definition of the position of a mass point. The choice of Cartesian coordinates is only one of many possibilities, such as cylindrical polar, spherical, etc., and in fact the actual choice of a coordinate system is a matter of convenience. In dealing with the motion of charged particles in electromagnetic fields, it is advisable to use a coordinate system matching the symmetry properties of the field distribution. Apart from any other possible advantages such a choice helps to reveal the applicability of conservation laws.

A large class of devices is based on the properties of electron motion in 'two-dimensional' fields whose distribution depends on two coordinates only. Electric fields of this kind are maintained between electrode surfaces generated by lines parallel to the x_3-axis of an orthogonal coordinate system. Such electric fields can be derived from a scalar potential $\Phi = \Phi(x_1, x_2)$. Similarly a magnetic field arising from currents carried by a system of conductors parallel to the x_3-axis can be derived from a vector potential with components $A_3 = A(x_1, x_2)$, $A_1 = A_2 = 0$. Considering for the sake of simplicity, the non-relativistic case only, one finds that equation **1.16a** reduces to

$$H = \frac{1}{2m}[\pi_1^2 + \pi_2^2 + \{\pi_3 - e\,A(x_1, x_2)\}^2] + e\,\Phi(x_1, x_2)$$

and hence from equation **1.17a**

$$\pi_3 = \text{constant}.$$

In other words, if the field distribution is independent of a coordinate, the corresponding component of the *canonical* momentum is conserved. To see the implications of this general law in terms of more conventional physical quantities the field distributions must be specified more closely.

1.2.1 *Plane fields*

Consider an electrostatic field maintained between electrodes of general cylindrical shape; all the cylinders being perpendicular to a common plane. If the inter-electrode distances are negligible compared with the length of the electrode, at distances sufficiently far from the ends the fringing fields can be disregarded and the electrodes considered as infinitely long. To describe the field distribution let us choose a Cartesian coordinate system, where $x_1 = x$, $x_2 = y$ are measured in the plane perpendicular to the electrodes and $x_3 = z$ along the electrodes. In this case the electric field can be derived from a scalar potential which is independent of z

$$\Phi = \Phi(x, y),$$

and the electric field $\mathbf{E} = \nabla\Phi$ has x- and y-components only, so $E_z = 0$.

A magnetic field of the same type is generated by a system of ('infinitely' long) current-carrying conductors parallel to the z-axis. In this case the vector potential $\mathbf{A}$ is independent of z and has a z-component only,

$$A_z = A(x, y), \qquad A_x = A_y = 0,$$

yielding a flux density $\mathbf{B} = \nabla \times \mathbf{A}$ which has x- and y-components only ($B_z = 0$).

Thus applying the Lagrangian **1.9**

$$\pi_3 \rightarrow \pi_z = p_z + eA_z.$$

In the absence of a magnetic field, $A = 0$, π_z is identical with the conventional momentum, and the relation π_z = constant expresses the conservation of the momentum component perpendicular to the electric field. In a magnetic field, however, $A(x, y)$ varies with the position of the particle, as does the momentum $p_z = \pi_z - e\,A(x, y)$.

1.2.2 *Fields of rotational symmetry*

Consider now electrodes which are rotationally symmetrical and are centred on a common axis. The electric field maintained by a system of such electrodes can be most easily described by using cylindrical polar coordinates. Let us choose as the common axis of symmetry the $x_1 = z$ axis, and in the plane perpendicular to this axis, let $x_2 = \rho$ be the radial distance and $x_3 = \theta$ the azimuth angle measured from an arbitrary direction in this plane. The electric field in this case can be derived from a scalar potential which is independent of the azimuth angle,

$$\Phi = \Phi(\rho, z),$$

giving a field $\mathbf{E} = -\nabla\Phi$, which has radial and axial components only ($E_\theta = 0$).

If a magnetic field is generated by currents through conductors which are rotationally symmetric and centred on the z-axis, the vector potential $\mathbf{A}$ is independent of the azimuth and has an azimuthal component only,

$$A_\theta = A_\theta(\rho, z), \quad A_\rho = A_z = 0.$$

Hence the magnetic flux density, $\mathbf{B} = \nabla \times \mathbf{A}$, has radial and axial components only, and $B_\theta = 0$.

The implications of the conservation law $\pi_3 \rightarrow \pi_\theta$ = constant, under the present conditions, are shown by recalling that in terms of the cylindrical polar coordinates introduced, the velocity vector takes the form

$$\mathbf{v} = \dot{z}\mathbf{e}_z + \dot{\rho}\mathbf{e}_\rho + \rho\dot{\theta}\mathbf{e}_\theta,$$

where $\mathbf{e}_z$, $\mathbf{e}_\rho$ and $\mathbf{e}_\theta$ are unit vectors parallel to the coordinate axes. Substituting this expression into the Lagrangian equation **1.9** and taking into account the symmetry properties of the potential and the vector potential discussed above, one finds

$$\pi_1 \to \pi_z = m\dot{z},$$
$$\pi_2 \to \pi_\rho = m\dot{\rho},$$
$$\pi_3 \to \pi_\theta = m\rho^2\dot{\theta}+e\rho A_\theta.$$

Note that $m\rho^2\dot{\theta} \equiv J_z$ is the z-component of the angular momentum **J** about a point on the symmetry axis. Now the Hamiltonian takes the form

$$H = \frac{1}{2m}\left\{\pi_z^2+\pi_\rho^2+\frac{1}{\rho^2}(\pi_\theta-e\rho A_\theta)^2\right\}+e\Phi \qquad \mathbf{1.19}$$

and hence $\partial H/\partial\theta = 0$. This yields the conservation law π_θ = constant for the azimuthal component of the canonical momentum. In the absence of a magnetic field, $A = 0$, $\pi_\theta \to m\rho^2\dot{\theta}$, that is, the azimuthal component of the canonical momentum is identical with the z-component of the angular momentum about a point on the symmetry axis of the field. In other words *in an electrostatic field of cylindrical symmetry the angular momentum component parallel to the symmetry axis is conserved.*

However, in the presence of a magnetic field, the z-component of the angular momentum is not conserved, but

$$J_z(\rho,\theta) \equiv m\rho^2\dot{\theta} = \pi_\theta - e\rho A_\theta. \qquad \mathbf{1.20}$$

In the Hamiltonian **1.19** the last two terms can be combined to give an 'effective potential'

$$U(\rho, z) = \frac{1}{2m}\left(\frac{J_z}{\rho}\right)^2+e\Phi \qquad \mathbf{1.21}$$

yielding $$H = \frac{1}{2m}(\pi_z^2+\pi_\rho^2)+U(\rho, z). \qquad \mathbf{1.22}$$

This expression indicates that the motion can be described as taking place in a plane containing the symmetry axis ('meridian plane') in the two-dimensional scalar potential U:

$$m\frac{d^2\rho}{dt^2} = -\frac{\partial U}{\partial\rho},$$
$$m\frac{d^2z}{dt^2} = -\frac{\partial U}{\partial z}. \qquad \mathbf{1.23}$$

Denoting the velocity of the electron in the meridian plane by u_m, we find that $\frac{1}{2}mu_m^2 = -U$,

if $u_m = 0$ where $U = 0$. The meridian plane, however, is not fixed but is rotating with an angular velocity $\dot{\theta}$ determined by equation **1.20** which relates the angular momentum of the particle and the vector potential.

Equation **1.20** is often quoted in another form. Stokes's theorem states that

$$\oint_C \mathbf{A}.d\mathbf{s} = \int_S (\nabla \times \mathbf{A}).d\mathbf{S},$$

where the integral on the right-hand side is extended to an arbitrary surface S, the boundary of which is the closed curve C over which the integral on the left-hand side is carried out; $d\mathbf{S} = \hat{\mathbf{n}}\, dS$, $\hat{\mathbf{n}}$ being the unit vector normal to the surface element dS. Therefore one can write

$$\int_0^{2\pi} A_\theta \rho\, d\theta = \int_0^{2\pi}\int_0^{\rho} (\nabla \times \mathbf{A})_z \rho'\, d\theta\, d\rho'$$

and hence $$A_\theta = \frac{1}{2\pi\rho}\int_0^{\rho} B_z(z, \rho')\rho'\, d\rho' = \frac{1}{2\pi\rho} f_\rho(z),$$

where $f_\rho(z)$ is the magnetic flux linked by a circle of radius ρ centred on the symmetry axis at z.

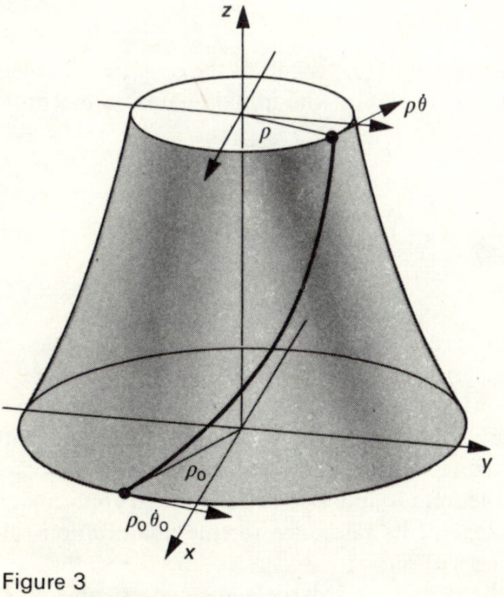

Figure 3

Now the electron proceeds from a point (ρ_0, z_0) where its angular velocity is $\dot{\theta}_0$, to a point (ρ, z) where its angular velocity is $\dot{\theta}$ (Figure 3). Since the canonical momentum π_0 is constant, equation **1.20**, with the above result, yields

$$J_z(\rho, z) - J_z(\rho_0, z_0) = -\frac{e}{2\pi}[f_\rho(z) - f_{\rho 0}(z_0)]. \quad \mathbf{1.24}$$

That is, the change in the angular momentum about the symmetry axis is proportional to the change in the flux linked by circles whose radii are the distances of the respective positions of the electron. This formulation of the angular momentum relations in a cylindrically symmetric field is frequently called *Busch's theorem*.

It can be shown (see Appendix, section A.3) that the z-component of a cylindrically symmetric magnetic field is described by the power series

$$B_z(\rho, z) = \sum_{n=0}^{\infty} \frac{(-1)^n}{(n!)^2} \frac{\partial^{2n} B_0(z)}{\partial z^{2n}} (\tfrac{1}{2}\rho)^{2n}, \qquad \textbf{1.25}$$

where $B_0(z)$ is the field on the axis at z. Consideration of the field so near to the z-axis that second- and higher-order terms in ρ are negligible shows that

$$f_\rho(z) = \pi\rho^2 B_0(z).$$

If the electron is moving in the vicinity of the symmetry axis, one can use this expression and, substituting the explicit form of J_z in equation **1.24**, one finds

$$\rho^2\dot{\theta} - \rho_0^2\dot{\theta}_0 = \rho^2\omega_L - \rho_0^2\omega_{L_0}, \qquad \textbf{1.26}$$

where $\quad \omega_L(z) = -\dfrac{e}{2m} B_0(z) = \tfrac{1}{2}\omega(z)$

is called the Larmor frequency; its magnitude is half of the cyclotron frequency. In a uniform magnetic field $\omega_L \equiv \omega_{L_0}$ and if the initial angular momentum is zero **1.26** reduces to

$$\dot{\theta} = \omega_{L_0}\left(1 - \frac{\rho^2}{\rho_0^2}\right). \qquad \textbf{1.26a}$$

1.2.3 *Adiabatic invariants*

In the study of the motion of charged particles one often meets a situation in which some parameters characteristic of the physical conditions change very slowly instead of remaining constant, and therefore some of the conservation laws discussed above do not apply. In such a situation it may be of great advantage to resort to an invariance principle, usually quoted as the 'adiabatic theorem'. Although purely classical this theorem found its most celebrated applications in the 'old' quantum theory because of its relevance to the 'quantization rules', problems which will be no concern of ours.

Let us start by considering the motion of a particle in a conservative field. In this case – as was seen – the Hamiltonian is a constant of motion:

$$H(x_i, \pi_i) = W = \text{constant}.$$

For simplicity we confine our first considerations to one-dimensional motion. Plotting the coordinate x and the conjugate momentum π along two mutually perpendicular coordinate axes, the conservation law $H(x, \pi) = W$ defines a curve in the (x, π) 'phase plane'. As the particle of constant energy W moves, its representative point traces out a curve in the phase plane. If the motion is confined to a finite interval in real space the representative point in the phase plane describes a closed or a periodic curve. Treating the motion of a simple pendulum in such terms will illustrate the point.

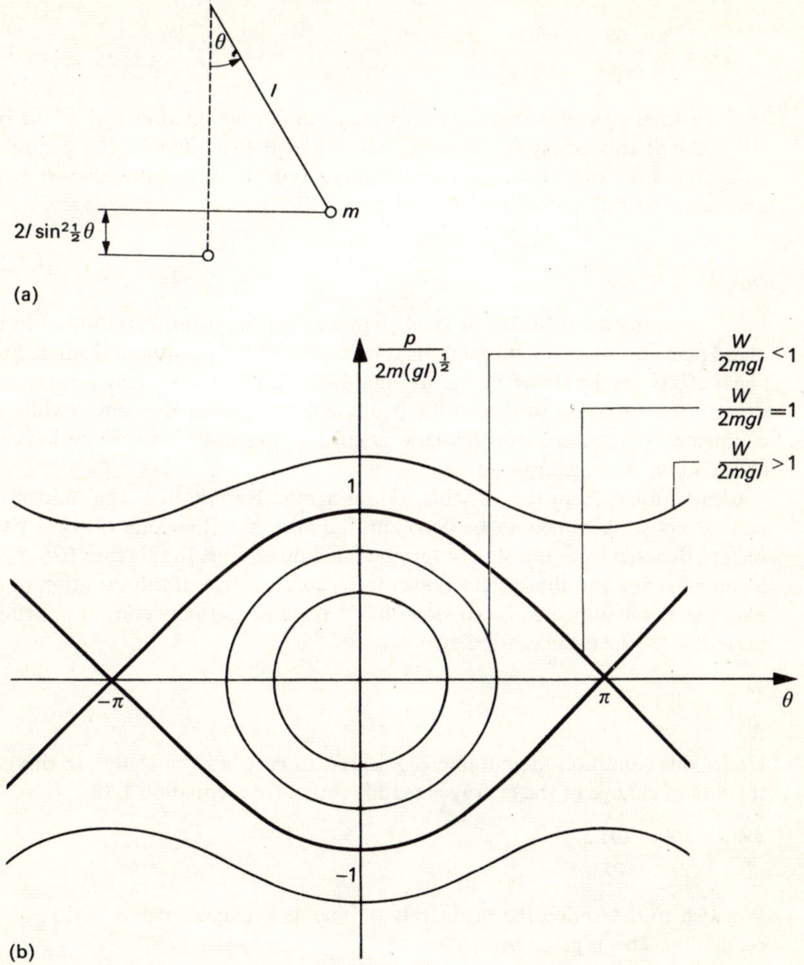

Figure 4

The velocity of a pendulum bob moving in a uniform gravitational field is a function of the angular deviation from its equilibrium position. If its potential energy U is taken as zero in the equilibrium position, it can be expressed in terms of the mass m of the bob, the gravitational acceleration g and the length of the pendulum l by $U = 2mgl\sin^2\frac{1}{2}\theta$ (see Figure 4). Hence the Hamiltonian is

$$H = \frac{1}{2m}\left(\frac{\pi_\theta}{l}\right)^2 + 2mgl\sin^2\tfrac{1}{2}\theta,$$

where θ and π_θ are canonically conjugate. Observe that π_θ is the angular momentum of the pendulum. For a given total energy W, the equation $H = W$ gives

$$\frac{\pi_\theta^2}{2ml^2W} + \frac{\sin^2\frac{1}{2}\theta}{W/2mgl} = 1. \qquad \textbf{1.27}$$

If the total energy is smaller than the maximum potential energy of the bob ($W < 2mgl$) this equation requires that $-\pi < \theta < \pi$, that is, the pendulum swings between two extreme values of elongation. If $W \ll 2mgl$, $\sin^2\frac{1}{2}\theta \simeq \frac{1}{4}\theta^2$ and equation **1.27** reduces to

$$\frac{\pi_\theta^2}{2ml^2W} + \frac{\theta^2}{2W/mgl} = 1. \qquad \textbf{1.27a}$$

In other words the orbit of the point representing the pendulum motion in the phase plane becomes an ellipse if the total energy of the pendulum is small. Such phase orbits are the closed curves in Figure 4(b). If $W > 2mgl$ there is no restriction on the variation of θ, and the pendulum bob moves in a circle while the conjugate momentum – in this case angular momentum – varies periodically about a non-zero mean value.

Clearly, apart from the variables θ and π_θ, the Hamiltonian is a function of parameters we assumed to be constant. Suppose now that one of these parameters, denoted by λ, is a slowly varying function of time. In this case $H(\theta, \pi_\theta; \lambda)$ is not constant and the system ceases to be conservative. If the variation of the parameter is assumed to be so slow that it remains virtually constant during a period, T, of the oscillation, then

$$\frac{d\lambda}{dt}T \ll \lambda.$$

Under this condition the parameter λ is said to vary 'adiabatically'. In this case the rate of change of the energy is readily found from equation **1.18**:

$$\frac{dW}{dt} = \frac{\partial H}{\partial t} = \frac{\partial H}{\partial \lambda}\frac{d\lambda}{dt}.$$

We wish to determine the mean rate of change averaged over a period of the oscillation. This is given by

$$\overline{\frac{dW}{dt}} = \frac{1}{T}\int_0^T \frac{\partial H}{d\lambda}\frac{d\lambda}{dt}\,dt. \qquad \textbf{1.28}$$

Since λ is assumed to vary slowly, the same applies to $d\lambda/dt$, and it can be considered constant during the period T, so that

$$\frac{\overline{dW}}{dt} = \frac{d\lambda}{dt}\frac{1}{T}\int_0^T \frac{\partial H}{\partial \lambda}\,dt. \qquad \textbf{1.29}$$

From the Hamiltonian equation of motion $d\theta/dt = \partial H/\partial \pi_\theta$ one obtains

$$dt = \frac{d\theta}{\partial H/\partial \pi_\theta}$$

and hence $$T = \int_0^T dt = \oint \frac{d\theta}{\partial H/\partial \pi_\theta}$$

where the symbol $\oint$ means integration over the full range of variation of the coordinate during a complete period of the motion. Thus equation **1.29** gives

$$\frac{\overline{dW}}{dt} = \frac{d\lambda}{dt}\frac{\oint \frac{\partial H/\partial \lambda}{\partial H/\partial \pi_\theta}\,d\theta}{\oint \frac{d\theta}{\partial H/\partial \pi_\theta}}. \qquad \textbf{1.30}$$

Since λ is considered constant during a period of the motion, the integration is carried out along a path on which the energy is conserved. Therefore, expressing the momentum as a function of the coordinate and the *now* constant parameters W and λ: $\pi_\theta(\theta;\, W, \lambda)$, both sides of the equation $H(\theta, \pi_\theta; \lambda) = W$ can be differentiated with respect to λ:

$$\frac{\partial H}{\partial \lambda} + \frac{\partial H}{\partial \pi_\theta}\frac{\partial \pi_\theta}{\partial \lambda} = 0$$

and hence $$\frac{\partial H/\partial \lambda}{\partial H/\partial \pi_\theta} = -\frac{\partial \pi_\theta}{\partial \lambda}.$$

Furthermore $$\left(\frac{\partial H}{\partial \pi_\theta}\right)^{-1} = \left(\frac{\partial W}{\partial \pi_\theta}\right)^{-1} = \frac{\partial \pi_\theta}{\partial W},$$

so that we find from equation **1.30**

$$\frac{\overline{dW}}{dt} = \frac{d\lambda}{dt}\frac{-\oint \partial \pi_\theta/\partial \lambda\, d\theta}{\oint d\pi_\theta/\partial W\, d\theta},$$

or $$\oint\left(\frac{\partial \pi_\theta}{\partial W}\frac{\overline{dW}}{dt} + \frac{\partial \pi_\theta}{\partial \lambda}\frac{d\lambda}{dt}\right) d\theta = 0. \qquad \textbf{1.31}$$

If we introduce, by definition,

$$I_\theta = \oint \pi_\theta\, d\theta, \qquad \textbf{1.32}$$

we find $$\overline{\frac{dI_\theta}{dt}} = \frac{1}{T}\int_0^T \frac{d}{dt}\left(\oint \pi_\theta \, d\theta\right) dt = \oint \left(\frac{1}{T}\int_0^T \frac{d\pi_\theta}{dt}\, dt\right) d\theta$$

$$= \oint \left(\frac{\partial \pi_\theta}{\partial W}\overline{\frac{dW}{dt}} + \frac{\partial \pi_\theta}{\partial \lambda}\frac{d\lambda}{dt}\right) d\theta. \qquad \mathbf{1.33}$$

Comparing equation **1.33** with equation **1.31**, we find

$$\overline{\frac{dI_\theta}{dt}} = 0$$

and we conclude that the quantity I_θ defined by equation **1.32** is an 'adiabatic invariant', meaning that it remains constant if the nominally constant parameter λ of the Hamiltonian varies slowly in the sense explained above.

The quantity I_θ, often called the 'phase integral', has a simple geometrical significance. If the one-dimensional motion of a particle is periodic and the representative point in the phase plane describes a closed orbit, I_θ is the area enclosed by this orbit. (Note that this area is dimensionally energy × time = 'action'.) Bearing this geometric interpretation in mind we can say that if the parameter λ varies 'slowly' the closed orbit in the phase plane changes its shape in such a manner that the area enclosed is conserved. This is the '*adiabatic theorem*'.

Applying this result, for example, to the pendulum motion we observe that the phase orbit for constant pendulum parameters and small oscillations is an ellipse with semi-axes $(2mW)^{\frac{1}{2}}l$ and $(2W/mgl)^{\frac{1}{2}}$ respectively (see equation **1.27a**). Therefore the phase integral is $I = 2\pi W(l/g)^{\frac{1}{2}} = WT$, and the adiabatic theorem requires that if any of the nominally constant parameters m, g or l undergoes a slow change, the energy of the pendulum remains inversely proportional to its period of oscillation. In Chapter 4, the adiabatic theorem will be used to clarify some properties of linear accelerators.

Although so far only one-dimensional motion has been considered, the arguments used can be readily extended to a case when a particle performs periodic or multiply periodic motion. For a rigorous treatment of the problem, a study of 'canonical transformations' and the properties of the solution of the 'Hamilton–Jacobi differential equation' is necessary. Suffice it to say that all periodic or multiply periodic motions have the common feature that they present problems which can be solved by the separation of variables.

Consider, for example, a particle of mass m moving in a plane subject to a force described in Cartesian coordinates by the components $F_x = -ax$ and $F_y = -by$. The position of such a particle in the plane is given by $x = A_1 \sin[(a/m)^{\frac{1}{2}}t + \phi_1]$ and $y = A_2 \sin[(b/m)^{\frac{1}{2}}t + \phi_2]$ where A_1, A_2, ϕ_1 and ϕ_2 are constants determined by the initial conditions. The important feature to note is that the motion is the superposition of two independent harmonic oscillations of different frequencies (multiply periodic). If $a = b$ the motion is simply periodic and the particle moves on a closed trajectory.

In conservative, central or axially symmetric fields the problem can always be solved by the separation of variables, and the same may be true even in more complicated situations. It should be noted, however, that the possibility of the separation of the variables often depends on the choice of the coordinate system; in this respect the symmetry properties of the field in question give useful guidance.

Suppose now that x_i and π_i ($i = 1, 2, 3$) are coordinates and canonically conjugate momenta respectively of a particle moving in three-dimensional space in a conservative field. As the particle moves its representative point will trace out a complicated curve on the multi-dimensional surface $H(x_i, \pi_i) = W$, in a six-dimensional phase space. The motion is periodic if the variables can be separated and the projection of the phase-orbit of the representative point in one or more of the phase planes (x_k, π_k) is a closed curve. For each of these pairs of canonical variables the phase integral

$$I_k = \oint \pi_k \, dx_k$$

is an adiabatic invariant. In the next subsection we shall see an example of practical significance for the application of the adiabatic theorem.

1.2.4 *Applications*

There are a number of problems which can be solved purely by the application of the conservation laws discussed above.

(a) If one considers a photocell consisting of a photocathode formed by the inside surface of a cylinder and a rod-shaped anode situated in the axis of the cylinder, all the electrons emitted by the cathode with a given initial energy can reach the anode only if the potential difference between the anode and the cathode is higher than a threshold value.

(b) For a straight filament-shaped cathode, a cylindrical anode whose axis coincides with the filament, and a uniform magnetic field parallel to the symmetry axis of the electrodes, there is a threshold value for the voltage difference between anode and cathode, below which electrons emitted by the filament cannot reach the anode.

The solution of both of these problems is left to the reader.

Let us apply the general ideas developed in this section to the problem of trapping charged particles in a '*magnetic bottle*'.

If electrons are emitted in a uniform magnetic field they will be confined to a cylindrical region. If the energy of the electrons is W, the momentum component perpendicular to the magnetic field is smaller than $(2mW)^{\frac{1}{2}}$ and, if the magnetic field is B_0, the diameter of the cylindrical region to which the electrons are confined is $4R = 4(2mW)^{\frac{1}{2}}(eB_0/m)^{-1}$. Thus the uniform magnetic field acts as a cylindrical 'bottle' of diameter $4R$, its axis being parallel to the direction of the magnetic field. This bottle, however, is open at both ends, since if a particle has a velocity component parallel to the magnetic field, this is conserved and nothing stops the particle from moving along the direction of the magnetic field.

If we consider a magnetic field of cylindrical symmetry instead of a uniform one, we can show that the electron motion can be confined to a finite region along the direction of increasing field intensity. One might say that the region of relatively high intensity acts as a 'plug' in the 'bottle'.

We shall assume that the field is strong but it varies slowly along the z-axis; more precisely, the electrons drift along the z-axis with a velocity such that the field seen by the electrons during a period of their transverse motion can be considered constant. Under these conditions we can regard the magnetic field as a slowly varying parameter and its effect on the motion of an electron in a nominally uniform field can be discussed in terms of the adiabatic theorem.

The Hamiltonian of an electron moving in a rotationally symmetric field is given by equation **1.19**. In the present case $\Phi = 0$ and $A_\theta = \frac{1}{2}\rho B$, where B is constant and parallel to the z-axis. The variables in this case can evidently be separated. The particle describes a helix whose projection on the plane perpendicular to the magnetic field is a circle. Therefore there are two adiabatic invariants:

$$I_\rho = \oint \pi_\rho \, d\rho \qquad \text{and} \qquad I_\theta = \oint \pi_\theta \, d\theta,$$

and the adiabatic theorem requires that each phase integral is constant. Therefore

$$I_\rho + I_\theta = \text{constant.} \qquad \textbf{1.34}$$

Since the transverse motion of the electron is circular, the representative points in both phase planes (ρ, π_ρ) and (θ, π_θ) trace out closed curves on which the representative point moves with the same period. Therefore, when the explicit expressions $\pi_\rho = m\dot{\rho}$ and $\pi_\theta = m\rho^2\dot{\theta} + e\rho A_\theta$ are substituted in equation **1.34**, the additive terms may be rearranged as follows:

$$I_\rho + I_\theta = \oint m(\dot{\rho}\, d\rho + \rho\dot{\theta}\, \rho\, d\theta) + \oint e\rho A_\theta \, d\theta = \text{constant.} \qquad \textbf{1.35}$$

Denoting the transverse component of the velocity by $\mathbf{v}_t$, and the circular element of arc of the projection of the electron orbit on the transverse plane by $d\mathbf{s}$ the first integral in equation **1.35** gives

$$\oint m\mathbf{v}_t . d\mathbf{s} = \int_0^{2\pi/\omega} m\omega^2 R^2 \, dt = 2\pi m R^2 \omega,$$

where R is the orbit radius and ω is the cyclotron frequency. Since A_θ is the only non-zero component of the vector potential, Stokes's theorem gives

$$e\oint \mathbf{A} . d\mathbf{s} = eBR^2\pi = -\pi m\omega R^2.$$

Collecting our results, and applying the adiabatic theorem:

$$\pi m R^2 \omega = \text{constant,} \qquad \textbf{1.36a}$$

that is, under the conditions considered, angular momentum is conserved.

This relation can be re-stated in various forms. Substituting the cyclotron frequency expressed in terms of the magnetic field, and incorporating the electronic charge in the 'constant', it follows that

$$\pi R^2 B = \text{constant.} \qquad \mathbf{1.36b}$$

That is, the motion takes place in such a manner that the magnetic flux linked by the orbit stays constant. If the transverse component of the orbital momentum is denoted by p_t, we have $p_t = eBR$ and therefore can replace equation **1.36b** by

$$\frac{p_t^2}{B} = \text{constant.} \qquad \mathbf{1.36c}$$

Finally, the rotating electron constitutes a circulating current $i = (\omega/2\pi)e$, and as such, it has a magnetic moment $M = \pi R^2 i = \frac{1}{2}eR^2\omega$. Hence, under the conditions considered, the magnetic moment arising from the orbital rotation of the electrons is conserved:

$$M = \text{constant.} \qquad \mathbf{1.36d}$$

The claim that the 'magnetic bottle' confines the electron motion in the direction of increasing field intensity follows from the fact that the field is conservative. Denoting the momentum component parallel to the magnetic field by p_b and distinguishing the quantities at the initial position of the electron by the subscript '0', energy conservation requires that

$$p_b^2 + p_t^2 = p_{b0}^2 + p_{t0}^2 = p^2,$$

where p is the total orbital momentum. But from equation **1.36c**

$$\frac{p_t^2}{B} = \frac{p_{t0}^2}{B_0}$$

and therefore $$p_b^2 = p^2 - p_{t0}^2 \frac{B}{B_0}.$$

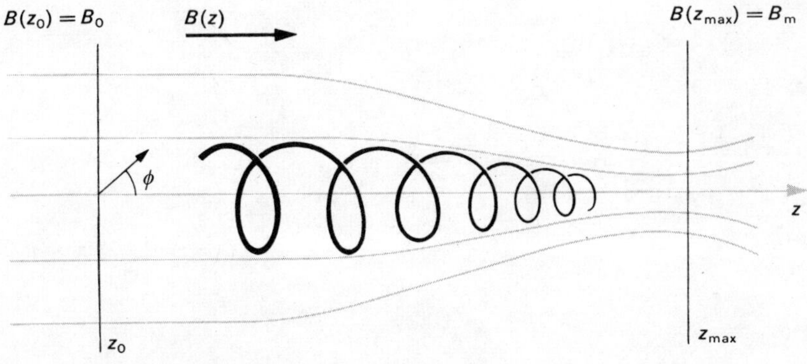

Figure 5

Since by necessity $p_b^2 \geqslant 0$, the electron can reach only regions where the field intensity

$$B \leqslant B_0 \left(\frac{p}{p_{t_0}}\right)^2.$$

For electron orbits centred on the symmetry axis of the field (Figure 5), starting from a region of uniform field B_0, the condition that the electrons will be trapped is

$$\frac{p_{t_0}}{p} = \sin \phi \leqslant \frac{B_0}{B_m},$$

where B_m is the maximum field and ϕ is the angle between the momentum of the electron and the symmetry axis in the uniform field B_0.

Further reading

T. W. KIBBLE, *Classical Mechanics*, McGraw-Hill, 1966.

An excellent introduction which gives a clear presentation of the fundamental concept and the conservation laws in particular. It emphasizes aspects of classical mechanics which are of outstanding importance in the theories of relativity and quantum mechanics.

H. GOLDSTEIN, *Classical Mechanics*, Addison-Wesley, 1950.

An advanced textbook which gives a thorough discussion of the mathematical formalism.

L. D. LANDAU and E. M. LIFSHITZ, *Mechanics*, Pergamon, 1960.

A very concise, advanced textbook; probably the most stimulating one in the modern literature of non-relativistic mechanics.

Chapter 2
Geometrical Electron Optics

In many cases the relations of the phenomena in two different physical questions have a certain similarity which enables us, when we have solved one of these questions, to make use of our solution in answering the other.

J. C. Maxwell

2.1 Introduction

There is a large class of devices whose role is to collect in one point a divergent beam of charged particles emitted by a point source. This is achieved by allowing the particles to pass through a suitably chosen electromagnetic field. All the charged-particle imaging devices ('lenses') and charged-particle spectrometers fall in this category. The examples mentioned indicate that the practical purpose of such devices can be very different, as are the design considerations which lead to the development of such devices. The fundamental physical principles, however, on which all the design considerations rest are essentially the same. The subsequent sections of this chapter are devoted to these basic ideas.

To express general principles in quantitative terms a flexible mathematical formalism is required which itself is sufficiently general, and yet at the same time can be easily adapted to the treatment of particular problems of practical significance. At present we wish to study the motion of charged particles in static fields so the development of the motion in time is of little interest. Therefore, instead of describing the position of the particles as a function of time, we shall consider the particle trajectories as curves in space. In mathematical terms, instead of the equations of motion where time is the independent variable, one requires differential equations of the trajectory in which the independent variable is a spatial coordinate and time does not occur explicitly.

If such considerations are carried out in complete generality, the results suggest that there must be a close similarity between the trajectories of charged particles in static electromagnetic fields and rays of light propagating in a refractive medium according to the laws of geometrical optics. We shall see in section 2.3 that the trajectories are determined by a variational principle (*principle of least action*) which can be written in the form

$$\delta \int_{(1)}^{(2)} [\{2m(W-e\Phi)\}^{\frac{1}{2}}+e\mathbf{A}.\hat{\mathbf{l}}]\, dl = 0,$$

where W is the total energy of the particle, Φ and $\mathbf{A}$ are the scalar and vector potentials, respectively, and $\hat{\mathbf{l}}$ is the unit vector tangential to the trajectory along which the integration is carried out between points (1) and (2). The integrand is a scalar function of the coordinates only, so one may introduce for brevity

$$[2m(W-e\Phi)]^{\frac{1}{2}}+e\mathbf{A}.\hat{\mathbf{l}} \equiv n(x_1, x_2, x_3)$$

when the variational principle takes the form

$$\delta \int_{(1)}^{(2)} n(x_1, x_2, x_3)\, dl = 0.$$

This equation is identical with the variational principle in geometrical optics, known as Fermat's principle, provided the function $n(x_1, x_2, x_3)$ is interpreted as the refractive index of an isotropic optical medium. Thus, to any static electromagnetic field there corresponds a refractive medium such that the trajectory of a charged particle in the field is the same as that of a light ray in the corresponding optical medium. This analogy suggests that the formalism to be adopted should be one which expresses the geometrical optical features of the particle trajectories. Besides, one can anticipate that the motion of charged particles will be particularly interesting in an electromagnetic field which has a spatial distribution with symmetry properties similar to those of the components of optical devices such as spherical and cylindrical lenses, prisms, and their combinations. For this reason the subsequent sections deal with problems characteristic of fields which have different kinds of symmetry properties. As a first step, however, it will be useful to become more familiar with some basic ideas and to obtain further guidance for tackling problems in general terms. Therefore we shall start by considering very simple physical conditions.

2.2 Electrons in magnetic fields

2.2.1 *Electron trajectories in uniform fields*

Consider a point source of charged particles on the z-axis of a cylindrical polar coordinate system situated in a uniform magnetic field B parallel to the z-axis. Particles of momentum $\mathbf{p}$ and energy $W = p^2/2m$ are emitted by the source at an angle $0 < \alpha < \frac{1}{2}\pi$ relative to the z-axis. The particles will describe a helical trajectory. The radial distance of a particle from the z-axis is

$$\rho = 2R \sin \tfrac{1}{2}\omega t, \qquad \textbf{2.1}$$

where $\omega = eB/m$ is the cyclotron frequency, $R = (p/eB) \sin \alpha$, and the velocity component parallel to the z-axis is conserved:

$$\frac{dz}{dt} = \frac{p}{m} \cos \alpha = \text{constant}$$

(see Figure 6). Thus one can eliminate time as an independent variable and from equation **2.1** one obtains

$$\rho = 2R \sin kz, \qquad \textbf{2.2}$$

where $$k = \frac{1}{2} \frac{eB}{p \cos \alpha} = \frac{1}{2} \frac{eB}{(2mW)^{\frac{1}{2}} \cos \alpha}. \qquad \textbf{2.3}$$

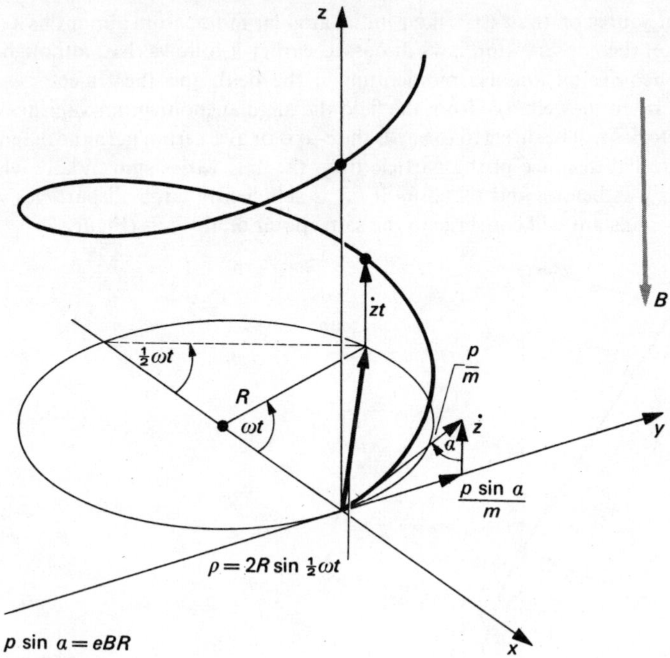

$p \sin \alpha = eBR$

Figure 6

Hence, all particles for which $p \cos \alpha$ = constant cross the z-axis at points where

$$z = \frac{n\pi}{k} = 2\pi n \frac{p \cos \alpha}{eB} \equiv Z_n \qquad (n = \text{integer}). \qquad \mathbf{2.4}$$

For particles emitted at a small angle α, such that $\cos \alpha \simeq 1$, one finds to a good approximation

$$k \simeq \frac{1}{2}\frac{eB}{p} = \tfrac{1}{2}eB(2mW)^{-\frac{1}{2}}$$

and $$Z_n \simeq \frac{2\pi np}{eB}.$$

In this case all the particles with p = constant emitted within the cone of semi-angle α are collected at $z = Z_n$. So a pinhole situated at a distance Z_n from the source will transmit only those particles within the narrow momentum range $p \pm \Delta p$. Some of the shortcomings of this system as a momentum analyser and the possibility of its improvement will be discussed in the next section.

Now one can raise the following question. Let the uniform magnetic field extend between the planes $z = z_0$ and $z = z_0 + Z$ only. Is it possible to focus particles diverging from a point source on the z-axis, at $z = 0$, outside the magnetic field? Qualitatively one can argue as follows. Since the particles are emitted by a point source on the z-axis, their initial angular momentum about this axis is zero. From the conservation laws discussed earlier it follows that, although the particles acquire an angular momentum in the field, and the trajectories get 'twisted', when they emerge from the field the angular momentum vanishes and the trajectories will be directed towards the z-axis or away from it. In the magnetic field the radial distance of the particle from the axis varies sinusoidally with a period $2\pi/k$ as before, and therefore if kZ is sufficiently large all particles with $p \cos \alpha =$ constant will converge to the same point on the axis (Figure 7).

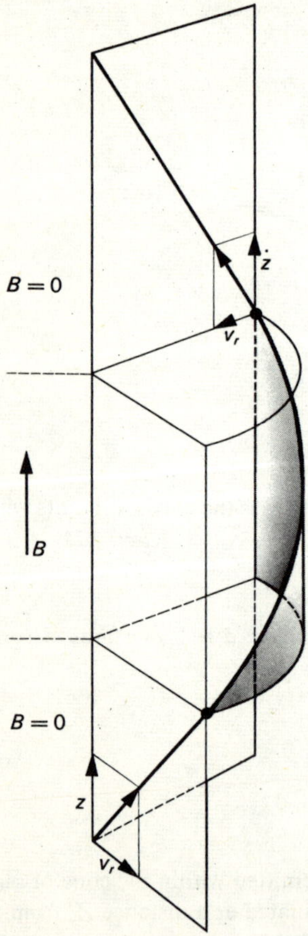

Figure 7

It is worthwhile pursuing this argument quantitatively. It follows from our previous considerations that, if the particle starts from a field-free region with vanishing angular momentum, then the canonical momentum component $\pi_\theta = 0$, and according to equation **1.20** the particle will move in a plane rotating about the z-axis with an angular velocity $\dot{\theta} = -eA_\theta mp\ (\rho \neq 0)$. The radial motion of the particle, as observed in the rotating plane, is determined by equation **1.23**. If the field is uniform,

$$A_\theta = \tfrac{1}{2}\rho B$$

so that the meridian plane rotates with the Larmor frequency. For the equivalent potential occurring in equation **1.23**, one finds from equations **1.20** and **1.21**

$$U = \frac{1}{2m}(eA_\theta)^2 = \frac{1}{2m}(\tfrac{1}{2}e\rho B)^2.$$

To obtain a differential equation for the trajectory in the meridian plane, one has to eliminate time by combining the equations of motion **1.23**. Let us choose z as the independent variable, and denote differentiation with respect to time by a dot and that with respect to z by a prime.

Thus $$\frac{d\rho}{dz} \equiv \rho' = \frac{\dot{\rho}}{\dot{z}}$$

and hence $$\frac{d^2\rho}{dz^2} \equiv \rho'' = \frac{1}{\dot{z}}\frac{\ddot{\rho}\dot{z}-\dot{\rho}\ddot{z}}{\dot{z}^2} = \frac{\ddot{\rho}}{\dot{z}^2},$$

since in the uniform magnetic field $\dot{z} = p_z/m =$ constant.

Thus we obtain $$\rho'' = -\frac{m}{p_z^2}\frac{\partial U}{\partial \rho}$$

and hence $$\rho'' = -k^2\rho, \qquad \textbf{2.5}$$

where $$k = \frac{eB}{2p_{z0}}$$

is the same as given by equation **2.3**.

The general solution of equation **2.5** is well known; it can be written as

$$\rho = a\sin(kz+\delta) \qquad \textbf{2.6}$$

and hence $$\rho' = ak\cos(kz+\delta),$$

where a and δ are constants determined by the initial conditions.

Figure 8(a) shows the trajectories of two electrons in the meridian plane. Both electrons are emitted at the same point P with the same axial momentum component, p_{z0}, but at different angles relative to the z-axis.

In a Cartesian coordinate system where ρ and ρ'/k are measured along the two axes, the radial position of the electron and the slope of its trajectory are represented by a point in this plane. As the electron proceeds on its trajectory in the field, the representative point describes a circle of radius a. In Figure 8(b), point A represents the radial position and the slope of an electron trajectory at the

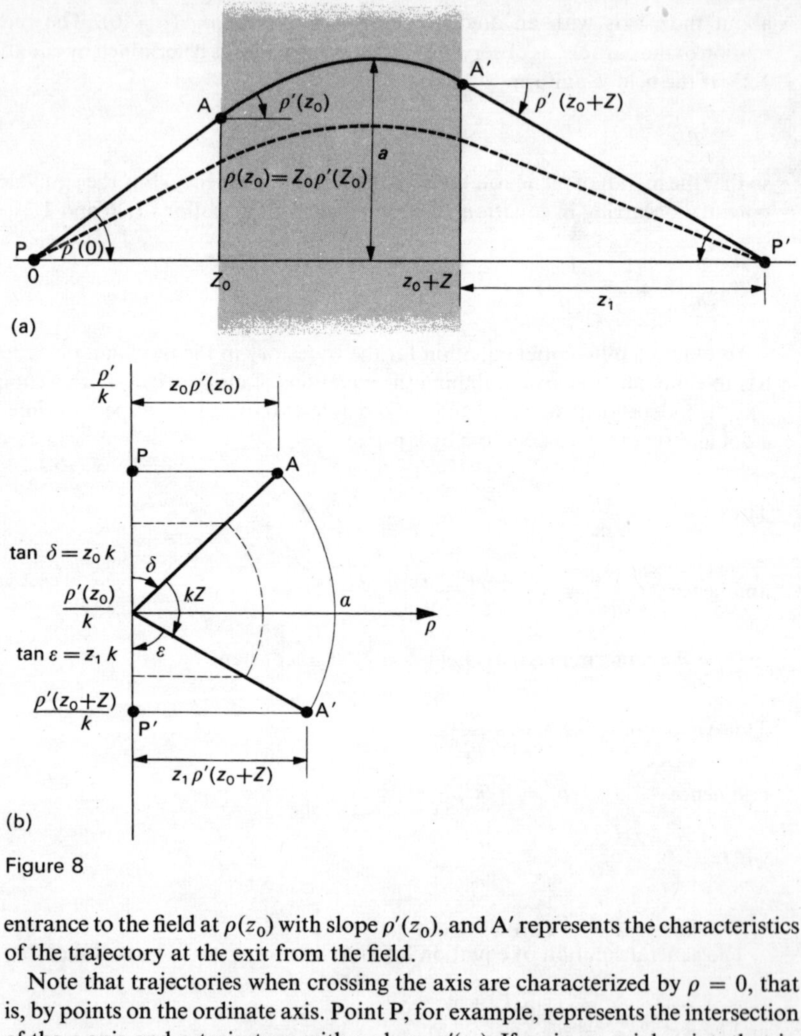

Figure 8

entrance to the field at $\rho(z_0)$ with slope $\rho'(z_0)$, and A′ represents the characteristics of the trajectory at the exit from the field.

Note that trajectories when crossing the axis are characterized by $\rho = 0$, that is, by points on the ordinate axis. Point P, for example, represents the intersection of the z-axis and a trajectory with a slope $\rho'(z_0)$. If z_0 is an axial point, that is, $\rho(z_0) = 0$, as in the example in Figure 8(a), the trajectory is a straight line until it enters the field. Thus, as the electron moves on this trajectory, point P in Figure 8(b) shifts to A and then to A′. Since $\rho(z_0) = \rho'(z_0)z_0$ one finds

$$\tan \delta = kz_0. \qquad \textbf{2.7a}$$

When the particle emerges from the field it moves on a rectilinear trajectory, that is, with a constant slope corresponding to the representative point A′. It moves towards the z-axis if

$$\delta + kZ < 0$$

and thus point A′ shifts towards P′, which again represents a point on the z-axis. Figure 8(a) shows that the distance of this point from the exit plane of the field, z_1, is such that

$$\rho'(z_0+Z+z_1) = \frac{\rho(z_1+Z)}{z_1}.$$

Hence according to Figure 8(b) one has

$$\tan \varepsilon = kz_1. \qquad \textbf{2.7b}$$

Figure 8(b) also shows that, to consider a trajectory crossing the z-axis at z_0 with different slopes, point P must move along the ordinate axis; point P′ shifts also, but equations **2.7a** and **b** remain valid. In other words, all the trajectories crossing the z-axis at z_0 will cross it again at z_1 if the condition

$$\delta+kZ+\varepsilon = \pi$$

is satisfied, that is, if $\tan(\delta+\varepsilon) = \tan kZ$. Hence, from equations **2.7a** and **b**, a relation between z_0 and z_1 is found for a given value of the parameter k and the field length Z:

$$\frac{kz_0+kz_1}{1-k^2z_0z_1} = \tan kZ. \qquad \textbf{2.8}$$

The point on the axis at a distance z_1 downstream from the exit plane of the field can be considered as the 'image' of the point source located on the axis at a distance z_0 in front of the entrance plane.

In order to study the problem of image formation by beams of charged particles in further detail, it will be advantageous to introduce a formalism familiar from geometrical optics. This is done in section 2.2.3. Further use will be made of diagrams in the (ρ, ρ') planes in section 2.9 and section 4.3.

2.2.2 *Ring focusing in long magnetic fields*

Let us return to the problem of the possible improvement of the momentum-selective properties of a 'long' *uniform* magnetic field discussed earlier. Equation **2.4** shows that if there is a point source of electrons and a pinhole at a distance Z_n, in a uniform magnetic field which is parallel to the straight line connecting them, the pinhole will transmit all the particles emitted by the source with the same axial velocity, that is, $p \cos \alpha$ = constant. To a first-order approximation this is compatible with a finite spread in momentum, δp, and in emission angle, $\delta\alpha$, provided

$$\delta Z_n = \delta p \cos \alpha - p \sin \alpha \, \delta\alpha = 0.$$

To obtain a transmitted beam of finite intensity, the particles must be emitted by the point source in a finite range of angles between, say, α and $\alpha+\delta\alpha$, that is, into a solid angle (Figure 9)

$$\Omega = 2\pi \sin \alpha \, \delta\alpha.$$

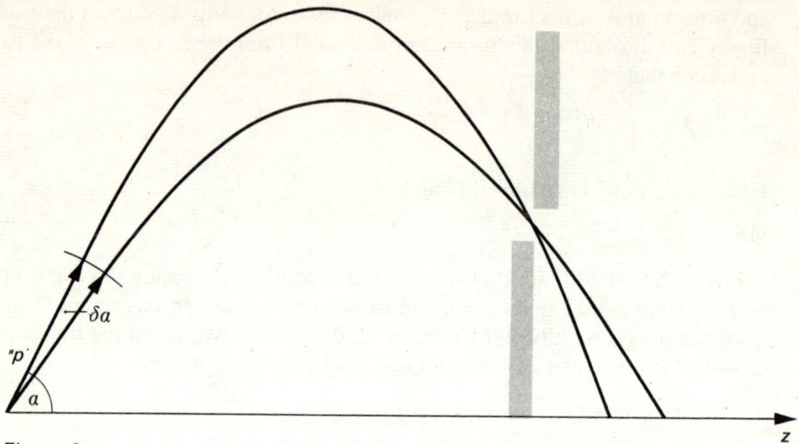

Figure 9

In this case the fractional spread of the momentum in the transmitted beam will be

$$\frac{\delta p}{p} = \frac{\sin \alpha\, \delta\alpha}{\cos \alpha} = \frac{\Omega}{2\pi \cos \alpha} \geqslant \frac{\Omega}{2\pi}.$$

If two particles of the same momentum p were emitted, one at an angle α and the other at an angle $\alpha+\delta\alpha$, from the same point on the axis, their trajectories will cross the axis at different distances from the source, but they will cross one another in a plane $z = z_0$ at some distance, $\rho(z_0)$, off the axis (Figure 9). Therefore an infinitesimally thin ring aperture of radius $\rho(z_0)$, situated in the $z = z_0$ plane, would transmit, to a first-order approximation, all the particles of the sharply defined momentum p emitted in the solid angle $\Omega = \sin \alpha\, \delta\alpha$, and would reject all particles which have different moments.

To express this argument in quantitative form one must consider the condition under which electrons emitted by the source will reach the same radial distance ρ in a plane $z =$ constant. This is achieved to a first-order approximation if the relation

$$\delta\rho = \frac{\partial \rho}{\partial p}\,\delta p + \frac{\delta \rho}{\partial \alpha}\,\delta\alpha = 0 \qquad \textbf{2.9}$$

is satisfied. On substitution of the explicit values for R and k into equation **2.2**, one finds from equation **2.9**

$$\frac{\partial \rho}{\partial \alpha} = \frac{2p}{eB}\left(\cos \alpha \sin kz + \frac{\sin^2 \alpha}{\cos \alpha}\, kz \cos kz\right)$$

and hence $\partial\rho/\partial\alpha = 0$, provided that z satisfies the equation

$$(kz)^{-1} \tan kz = -\tan^2 \alpha. \qquad \textbf{2.10}$$

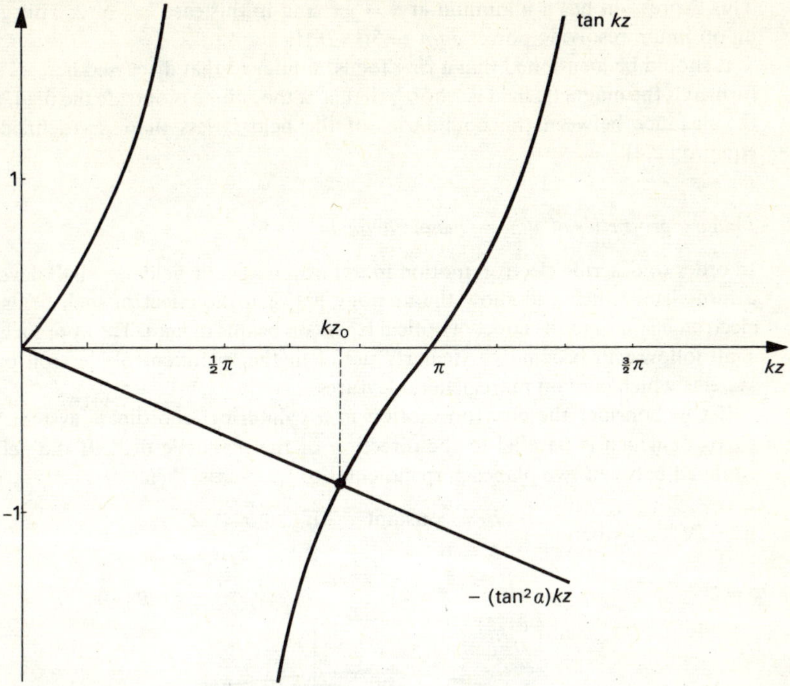

Figure 10

Denoting this value of z by z_0, we can easily convince ourselves that $(\partial\rho/\partial p)_{z=z_0} \neq 0$ and therefore, at $z = z_0$, equation **2.9** is satisfied only if $\delta p = 0$. Thus, as expected, if an infinitely narrow ring aperture of radius $\rho(z_0)$ is placed perpendicularly to the magnetic field at a distance z_0 from the source, it will transmit electrons of a sharply defined momentum ($\delta p = 0$), emitted in a small but finite range of angles between α and $\alpha+\delta\alpha$. This effect is called 'ring focusing' and it is widely used in the design of beta-ray spectrographs. The graphical solution of equation **2.10** is indicated in Figure 10.

The actual 'resolving power', that is, the fractional momentum spread, $\delta p/p$ of the transmitted beam is obtained by taking into account the *second-order* term in $\delta\alpha$. Since $(\partial\rho/\partial\alpha)_{z_0} = 0$,

$$(\delta\rho)_{z_0} = \left(\frac{\partial\rho}{\partial p}\right)_{z_0} \delta p + \frac{1}{2}\left(\frac{\partial^2\rho}{\partial\alpha^2}\right)_{z_0} (\delta\alpha)^2 = 0$$

and hence $$\frac{\delta p}{p} = \frac{1}{2}(3+\cos^2\alpha \tan^2 kz_0)\delta\alpha^2$$

$$= \frac{1}{2}\left(\frac{\Omega}{2\pi}\right)^2 \left[\frac{3}{\sin\alpha} + \left(\frac{\tan kz_0}{\tan\alpha}\right)^2\right].$$

This expression has a minimum at $\alpha = \frac{1}{4}\pi$, and in this case $kz_0 \simeq 2$. This gives an optimum resolving power $\delta p/p \simeq 5(\Omega/2\pi)^2$.

It should be mentioned that a ring focus, similar to that discussed here, is also formed if the magnetic field is 'short', that is, if the source is outside the field, and the distance between the boundaries of the field is less than Z_n (defined by equation **2.4**).

2.2.3 *Imaging properties of uniform magnetic fields*

In order to describe electron motion in uniform magnetic fields we shall develop a formalism which will show the analogy between the effect of such fields on electron beams and the effect of optical lenses on beams of light. The approach we shall follow will become particularly useful in the treatment of electron optic systems which contain more general features.

Let us consider the electron motion in a cylindrical coordinate system, the z-axis of which is parallel to the direction of the magnetic field. If the field is confined between two planes perpendicular to the z-axis, then

$$\mathbf{B} = B\mathbf{e}_z, \quad \text{where} \quad \begin{array}{ll} B = \text{constant} & \text{if } 0 \leqslant z \leqslant Z, \\ B = 0 & \text{if } z < 0 \text{ or } z > Z. \end{array}$$

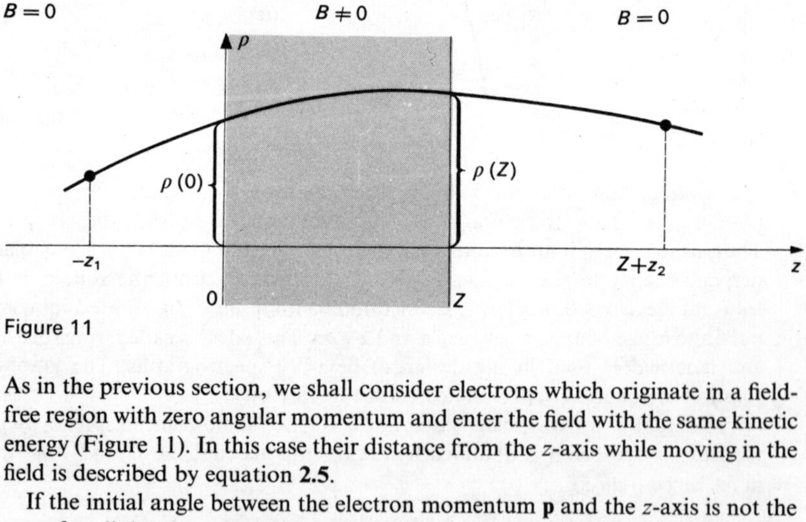

Figure 11

As in the previous section, we shall consider electrons which originate in a field-free region with zero angular momentum and enter the field with the same kinetic energy (Figure 11). In this case their distance from the z-axis while moving in the field is described by equation **2.5**.

If the initial angle between the electron momentum $\mathbf{p}$ and the z-axis is not the same for all the electrons in a beam but is always small ($\mathbf{p} \cdot \mathbf{e}_z \simeq p$), the quantity k is constant and is determined by the field intensity and the magnitude of the momentum ($k \simeq eB/2p$). This condition is characteristic of the 'paraxial' approximation.

Instead of using equation **2.6**, let us write the solution of the differential equation of the trajectory in the form

$$\rho(z) = a \cos kz + b \sin kz \qquad (0 \leqslant z \leqslant Z),$$

where a and b are constants dependent on the initial conditions. At entry $z = 0$, which gives

$$\rho(0) = a,$$
$$\left(\frac{d\rho}{dz}\right)_{z=0} \equiv \rho'(0) = kb$$

and hence the equation of the trajectory and its slope can be expressed in terms of the initial conditions:

$$\rho(z) = \rho(0)\cos kz + k^{-1}\rho'(0)\sin kz,$$
$$\rho'(z) = -k\rho(0)\sin kz + \rho'(0)\cos kz.$$

These equations show that the distance of the electron from the z-axis and the slope of its trajectory are everywhere linearly related to the initial values of these parameters. The two equations above can be summarized in a matrix equation

$$\begin{bmatrix}\rho(z)\\ \rho'(z)\end{bmatrix} = \mathrm{M}\begin{bmatrix}\rho(0)\\ \rho'(0)\end{bmatrix}, \qquad \textbf{2.11}$$

where M is a quadratic matrix

$$\mathrm{M}(z) \equiv \begin{bmatrix} m_{11} & m_{12}\\ m_{21} & m_{22}\end{bmatrix} = \begin{bmatrix}\cos kz & k^{-1}\sin kz\\ -k\sin kz & \cos kz\end{bmatrix}. \qquad \textbf{2.12}$$

Note that the determinant of the matrix is unity:

$$\det \mathrm{M} = 1.$$

The introduction of a matrix formalism enables an electron trajectory to be very easily followed through adjoining regions of different field intensity. In particular, in a field-free region $k = 0$ and if the length of the region is s, then $k^{-1}\sin ks \to s$, and the matrix becomes

$$\mathrm{M}_0(s) = \begin{bmatrix}1 & s\\ 0 & 1\end{bmatrix}. \qquad \textbf{2.13}$$

This gives $\rho(s) = \rho(0) + s\rho'(0)$ and $\rho'(s) = \rho'(0)$, as it should do, since in the field-free region the electron trajectory is a straight line.

If the electron is emitted by a point source in the $z = -z_1$ plane at a distance $\rho(-z_1)$ from the axis at an angle $\rho'(-z_1)$ with respect to the z-axis, the corresponding parameters at the entry into the field are given by the equation

$$\begin{bmatrix}\rho(0)\\ \rho'(0)\end{bmatrix} = \mathrm{M}_0(-z_1)\begin{bmatrix}\rho(-z_1)\\ \rho'(-z_1)\end{bmatrix}.$$

It leaves the field with parameter values

$$\begin{bmatrix}\rho(Z)\\ \rho'(Z)\end{bmatrix} = \mathrm{M}(Z)\,\mathrm{M}_0(-z_1)\begin{bmatrix}\rho(-z_1)\\ \rho'(-z_1)\end{bmatrix}$$

and, for a distance z_2 beyond the exit plane, one finds

$$\begin{bmatrix}\rho(Z+z_2)\\ \rho'(Z+z_2)\end{bmatrix} = \mathrm{M}_0(z_2)\,\mathrm{M}(Z)\,\mathrm{M}_0(-z_1)\begin{bmatrix}\rho(-z_1)\\ \rho'(-z_1)\end{bmatrix}. \qquad \mathbf{2.14}$$

The product of the three matrices is, of course, again a quadratic matrix, and denoting its elements by M_{ij} $(i, j = 1, 2, 3)$ equation **2.14** is equivalent to the following:

$$\rho(Z+z_2) = M_{11}\,\rho(-z_1)+M_{12}\,\rho'(-z_1),$$
$$\rho'(Z+z_2) = M_{21}\,\rho(-z_1)+M_{22}\,\rho'(-z_1).$$

These equations show that the trajectories of all electrons, emitted by a point source at $\rho(-z_1)$ in a meridian plane within a small range of the direction of emission, cross at a point $\rho(Z+z_2)$ of the same plane, independently of the initial direction of emission ρ' if

$$M_{12} = 0. \qquad \mathbf{2.15}$$

Provided this condition is satisfied, to any point of an extended source in the plane $z = -z_1$ there corresponds an 'image point' in the plane $z = Z+z_2$. Thus equation **2.15** gives the condition of 'stigmatic imaging' by means of a magnetic 'lens'. If this condition is satisfied, the 'linear magnification' is given by

$$\frac{\rho(Z+z_2)}{\rho(-z_1)} = M_{11}. \qquad \mathbf{2.16}$$

If the source is on the axis $\rho(-z_1) = 0$, the image point is also on the axis $\rho(Z+z_2) = 0$, and

$$\rho'(Z+z_2) = M_{22}\,\rho'(-z_1).$$

Hence $M_{22} = \dfrac{\rho'(Z+z_2)}{\rho'(-z_1)}$

gives the angular magnification, and since det M $= 1$ and $M_{12} = 0$, one finds

$$M_{11}M_{22} = 1,$$

that is, *the product of the linear and angular magnifications is unity*.

It should be remembered that the meridian plane, in which the imaging properties of the lens have been investigated, is rotating about the z-axis with an angular velocity determined by equation **1.20**, and as a result the electron optic image, produced by the uniform field, is rotated about the z-axis with respect to the source. The amount of the rotation can be readily determined. Since the electrons originate outside the magnetic field with zero initial angular momentum, and the magnetic field is uniform, the meridian plane rotates at a rate given by the Larmor frequency while the electron is in the magnetic field. Hence the rotation of the image is given by

$$\theta_i = \frac{mZ}{p}\,\omega_L = \frac{eBZ}{2p}.$$

Further implications of equation **2.15** can be seen by calculating the matrix product $M_0(z_2)\, M(Z)\, M_0(-z_1)$, and this is found to be:

$$\begin{bmatrix} m_{11}+z_2m_{21} & -z_1m_{11}+m_{12}-z_2z_1m_{21}+z_2m_{22} \\ m_{21} & -z_1m_{21}+m_{22} \end{bmatrix}. \qquad \textbf{2.17}$$

Hence equation **2.15** gives a relation between the position of the source ('object') plane $z = -z_1$ and the image plane $z = Z+z_2$,

$$\frac{m_{22}}{z_1}-\frac{m_{11}}{z_2} = m_{21}-\frac{m_{12}}{z_1z_2}. \qquad \textbf{2.18}$$

Substitution of the matrix elements from equation **2.12** shows that this equation is equivalent to equation **2.8**, which was obtained by geometrical arguments in the $(\rho, \rho'/k)$ plane.

In geometrical optics, the focal planes are defined as the planes whose conjugates are at infinity. Thus the position of the focal plane in the image space, z_{f2}, is found from equation **2.18** by making $z_1 \to \infty$; the position of the focal plane in the object space, z_{f1}, is found by making $z_2 \to \infty$:

$$\frac{1}{z_{f2}} = -\frac{m_{21}}{m_{11}}, \qquad \frac{1}{z_{f1}} = \frac{m_{21}}{m_{22}}. \qquad \textbf{2.19}$$

Note that the quantities z_{f1} and z_{f2} give the position of the foci with respect to the boundaries of the magnetic field (Figure 12). The foci are located symmetrically if $m_{11} = m_{22}$. From equation **2.12**, which gives the definition of the matrix elements m_{ij}, we find that

$$\frac{Z}{z_{f2}} = -\frac{Z}{z_{f1}} = kZ \tan kZ. \qquad \textbf{2.19a}$$

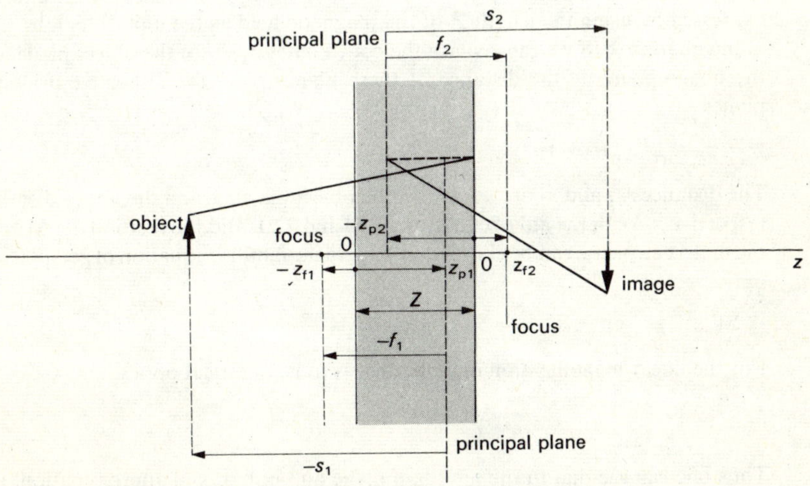

Figure 12

The principal planes of a lens in geometrical optics are defined as conjugate planes in which the linear magnification is unity. Thus the positions of the principal planes, z_{p1} and z_{p2}, must satisfy two conditions simultaneously: $M_{12} = 0$, and $M_{11} = 1$. Taking the appropriate matrix elements from **2.17**, and solving the equations for z_1 and z_2, one obtains

$$z_{p1} = -\frac{1-m_{22}}{m_{21}}, \qquad z_{p2} = \frac{1-m_{11}}{m_{21}}. \qquad \textbf{2.20}$$

For the uniform magnetic field

$$\frac{z_{p1}}{Z} = -\frac{z_{p2}}{Z} = \frac{1}{2}\frac{\tan \frac{1}{2}kZ}{\frac{1}{2}kZ}. \qquad \textbf{2.20a}$$

If the magnetic field is so short that $kZ \ll 1$ then $z_{p1} = -z_{p2} \simeq \frac{1}{2}Z$, that is, the two principal planes coincide in the mid-plane of the magnetic field, and the magnetic field has the properties of a 'thin' lens.

As in geometrical optics, the focal length of the lens is defined as the distance of the focal plane from the corresponding principal plane

$$f_i = z_{fi} - z_{pi} \qquad (i = 1, 2).$$

This is found to have the same magnitude for the image space and for the object space:

$$f_1^{-1} = -f_2^{-1} = m_{21}, \qquad \textbf{2.21}$$

giving, for the uniform magnetic field,

$$-\frac{Z}{f_1} = \frac{Z}{f_2} = kZ \sin kZ. \qquad \textbf{2.21a}$$

In equations **2.19a**, **2.20a** and **2.21a** the positions of the cardinal points are expressed by using the length Z of the magnetic field as the unit of length.

In equation **2.18** we can replace the coordinates z_1, z_2 of the object plane and the image plane by the distance of these planes from the respective principal planes

$$s_i = z_i - z_{pi} \qquad (i = 1, 2).$$

The distances s_1 and s_2 are known as the object distance and the image distance respectively. With the aid of equations **2.20** and **2.21**, and the relation det M = 1, the matrix elements can be eliminated, to give the familiar equation of geometrical optics:

$$s_2^{-1} + s_1^{-1} = |f|^{-1}.$$

For the linear magnification one obtains, as in geometrical optics,

$$M_{11} = 1 - \frac{s_2}{f_2}.$$

Thus one can see that in the meridian plane and in paraxial approximation, the imaging properties of a uniform magnetic field are the same as those of a spherical lens in an optically uniform environment (that is, the refractive index outside the lens is the same on both sides). The conventional methods of image construction are applicable as indicated in Figure 12. The positions of the cardinal points are defined by the parameter $kZ = \frac{1}{2}ZeB/p$: for a magnetic field of given intensity and given length their position depends on the momentum of the electrons. Therefore if the electron beam is not mono-energetic, there is an aberration in the imaging analogous to the chromatic aberration in optics. The analogy becomes closer if one considers that the de Broglie wavelength of a particle is uniquely determined by its momentum $\lambda = h/p$ (h is Planck's constant). If the initial angle maintained by the electron beam is not small, 'spherical aberration' arises, because in this case, as pointed out earlier, the angular dependence of the parameter $kZ = ZeB/2p \cos \alpha$ is not negligible.

We shall see later that the general formalism just developed can be generalized for the treatment of the imaging properties of any static electromagnetic fields provided that the position of the electron and the slope of its trajectory can be linearly related to the initial values, as in equation **2.16**, where the elements of the quadratic matrix can be arbitrary functions of z subject to the only restriction that det M = 1.

Regarding the applications of the description of the trajectories in optical terms some warning is appropriate. As in conventional optics, the imaging laws are designed to relate the positions of source points and image points. In plotting rays to determine the performance of a lens, one should remember that the straight sections one draws are not actual trajectories unless the region considered is outside the field of force, so the construction gives no information about the trajectories between the principal planes. This fact should be particularly noted if the lens is 'thick', which is certainly the case if the crucial parameter $kZ > \frac{1}{2}\pi$. It is somewhat misleading, for example, to judge the focusing or defocusing effect of such a thick lens purely by the criterion usually applied to thin lenses, namely that it is considered as a focusing lens if $-1/f_1 = 1/f_2 > 0$. We shall return to this problem in section 2.8.

2.3 Imaging properties of static fields of rotational symmetry.

Electron-optic imaging devices, which are analogous to spherical lenses, consist of electric or magnetic fields of rotational symmetry. Such electric fields are produced by rotationally symmetric electrode systems centred on a common axis, for example, circular apertures and/or cylindrical electrodes. In the case of magnetic fields, rotational symmetry is obtained if the current-carrying conductors, or magnetic pole faces, have a common axis of rotational symmetry.

If the fields are described in terms of cylindrical polar coordinates ρ, θ, z, then the electrostatic field can be derived from a scalar potential Φ which is a function of r and z only. If the potential on the symmetry axis is denoted by $V(z)$, the potential can be written as a power series (see Appendix):

$$\Phi(\rho, z) = \sum_{n=0}^{\infty} (-1)^n (n!)^{-2} \frac{d^{2n}V}{dz^{2n}} \left(\frac{\rho}{2}\right)^{2n}. \qquad \textbf{2.22a}$$

Thus, the potential is an even function of ρ. In the vicinity of the symmetry axis it is sufficient to consider only the lowest term in ρ.

A magnetic field of rotational symmetry can be derived from a vector potential **A**, for which $A_\rho = A_z = 0$, and $A_\theta = A(\rho, z)$ depends on ρ and z only. For this component there is a power-series expansion similar to the one above. It follows from the symmetry conditions that the flux density on the axis is parallel (or anti-parallel) to the z-axis. If this is denoted by $B_0(z)$, the non-vanishing component of the vector potential is given by the series:

$$A_\theta(\rho, z) = \sum_{n=0}^{\infty} \frac{(-1)^n}{n!\,(n+1)!} \frac{d^{2n}B_0}{dz^{2n}} \left(\frac{\rho}{2}\right)^{2n+1}. \qquad \textbf{2.22b}$$

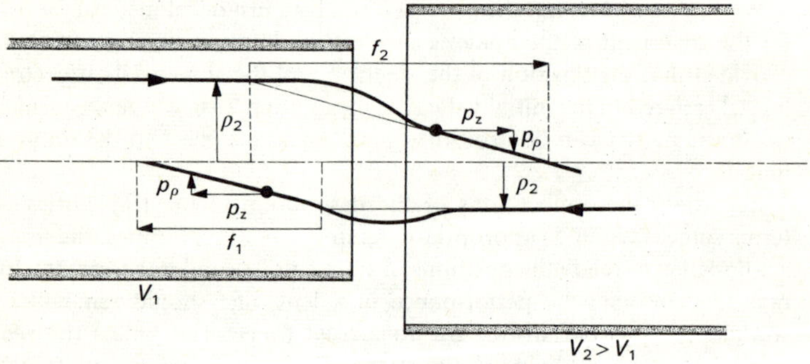

Figure 13

In some simple situations qualitative considerations permit a good guess to be made regarding the focusing action of rotationally symmetric fields. Consider for example the motion of an electron through a rotationally symmetric electric field between two field-free regions, that is, between two regions where the potential is constant, say V_1 and V_2 (see Figure 13). If the electron is moving along the symmetry axis, the force acting on it has no radial component, and therefore the electron keeps moving along the axis with an increasing or decreasing kinetic energy, depending on the direction of its motion. But if the electron reaches the field in a point off the axis, it will be acted upon by a radial force whose magnitude – to a first-order approximation – is proportional to the radial distance of the electron from the axis. If the electron is traversing the field in the direction of axial acceleration, it is accelerated, at first, towards the axis, and later in the opposite direction. The net deflection is directed towards the axis because (a) the deflection produced by a given radial force decreases as the velocity of the particle increases, and (b) as the particle approaches the axis the radial force is decreasing. For a particle traversing the field in the opposite direction the sequence of events is reversed, but the same argument shows that the net effect is again a deflection towards the axis. Considering the projection of the straight sections of the trajectory in the field-free regions, and assuming that the net change of radial momentum of a particle, p_ρ; is independent of the direction in which it crosses the field, one expects that – in the notation of Figure 13 –

$$\frac{\rho_i}{f_i} = \left(\frac{p_\rho}{p_z}\right)_i \quad (i = 1, 2).$$

Taking the axial distance of the incident electrons to be the same, that is, $\rho_1 = \rho_2$, it follows that

$$\frac{f_2}{f_1} = \frac{p_{z_2}}{p_{z_1}} = \left(\frac{V_2}{V_1}\right)^{\frac{1}{2}}.$$

Let us consider now the passage of an electron through an axially symmetric magnetic field. If the particle is incident along the symmetry axis, there is no force acting on it, and it stays on the axis. If it is incident parallel to the axis, the radial component of the magnetic field deflects it in a direction perpendicular to the plane defined by the symmetry axis and the point of incidence. The electron acquires an azimuthal velocity $\rho\dot{\theta}$.

An electron moving in a plane which is rotating with the electron about the symmetry axis, that is, with the Larmor frequency $\dot{\theta} = \omega_L = (e/2m)B(z)$, will be acted upon by the centrifugal force $m\rho\omega_L^2$ and the radial component of the Lorentz force, $e\rho\omega_L B(z)$. Thus, in this plane, the electron motion is determined by the equation $\ddot{\rho} = -\omega_L^2\rho$, and the electron oscillates about the symmetry axis. The difference between this motion and that in a uniform magnetic field is that now the oscillation is not simple harmonic, because ω_L is not a constant, but a function of z. All the same, the electron is confined to a region in the vicinity of the symmetry axis, and if it performs less than half a cycle of its oscillation in the magnetic field during its transit, it leaves the field deflected towards the symmetry axis.

For a quantitative description of the formation of an electron-optical image, the properties of the electron *trajectories* must be analysed in further detail. One of the most general methods of obtaining a differential equation for the trajectory of any moving particle is based on a variational principle known as the principle of least action.

In mechanics, the quantity *action* is defined by the integral

$$\int_{t_1}^{t_2} \boldsymbol{\pi}.\mathbf{v}\,dt = \int_{r_1}^{r_2} \boldsymbol{\pi}.d\mathbf{r},$$

where $\boldsymbol{\pi}$, $\mathbf{v}$ and $\mathbf{r}$ are the canonical momentum, the velocity and the position vector respectively. The principle of least action states that a particle follows a trajectory on which the action has an extreme value, that is, the variation of the integral vanishes,

BIRMINGHAM UNIVERSITY LIBRARY

$$\delta \int_{r_1}^{r_2} \boldsymbol{\pi}.d\mathbf{r} = 0, \qquad 2.23$$

if the end points of the path are fixed in space, but time as well as coordinates are varied in such a manner that the total energy is conserved on the varied paths.*

In section 2.1 this principle was quoted in a form which is applicable if charged particles are moving in a static electromagnetic field. In this case, from equation **1.10a**: $\pi = \mathbf{p}+e\mathbf{A}$. Denoting the unit vector tangential to the trajectory by $\mathbf{l}$ and the path element by dl, that is, $d\mathbf{r} = \mathbf{l}\, dl$, we find that

$$\pi . d\mathbf{r} = (\mathbf{p}+e\mathbf{A}).\mathbf{l}\, dl.$$

Since the total energy W is conserved in static fields, one can substitute

$$p = [2m(W-e\Phi]^{\frac{1}{2}},$$

leading to the equation

$$\delta \int_{(1)}^{(2)} \{[2m(W-e\Phi)]^{\frac{1}{2}}+e\mathbf{A}.\mathbf{l}\}\, dl = 0. \qquad \mathbf{2.23a}$$

Consider the integrand in equation **2.23**. Using a cylindrical polar coordinate system, as before, where the z-axis is the symmetry axis of the fields,

$$\pi . d\mathbf{r} = \pi_z\, dz+\pi_\rho\, d\rho+\pi_\theta\, d\theta.$$

Let us recall that the particle moves in a plane rotating about the z-axis, with an angular velocity $\dot{\theta}$ given by equation **1.20**, the motion of the particle in this plane being determined by the equivalent potential $U(\rho, z)$, which is defined by equation **1.21**. The velocity u_m of the particle in this plane is related to the equivalent potential U by $\frac{1}{2}mu_m^2 = -U$. Denoting the path element in the rotating plane by ds, we have $ds = (1+\rho'^2)^{\frac{1}{2}}\, dz$, and hence

$$\pi_z\, dz+\pi_\rho\, d\rho = mu_m\, ds = (-2mU)^{\frac{1}{2}}(1+\rho'^2)^{\frac{1}{2}}\, dz.$$

It was seen in section 1.2 that π_θ = constant, and therefore equation **2.23** reduces to

$$\delta \int_{z_1}^{z_2} F(\rho, \rho', z)\, dz = 0, \qquad \mathbf{2.24}$$

where $\quad F(\rho, \rho', z) = (-2mU)^{\frac{1}{2}}(1+\rho'^2)^{\frac{1}{2}}.$

*This amounts to saying that the varied paths connect the fixed terminal points in space but the particle moves between these points at a rate, and over a total transit time, which is different from that needed to cover the unvaried path – the condition of energy conservation is satisfied in each case. Note that in its physical content this principle is distinctly different from a more often quoted variational principle – Hamilton's principle:

$$\delta \int_{t_1}^{t_2} L\, dt = 0.$$

Here L is the Lagrangian and both the end points of the trajectory and the time stay unvaried. In this case corresponding points of the unvaried, and of the varied, paths are reached by the moving point at the same instant, but energy is not conserved on the varied paths.

It is well known that equation **2.24** yields, for F, the Euler–Lagrange differential equation:

$$\frac{d}{dz}\frac{\partial F}{\partial \rho'} - \frac{\partial F}{\partial \rho} = 0$$

and hence $$\frac{d^2\rho}{dz^2} \equiv \rho'' = \frac{1+\rho'^2}{2U}\left[(1+\rho'^2)\frac{\partial U}{\partial \rho} - \rho'\frac{\partial U}{\partial z}\right]. \qquad \textbf{2.25}$$

Equation **2.25** determines the electron trajectory in the rotating meridian plane. The time rate of rotation of the meridian plane is determined by equation **1.20**, and hence the rate of rotation in terms of the displacement of the electron along the z-axis can be readily obtained. Since $\theta' = \dot{\theta}/\dot{z}$ and $\frac{1}{2}mu_m^2 = \frac{1}{2}m\dot{z}^2(1+\rho'^2) = -U$, from equations **1.20** and **1.21**

$$\theta' = \frac{1}{\rho}(1+\rho'^2)^{\frac{1}{2}}\left(\frac{U-e\Phi}{-U}\right)^{\frac{1}{2}}. \qquad \textbf{2.26}$$

Equations **2.25** and **2.26** are rigorously valid. In order to investigate *paraxial* imaging, we shall confine our attention to trajectories whose inclinations to the symmetry axis are always small, $\rho' \ll 1$, and which run close to the axis. These conditions allow the use of a first-order approximation. From equations **1.20** and **1.21**

$$U = e\Phi + \frac{1}{2m}\left(\frac{\pi_\theta}{\rho} - eA_\theta\right)^2.$$

In the power series in equation **2.22b** the first-order term in ρ gives

$$A_\theta = \tfrac{1}{2}B_0\,\rho.$$

From **2.22a** we retain $\Phi = V$ and, when needed, $\Phi' = -\frac{1}{2}V''$. The flux density and the electric potential on the symmetry axis are $B_0 = B_0(z)$ and $V = V(z)$, respectively. The partial derivatives $\partial U/\partial \rho$ and $\partial U/\partial \rho'$ in equations **2.25** and **2.26** are calculated with these assumptions. Furthermore, it is assumed that the second term in U, representing the rotational energy, is small compared with the energy gained from the accelerating field, and therefore $U \simeq eV$. Finally to emphasize that π_θ is a constant of motion, the notation $\pi_\theta = C$ is introduced. The equations thus obtained are called the *paraxial ray equations.*

$$\rho'' = -\frac{1}{2}\frac{V'}{V}\rho' - \left\{\frac{1}{4}\frac{V''}{V} + \frac{1}{8}\frac{eB_0^2}{mV} + \frac{1}{2emV}\left(\frac{C}{\rho^2}\right)^2\right\}\rho \qquad \textbf{2.27}$$

and $$\theta' = -\frac{1}{2}\left(\frac{e}{2m}\right)^{\frac{1}{2}} V^{-\frac{1}{2}}B_0 + (2emV)^{-\frac{1}{2}}\frac{C}{\rho^2}. \qquad \textbf{2.28}$$

It is easy to show that equation **2.27** can be written in the following form

$$\frac{d}{dz}(V^{\frac{1}{2}}\rho') = -\left\{P(z) - \frac{V^{-\frac{1}{2}}}{2em}\left(\frac{C}{\rho^2}\right)^2\right\}\rho, \qquad \textbf{2.29}$$

where $$P(z) = \tfrac{1}{4}V^{-\frac{1}{2}}\left(V'' + \frac{e}{2m}B_0^2\right). \qquad \textbf{2.30}$$

The trajectories in a uniform magnetic field were discussed previously. It was assumed that the canonical momentum component $\pi_\theta \equiv C = 0$, and that in the absence of an electric field V was constant. In this case equation **2.27** reduced to equation **2.26**, with $k = (eB_0^2/8mV)^{\frac{1}{2}}$, where B_0 is constant. But if the magnetic field is not uniform, k is a function of z, and the equation of the trajectory resembles that of a non-linear oscillator.

Still assuming that $C = 0$, we see from equation **2.29** that the presence of an electric field, $V \neq$ constant, leads to additional complications whether or not a magnetic field is present. Not only is the term $P(z)$ analogous to a variable restoring force, but the quantity playing the part of the mass, the coefficient of ρ', is also z-dependent. The term which appears if $C \neq 0$ plays the part of a repulsive potential proportional to the inverse square of the electron's distance from the symmetry axis. Therefore electrons which enter the field with an initial angular momentum cannot cross the symmetry axis.

We shall now consider quantitatively the imaging properties of axially symmetric fields if the condition $C = 0$ is satisfied. This means that the results will be applicable only if the electrons in question leave their source with an initial velocity which is co-planar with the z-axis. Under this restriction, the electron trajectories in the meridian plane are described by the equation

$$\frac{d}{dz}(V^{\frac{1}{2}}\rho') = -P(z)\rho. \qquad \textbf{2.31}$$

Equation **2.31** is a linear second-order differential equation, so its solution $\rho(z)$ and its derivative $\rho'(z)$ must be linearly related to the initial values of these quantities:

$$\begin{bmatrix}\rho(z)\\ \rho'(z)\end{bmatrix} = \mathrm{M}\begin{bmatrix}\rho(0)\\ \rho'(0)\end{bmatrix},$$

where M is a quadratic matrix, the elements of which, m_{ij}, are functions of z. In general, however, the condition det M = 1 is not satisfied and therefore this matrix is not suitable for the introduction of geometrical optical concepts. But given the above relation, there must exist a quadratic matrix N such that

$$\begin{bmatrix}\rho(z)\\ V^{\frac{1}{2}}(z)\,\rho'(z)\end{bmatrix} = \mathrm{N}\begin{bmatrix}\rho(0)\\ V^{\frac{1}{2}}(0)\,\rho'(0)\end{bmatrix} \qquad \textbf{2.32}$$

provided that $V(0) \neq 0$, and it can be shown (see Appendix, section A.2.1) that the matrix N satisfies the relation

det N = 1.

Consequently, the elements n_{ij} of the matrix N can be used to describe the properties of rotationally symmetric fields, in geometrical optic terms, using the same arguments as those developed in the preceding section. The expressions which define the cardinal elements in terms of the matrix element n_{ij} will be slightly different from those found earlier. In a field-free region of length s the electron trajectories are straight lines, that is,

$$\rho(s) = \rho(0)+s\,\rho'(0),$$
$$\rho'(s) = \rho'(0),$$

and since V = constant = V_0, one obtains

$$\mathrm{N}_0(s) = \begin{bmatrix} 1 & V_0^{-\frac{1}{2}}s \\ 0 & 1 \end{bmatrix}.$$

Let us consider now an electron source at $z = -z_1$ in a field-free region of constant potential V_1, and follow the electron through a rotationally symmetrical field of length Z into a field-free region of constant potential V_2. The matrix relating the positions of the electrons and the slope of the trajectories is

$$\mathrm{N}_0(z_2)\,\mathrm{N}(Z)\,\mathrm{N}_0(z_1)$$
$$= \begin{bmatrix} n_{11}+V_2^{-\frac{1}{2}}z_2 n_{11} & -V_1^{-\frac{1}{2}}z_1 n_{11}+n_{12}- \\ & -V_2^{-\frac{1}{2}}V_1^{-\frac{1}{2}}z_2 z_1 n_{21}+V_2^{-\frac{1}{2}}z_2 n_{22} \\ n_{21} & -V_1^{-\frac{1}{2}}z_1 n_{21}+n_{22} \end{bmatrix}. \qquad 2.33$$

Denoting the elements of the matrix above by N_{ij}, the condition for stigmatic imaging is $N_{12} = 0$. Hence the relation between conjugate planes becomes

$$\frac{V_1^{\frac{1}{2}}n_{22}}{z_1}-\frac{V_2^{\frac{1}{2}}n_{11}}{z_2} = n_{21}-\frac{V_1^{\frac{1}{2}}V_2^{\frac{1}{2}}}{z_1 z_2}n_{12}. \qquad 2.34$$

The positions of the foci and the principal planes are given by

$$z_{f_1} = V_1^{\frac{1}{2}}\frac{n_{22}}{n_{21}} \qquad z_{f_2} = -V_2^{\frac{1}{2}}\frac{n_{11}}{n_{21}} \qquad 2.35$$

$$z_{p_1} = -V_1^{\frac{1}{2}}\frac{1-n_{22}}{n_{21}} \qquad z_{p_2} = V_2^{\frac{1}{2}}\frac{1-n_{11}}{n_{21}} \qquad 2.36$$

and hence the focal distances are

$$f_1 = \frac{V_1^{\frac{1}{2}}}{n_{21}} \qquad f_2 = -\frac{V_2^{\frac{1}{2}}}{n_{21}}. \qquad 2.37$$

It should be noted that

$$\frac{f_2}{f_1} = -\left(\frac{V_2}{V_1}\right)^{\frac{1}{2}},$$

as anticipated (the sign is a matter of convention). The relation between source distance, image distance, and focal lengths (each measured relative to the appropriate principal plane) is

$$\frac{V_1^{\frac{1}{2}}}{s_1}-\frac{V_2^{\frac{1}{2}}}{s_2}=\left(-\frac{V_1^{\frac{1}{2}}V_2^{\frac{1}{2}}}{f_1 f_2}\right)^{\frac{1}{2}}. \tag{2.38}$$

This relation is analogous to the lens formula, which is applicable if the media in the object space and the image space have different refractive indices; $V_i^{\frac{1}{2}}$ plays the role of the refractive index.

The linear magnification of the image is

$$\frac{\rho(z_2)}{\rho(-z_1)}=N_{11}$$

and if the source is on the axis, $\rho(-z_1)=0$, the image is also on the axis, therefore $\rho(z_2)=0$. For a source point on the axis the angular magnification is

$$\frac{\rho'(z_2)}{\rho'(-z_1)}=N_{22}\left[\frac{V(-z_1)}{V(z_2)}\right]^{\frac{1}{2}}.$$

Since $N_{12}=0$ and det N $=1$, one obtains $N_{11}N_{22}=1$ and hence the product of the linear and angular magnification is

$$\frac{\rho(z_2)}{\rho(-z_1)}\frac{\rho'(z_2)}{\rho'(-z_1)}=\left[\frac{V(-z_1)}{V(z_2)}\right]^{\frac{1}{2}}, \tag{2.39}$$

that is, the product is determined by the ratio of the potentials at the source and image planes. This relation is analogous to the Helmholtz–Lagrange relation of geometrical optics, where the refractive indices take the place of the square roots of the potentials.

Equation **2.39** gives the special form of a more general and very important invariance law derived in the Appendix (A.3). If $\rho_1(z)$ and $\rho_2(z)$ describe two trajectories in the same field $V(z)$, then in general

$$V^{\frac{1}{2}}(z)[\rho_1'(z)\,\rho_2(z)-\rho_1(z)\,\rho_2'(z)]=\text{constant} \tag{2.40}$$

at any position z. This constant quantity is known as the *Lagrange invariant* of geometrical electron optics.

If $z=-z_1$ and $z=z_2$ are the positions of conjugate planes, and the trajectories ρ_1 and ρ_2 connect a pair of conjugate points, then equation **2.40** reduces to an equivalent of equation **2.39**. In this case $\rho_1(-z_1)=\rho_2(-z_1)\equiv\rho(-z_1)$, $\rho_1(z_2)=\rho_2(z_2)\equiv\rho(z_2)$ and therefore

$$\frac{\rho(z_2)}{\rho(-z_1)}\frac{\rho_1'(z_2)-\rho_2'(z_2)}{\rho_1'(-z_1)-\rho_2'(-z_1)}=\left[\frac{V(-z_1)}{V(z_2)}\right]^{\frac{1}{2}},$$

where the second fraction on the left-hand side gives the ratio of the angles maintained by the two trajectories at the source and at the image point respectively.

The Lagrange invariance can be given a simple physical meaning. Consider a Cartesian coordinate system, and plot along its axes the quantities ρ and $V^{\frac{1}{2}}\rho'$.

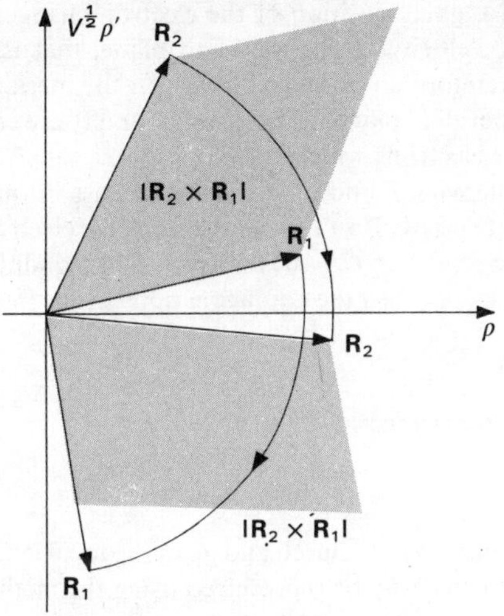

Figure 14

As an electron moves along its trajectory, ρ and $V^{\frac{1}{2}}\rho'$ vary as a function of z. Thus each point of the trajectory can be represented by a point in the $(\rho, V^{\frac{1}{2}}\rho')$ plane; so each point defines a vector **R** with components $(\rho, V^{\frac{1}{2}}\rho')$ (see Figure 14). As the particle moves along its trajectory, the endpoint of the vector **R** traces out a curve in the $(\rho, V^{\frac{1}{2}}\rho')$ plane, which plays the part of a phase plane.

If two trajectories are considered in the same field, for a given value of z the representative points in the phase plane will define two vectors $\mathbf{R}_1$ and $\mathbf{R}_2$, with components $\rho_1(z)$, $V^{\frac{1}{2}}(z)\,\rho_1'(z)$ and $\rho_2(z)$, $V^{\frac{1}{2}}(z)\,\rho_2'(z)$ respectively. In this representation the Lagrange invariant takes the form

$$V^{\frac{1}{2}}(z)[\rho_1'(z)\,\rho_2(z)-\rho_1(z)\,\rho_2'(z)] = |\mathbf{R}_2\times\mathbf{R}_1| = \text{constant.} \qquad \textbf{2.40a}$$

Since $|\mathbf{R}_2\times\mathbf{R}_1|$ gives the area of a vector parallelogram, equation **2.40a** states that the representative points corresponding to two different trajectories move in the phase plane in such a manner that the area of the vector parallelogram, $|\mathbf{R}_2\times\mathbf{R}_1|$, is conserved.

To establish the imaging properties of rotationally symmetric fields, we have been considering the properties of paraxial trajectories of electrons for which the azimuthal component of the canonical momentum π_θ vanished. It was mentioned above that this is the case if the electron source is a point on the z-axis, or if the electron source is outside the field and the initial electron velocity is coplanar with the z-axis. Thus the assumption $\pi_\theta \equiv C = 0$ imposes a severe restriction on the initial conditions of the electron emission, and it is important to realize that the imaging laws derived previously remain valid even if this restriction is removed.

A quantitative analysis of this problem is given in the Appendix (A.4). Qualitatively, equation **2.28** shows that one consequence of lifting the restriction

$C = 0$ is that the rate at which the angular position of the electron changes is different from the assumed angular velocity of the meridian plane, that is, if $C \neq 0$ then $\theta' \neq -\frac{1}{2}(e/2mV)^{\frac{1}{2}}B$. Therefore an observer at rest on the meridian plane will see the electron leave the meridian plane. If, however, P and P′ are conjugate points in a meridian plane for electrons which leave the source satisfying the condition $C = 0$, then electrons leaving P and $C \neq 0$ will also pass through point P′. In other words, the conjugate points P and P′ are the same for electrons (of a given energy) whether or not the condition $C = 0$ is satisfied. This condition determines only whether the trajectory between the conjugate points runs in the meridian plane or not.

2.4 Examples of rotationally symmetric electron lenses

2.4.1 *Weak magnetic lens*

The first electron-optical formula was derived by Busch and gives the focal length of a '*weak*' *magnetic lens*. This result can easily be reproduced using the method developed above.

In the absence of an electric field $V' = V'' = 0$, $V = V_0$ = constant, and without loss of generality we may assume that $C = 0$. In this case equation **2.31** reduces to

$$\frac{d}{dz}(V_0^{\frac{1}{2}}\rho') = -\frac{eB_0^2}{8mV_0^{\frac{1}{2}}}\rho,$$

where $B_0 = B_0(z)$ the field is along the axis of rotational symmetry and it is significantly different from zero only over the range $0 \leqslant z \leqslant Z$.

The lens being 'weak' means that the electron's distance from the axis does not vary significantly while it is in the field of the lens, that is,

$$\rho(Z) \simeq \rho(0).$$

Thus the equation of the trajectory can be integrated to give

$$V_0^{\frac{1}{2}}\rho'(Z) = -\rho(0)\frac{e}{8mV_0^{\frac{1}{2}}}\int_0^z B_0^2\,dz + V_0^{\frac{1}{2}}\rho'(0)$$

and hence $$\begin{bmatrix} \rho(Z) \\ V_0^{\frac{1}{2}}\rho'(Z) \end{bmatrix} = \begin{bmatrix} 1 & 0 \\ -\frac{e}{8mV_0^{\frac{1}{2}}}\int B_0^2\,dz & 1 \end{bmatrix} \begin{bmatrix} \rho(0) \\ V_0^{\frac{1}{2}}\rho'(0) \end{bmatrix}.$$

Thus from equation **2.36**, $z_{p1} = z_{p2} = 0$, and from equation **2.37**

$$\frac{1}{|f|} = -\frac{1}{f_1} = \frac{1}{f_2} = \frac{e}{8mV_0}\int_{-\infty}^{+\infty} B_0^2\,dz. \qquad \textbf{2.41}$$

Equation **2.41** shows that the lens is symmetrical, and that it is always a focusing lens giving a real focus ($f_2 = -f_1 > 0$), independently of the direction of the magnetic field.

It is important to have some indication of just how good the above approximation is, or in other words, what conditions make the lens 'weak'. If the magnetic field is uniform, that is, B_0 = constant in the region $0 \leqslant z \leqslant Z$ and $B_0 = 0$ outside this region, the matrix element determining the focal length is, in the above approximation,

$$n_{21} = -\frac{e}{8mV_0} B_0^2 Z.$$

For a uniform field of length Z, however, we know the exact expression of this matrix element. Substituting

$$k = \frac{eB_0}{2p} = \left(\frac{e}{8mV_0}\right)^{\frac{1}{2}} B_0$$

we have, from equation **2.12**,

$$m_{21} = -\left(\frac{e}{8mV_0}\right)^{\frac{1}{2}} B_0 \sin\left[\left(\frac{e}{8mV_0}\right)^{\frac{1}{2}} BZ\right]$$

and therefore $\quad n_{21} \simeq m_{21},$

$$\text{if} \left(\frac{e}{8mV_0}\right)^{\frac{1}{2}} B_0 Z \ll \frac{\pi}{2} \qquad \text{or} \qquad \frac{eB_0 Z}{p} \ll \pi.$$

Thus, whether a magnetic lens is 'weak' or not depends on the field intensity and its length along the axis, as well as on the momentum of the electrons.

2.4.2 *A single aperture as an electric lens*

Consider a small circular aperture in a thin conducting sheet of thickness ε located at $z = 0$. On the two sides of the aperture let there be uniform fields of different intensities, $E_1 = V_1'$ if $z < -\frac{1}{2}\varepsilon$ and $E_2 = V_2'$ if $z > \frac{1}{2}\varepsilon$. In the immediate vicinity of the aperture the potential on the axis can be considered constant (see Figure 15):

$$V = V_0 \qquad \text{if } -\tfrac{1}{2}\varepsilon < z < \tfrac{1}{2}\varepsilon.$$

In this region the distance of the electron from the axis does not change significantly:

$$\rho(z) \simeq \rho(-\tfrac{1}{2}\varepsilon) \qquad \text{if } -\tfrac{1}{2}\varepsilon < z < \tfrac{1}{2}\varepsilon.$$

In the absence of magnetic fields $B = 0$ and, assuming $C = 0$, equation **2.31** can be integrated to give

$$V_0^{\frac{1}{2}}[\rho'(\tfrac{1}{2}\varepsilon)-\rho'(-\tfrac{1}{2}\varepsilon)] = -\tfrac{1}{4}V_0^{-\frac{1}{2}}(V_2'-V_1')\rho(-\tfrac{1}{2}\varepsilon)$$

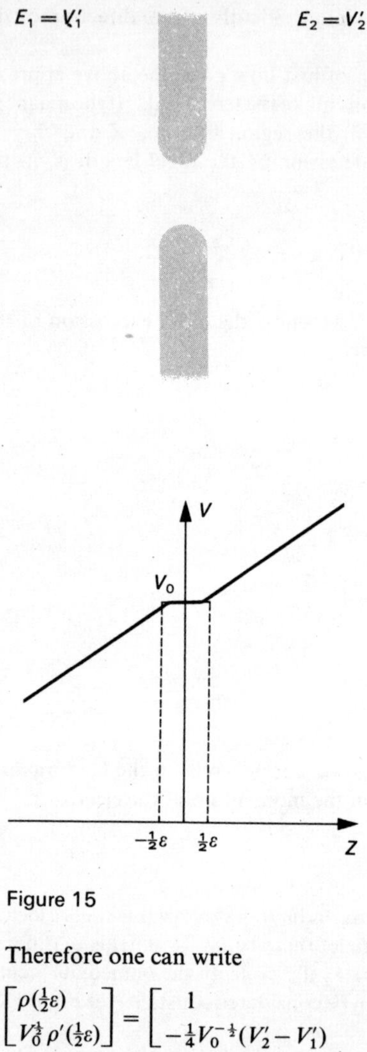

Figure 15

Therefore one can write

$$\begin{bmatrix} \rho(\frac{1}{2}\varepsilon) \\ V_0^{\frac{1}{2}}\rho'(\frac{1}{2}\varepsilon) \end{bmatrix} = \begin{bmatrix} 1 & 0 \\ -\frac{1}{4}V_0^{-\frac{1}{2}}(V_2' - V_1') & 1 \end{bmatrix} \begin{bmatrix} \rho(-\frac{1}{2}\varepsilon) \\ V_0^{\frac{1}{2}}\rho'(-\frac{1}{2}\varepsilon) \end{bmatrix}. \qquad \textbf{2.42}$$

Equation **2.36** shows that the principal planes coincide with the mid-plane of the aperture, and the focal lengths are given by

$$-f_1^{-1} = f_2^{-1} = \tfrac{1}{4}V_0^{-1}(V_2' - V_1'). \qquad \textbf{2.43}$$

This lens, therefore, can focus or defocus depending on the relative field intensities on the two sides of the aperture, that is, whether $V_2' - V_1' > 0$, or *vice versa*.

It must be emphasized that the results obtained describe the optical properties of the field in the immediate neighbourhood of the plane of the aperture. Since there is hardly any practical situation when a beam of particles is subject to such a field in isolation, the significance of the foregoing example arises from the fact that it presents an approximate description of elements of more complicated lenses.

2.4.3 *Bi-potential lenses*

Electric lenses of rotational symmetry usually consist of a series of circular apertures or a series of cylinders centred on a common axis and held at different potentials. As a simple example let us consider the combination of two small apertures at a distance d apart and at potentials V_1 and V_2 respectively. For simplicity's sake let us assume that outside the lens there is no electric field, and inside the lens there is a uniform field due to the potential difference between the parallel planes of the apertures (see Figure 16).

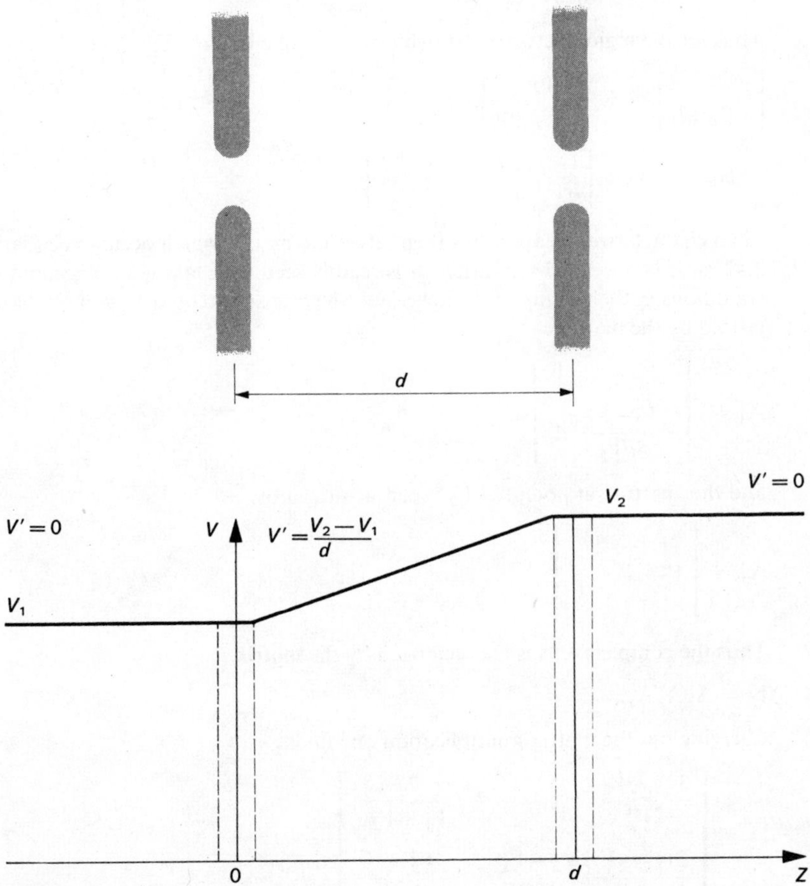

Figure 16

The optical properties of such an arrangement can be described by the product of three matrices: one for each aperture, similar to that given by equation **2.42**, and one to describe the region between the apertures. In the absence of a magnetic field, with the assumption $C = 0$ and, since the electric field between the apertures ($0 < z < d$) is uniform, $V'' = 0$, the paraxial trajectory equation **2.31** is readily integrated

$$V^{\frac{1}{2}}(z)\,\rho'(z) = \text{constant} = V_1^{\frac{1}{2}}\,\rho'(0).$$

In the region $0 < z < d$ we have

$$V(z) = V_1 + (V_2 - V_1)\frac{z}{d}$$

and therefore a second integration gives

$$\rho(d) = \rho(0) + \rho'(0)2dV_1^{\frac{1}{2}}(V_2^{\frac{1}{2}} + V_1^{\frac{1}{2}})^{-1}.$$

Thus for the region between the two apertures one finds

$$\begin{bmatrix} \rho(d) \\ V_2^{\frac{1}{2}}\,\rho'(d) \end{bmatrix} = \mathrm{A}_2 \begin{bmatrix} \rho(0) \\ V_1^{\frac{1}{2}}\,\rho'(0) \end{bmatrix},$$

where $$\mathrm{A}_2 = \begin{bmatrix} 1 & 2d(V_2^{\frac{1}{2}} + V_1^{\frac{1}{2}})^{-1} \\ 0 & 1 \end{bmatrix}.$$

To characterize the apertures themselves the matrix which occurs in equation **2.42** must be modified somewhat. It is readily seen that, taking into account the conditions in the vicinity of the apertures, the one at potential V_1 will be characterized by the matrix

$$\mathrm{A}_1 = \begin{bmatrix} 1 & 0 \\ -\dfrac{V_2 - V_1}{4dV_1^{\frac{1}{2}}} & 1 \end{bmatrix}$$

and the aperture at potential V_2 is characterized by

$$\mathrm{A}_3 = \begin{bmatrix} 1 & 0 \\ \dfrac{V_2 - V_1}{4dV_2^{\frac{1}{2}}} & 1 \end{bmatrix}.$$

Thus the complete lens is characterized by the matrix

$$\mathrm{N} = \mathrm{A}_3\mathrm{A}_2\mathrm{A}_1.$$

Carrying out the matrix multiplication one finds:

$$\mathrm{N} = \begin{bmatrix} \dfrac{3V_1^{\frac{1}{2}} - V_2^{\frac{1}{2}}}{2V_1^{\frac{1}{2}}} & \dfrac{2d}{V_1^{\frac{1}{2}} + V_2^{\frac{1}{2}}} \\ \dfrac{3(V_2 - V_1)}{8d}\,\dfrac{V_2^{\frac{1}{2}} - V_1^{\frac{1}{2}}}{V_1^{\frac{1}{2}}V_2^{\frac{1}{2}}} & \dfrac{3V_2^{\frac{1}{2}} - V_1^{\frac{1}{2}}}{2V_2^{\frac{1}{2}}} \end{bmatrix}. \qquad \mathbf{2.44}$$

Hence, by equation **2.37**, the focal lengths are

$$f_2^{-1} = \frac{3}{8d}\left(1-\frac{V_1}{V_2}\right)\left[\left(\frac{V_2}{V_1}\right)^{\frac{1}{2}}-1\right]$$

and $$f_1^{-1} = -\frac{3}{8d}\left(\frac{V_2}{V_1}-1\right)\left[1-\left(\frac{V_1}{V_2}\right)^{\frac{1}{2}}\right].$$

Note that the bi-potential lens described is always a focusing lens, since, if $V_1 < V_2$, or *vice versa*, one always finds $f_1 > 0$ and $f_1 < 0$.

For the mutual distance of the principal planes, $D = d+z_{p2}-z_{p1}$, one finds that, if the appropriate matrix elements given by equation **2.36** are taken, $D = -\frac{1}{3}d$. The negative sign indicates that the positions of the principal planes are 'reversed'. It can also be seen that the principal planes are *not* symmetrically located in the field, but both are shifted to the low potential side of the lens. The bi-potential lens is analogous to the 'immersion objective' in optics. The matrix element N_{11} gives the linear magnification and $(V_1/V_2)^{\frac{1}{2}}N_{22}$ gives the angular magnification, as seen previously.

2.4.4 *Single aperture in front of a plane cathode*

A simple example which shows the limitations of the above considerations merits special attention. Consider a *plane cathode* at potential $V_1 = 0$ and $z = 0$ and a small *circular aperture* at a potential $V_2 > 0$ and $z = d$ (see Figure 17a). Suppose that the region beyond the aperture is field free. Where is the image of the cathode formed?

The assumption $V_1 = 0$ precludes the use of the method developed, but one can argue as follows. In the region of uniform field in front of the cathode the electrons emitted with *zero* initial velocity form a parallel beam, and only the effect of the aperture need be taken into account. To apply equation **2.43** one must substitute $V_0 = V_2$, $V_2' = 0$ and $V_1' = V_2/d$. Hence $1/f_2 = -1/4d$. Since the incident beam is parallel to the z-axis, beyond the aperture the electrons move as if they had originated from a point on the axis at a distance $4d$ *behind* the aperture (see Figure 17b). This argument is correct but it contains an inadmissible assumption.

Electrons emitted by a cathode always have a small initial energy, say eU, and the initial direction of emission is distributed at random. Therefore, if the electrons are accelerated by a uniform field to an energy eV, those which were emitted by one point fill a cone about the direction of the normal to the cathode; the semi-angle of the cone is, for small initial energies, $(U/V)^{\frac{1}{2}}$. However small this angle may be, one must realize that the electron trajectories in front of the cathode are parabolae and not straight lines, and the slopes of the trajectories near the point of emission are very large. Thus in this region the paraxial approximation is invalid. From the properties of the parabola, it follows that the electrons reach the plane of the aperture as if they had been emitted by a 'virtual source', located at a distance $2d$ behind the aperture in a field-free region, and the aperture lens

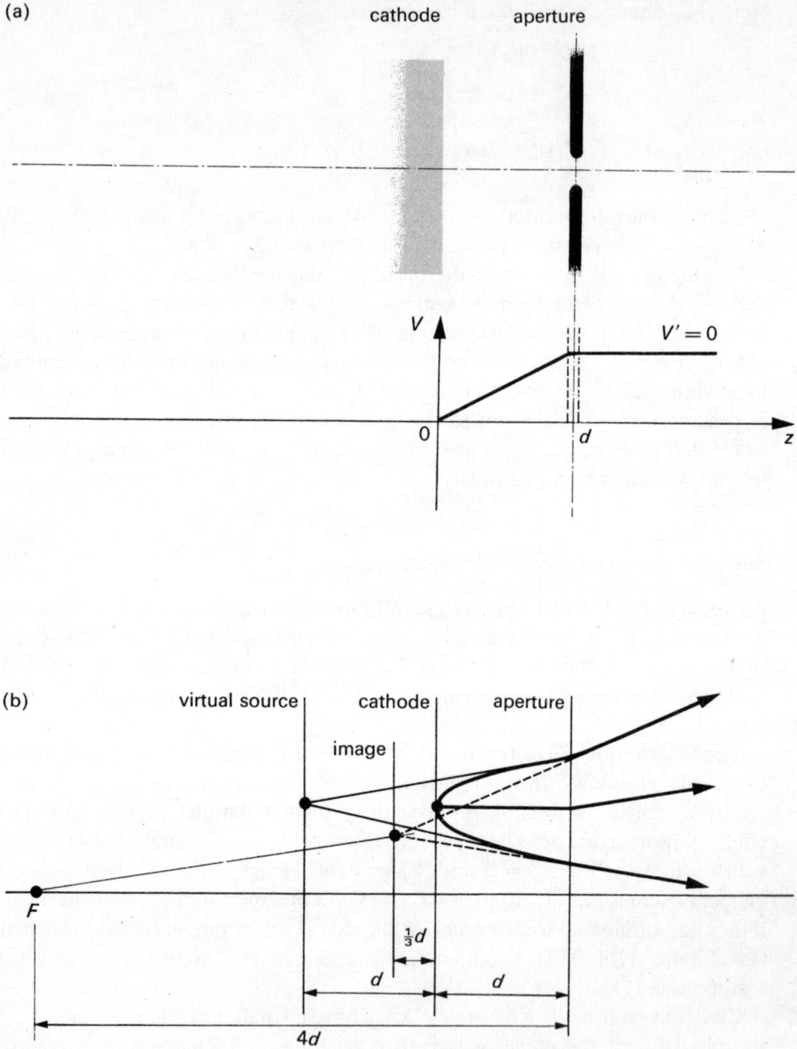

Figure 17

produces an image of this virtual source (see Figure 17b). Since the principal plane coincides with the plane of the aperture and its focal length is $-4d$, the image of the cathode and the linear magnification of the system can now be determined by elementary considerations.

2.5 Quadrupole lenses

Let us consider the simplest possible plane electrostatic field, which has two planes of mirror symmetry perpendicular to one another. These requirements mean that if the potential is described in terms of Cartesian coordinates, X_1, X_2, Z, it will depend on only two of these, say $\Phi(X_1, X_2)$, and that only even powers of X_1 and X_2 occur. Thus the simplest field distribution of this kind arises from a potential of the form

$$\Phi = a_1 X_1^2 + a_2 X_2^2,$$

where a_1 and a_2 are constants. If this function describes a scalar potential *in vacuo* in the absence of charges, it must satisfy Laplace's equation $\nabla^2\Phi = 0$, and this implies the condition $a_1 = -a_2 \equiv \frac{1}{2}a^2$.

$$\text{Thus} \quad \Phi = \tfrac{1}{2}a^2(X_1^2 - X_2^2). \qquad \textbf{2.45}$$

The potential is obviously zero on the Z-axis and where $X_1 = X_2$. It is easy to see that the equipotential surfaces are equilateral hyperbolic cylinders perpendicular to the (X_1, X_2) plane. If a pair of conductors at a potential U has a profile $(X_1/\frac{1}{2}d)^2 - (X_2/\frac{1}{2}d)^2 = 1$, and another pair of conductors at the potential $-U$ has a profile $(X_1/\frac{1}{2}d)^2 - (X_2/\frac{1}{2}d)^2 = -1$, the potential distribution between these electrodes is of the type given by equation **2.45**, namely

$$\Phi = \frac{U}{(\frac{1}{2}d)^2}(X_1^2 - X_2^2).$$

From equation **2.45**, the components of the field intensity are

$$E_1 = -\frac{\partial\Phi}{\partial X_1} = -a^2 X_1 \qquad \text{and} \qquad E_2 = \frac{\partial\Phi}{\partial X_2} = a^2 X_2$$

and hence the magnitude of the field

$$E = (E_1^2 + E_2^2)^{\frac{1}{2}} = a^2(X_1^2 + X_2^2)^{\frac{1}{2}}$$

is everywhere proportional to the distance $r = (X_1^2 + X_2^2)^{\frac{1}{2}}$ measured from the Z-axis. The gradient of the field is a constant:

$$\frac{\partial E}{\partial r} = a^2.$$

Hence this type of field is often called a *constant-gradient* field; from the way it can be produced it is usually referred to as an electrostatic quadrupole field (Figure 18).

Certain features of the electron motion follow directly from the symmetry properties of the field. In a symmetry plane the field component normal to it vanishes, and therefore, if at any instant an electron is in a symmetry plane and its momentum has no perpendicular component, the electron will stay in the symmetry plane. Along the Z-axis, which is the intersection of the symmetry planes, the total field vanishes, and therefore an electron moving along the Z-axis at any instant will remain on it.

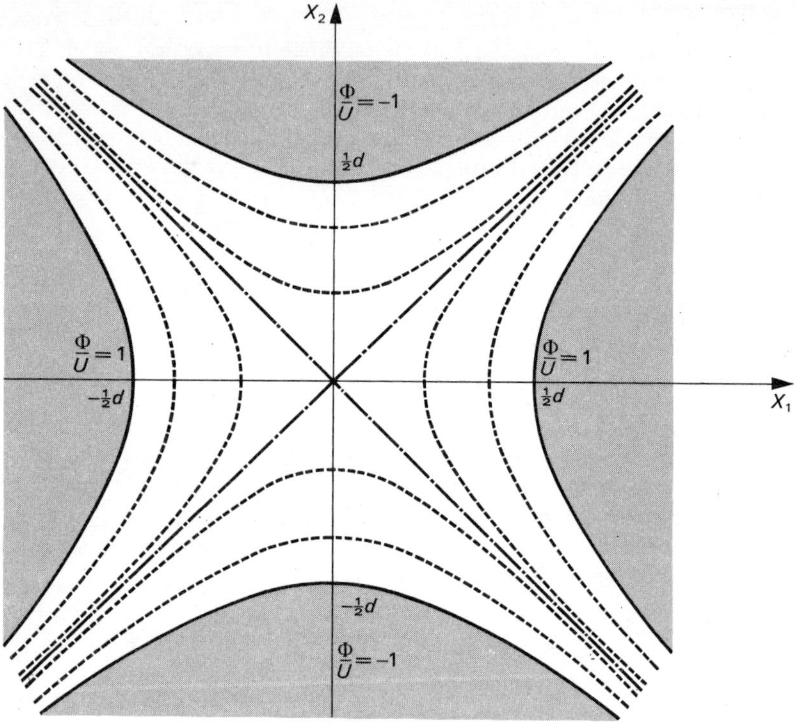

Figure 18

Some further properties of the electron motion becomes apparent from the equations of motion. From $m\ddot{\mathbf{r}} = -e\nabla\Phi$:

$$\ddot{X}_1 = -a^2\left(\frac{e}{m}\right)X_1,$$

$$\ddot{X}_2 = a^2\left(\frac{e}{m}\right)X_2$$

and hence

$$X_1 = X_{10}\sin(\omega t+\delta_1), \qquad X_2 = X_{20}\sinh(\omega t+\delta_2), \qquad \textbf{2.46}$$

where $\omega = \left(\frac{e}{m}\right)^{\frac{1}{2}}a,$

and X_{10}, X_{20}, δ_1 and δ_2 are constants which depend on the initial conditions.

Thus, in a direction parallel to the X_1-axis, the particle performs simple harmonic motion and therefore stays in the vicinity of the (X_2, Z) plane. In the direction parallel to the X_2-axis, however, its motion is not bounded at all, that is, the particle flies away from the (X_1, Z) plane. In other words the quadrupole lens is focusing in one plane and defocusing in the plane perpendicular to it. We shall see that – in optical terms – the quadrupole lens acts as an *astigmatic lens.*

A magnetic field of analogous properties can be derived from a vector potential

$$\mathbf{A} = \tfrac{1}{2}b^2(X_1^2 - X_2^2)\mathbf{e}_z \qquad (b = \text{constant}), \qquad \mathbf{2.47}$$

from which one obtains the components of the field $\mathbf{B} = \nabla \times \mathbf{A}$

$$B_1 = -b^2 X_2, \qquad B_2 = +b^2 X_1, \qquad B_2 = 0.$$

As in the case of the electric field discussed previously

$$B = (B_1^2 + B_2^2)^{\frac{1}{2}} = b^2 r \quad (r^2 = X_1^2 + X_2^2)$$

and $$\frac{\partial B}{\partial r} = b^2.$$

In a region where there are no currents, the magnetic field can also be derived from a static potential Φ_m, and the field distribution is obtained from $\Phi_m = b^2 X_1 X_2$. The pole faces of magnets can be considered magnetic equipotential surfaces if the magnetic materials involved are far from saturation. If the profiles of two pairs of pole faces are described by the equations $X_1 X_2 = \frac{1}{2}d$ and $X_1 X_2 = -\frac{1}{2}d$ respectively, and they are arranged with alternating polarity, the field obtained is of the type just described. If the flux density at the pole tips is $\pm B_0$, then

$$b^2 = \frac{2B_0}{d}.$$

The vector potential from which the magnetic field is derived has the same form as the scalar potential used to derive the electric field previously. The symmetry properties of the magnetic field, however, are different: in the symmetry planes the magnetic field is everywhere perpendicular to them. Consequently, if at any instant an electron is moving *in* the symmetry plane, the force will have no component perpendicular to it, and the electron stays in this plane, as in the case of the electric quadrupole field.

Further properties of the motion are also similar in the two cases. From the equation $m\ddot{\mathbf{r}} = e(\dot{\mathbf{r}} \times \mathbf{B})$, one finds

$$\ddot{X}_1 = -b^2\left(\frac{e}{m}\right)\dot{Z}X_1 \qquad \ddot{X}_2 = b^2\left(\frac{e}{m}\right)\dot{Z}X_2.$$

Assuming that $\dot{Z}$ = constant, the solutions are of the form of equations **2.46**, but here $\omega = b(e/m)^{\frac{1}{2}}\dot{Z}^{\frac{1}{2}}$. Thus, the main features of the electron motion in magnetic quadrupole fields are the same as in electric ones.

In order to develop a description of the properties of quadrupole fields in geometrical optic terms, one requires the equation of the trajectories. Just as in the case of rotationally symmetric fields, the treatment will be confined to trajectories adjacent to the Z-axis, and in this way we shall be led to the paraxial imaging properties of quadrupole lenses.

In the case of electric quadrupoles, there is no force acting parallel to the Z-axis and therefore $\dot{Z}$ = constant. For magnetic quadrupoles this condition applies only as a first-order approximation, valid when only *paraxial trajectories* are contemplated. Under these conditions $x_j'' = \ddot{x}_j/\dot{Z}^2$, where x_j, $\dot{x}_j$, x_j', etc. denote first-order small quantities, relevant in the description of paraxial trajectories. Thus the differential equation of the trajectories takes the form

$$x_j'' = -k_j^2 x_j \qquad (j = 1, 2), \tag{2.48}$$

where k_j^2 is a constant dependent on the particle's kinetic energy and on the distribution of the field. For electric quadrupoles

$$k_{1,2}^2 = \pm\frac{e}{2W}\frac{\partial E}{\partial r} \tag{2.49}$$

and for magnetic quadrupoles

$$k_{1,2}^2 = \pm\frac{e}{p}\frac{\partial B}{\partial r}. \tag{2.50}$$

The solution of equation **2.48** is familiar. If $k_j^2 > 0$, it is given in terms of circular functions, and if $k_j^2 < 0$ it is given in terms of hyperbolic functions. In other words in the first case the trajectories oscillate about the Z-axis within a finite region, and in the second case the trajectories run away from the Z-axis. The position of the electron x_j and the slope of its trajectory x_j' is in any case related linearly to the initial values of these quantities,

i.e. $$\begin{bmatrix} x_j \\ x_j' \end{bmatrix} = \mathrm{M}\begin{bmatrix} x_j(0) \\ x_j'(0) \end{bmatrix} \tag{2.51}$$

where $$\mathrm{M} = \begin{bmatrix} \cos k_j Z & k_j^{-1}\sin k_j Z \\ -k_j \sin k_j Z & \cos k_j Z \end{bmatrix} \equiv \mathrm{M}^+. \tag{2.52a}$$

This expression is directly applicable if $k_j^2 > 0$. If $k_j^2 < 0$, then $k_j = i|k_j|$, and therefore

$$\mathrm{M} = \begin{bmatrix} \cosh|k_j|Z & |k_j|^{-1}\sinh|k_j|Z \\ |k_j|\sinh|k_j|Z & \cosh|k_j|Z \end{bmatrix} \equiv \mathrm{M}^-. \tag{2.52b}$$

Note that the condition $\det \mathrm{M} = 1$ is satisfied.

From these matrices one can obtain the cardinal elements of the quadrupole lens following the arguments and results of section 2.2.3 (see equations **2.19** and **2.21**). If the quadrupole lens is of length Z_0 and this quantity is used as the unit of length, it is found, for example, that the focal length is given by

$$-\frac{Z_0}{f_1} = \frac{Z_0}{f_2} = k_j Z_0 \sin k_j Z_0 \qquad \text{for } k_j^2 > 0,$$

$$-\frac{Z_0}{f_1} = \frac{Z_0}{f_2} = -|k_j| Z_0 \sinh|k_j| Z_0 \qquad \text{for } k_j^2 < 0.$$

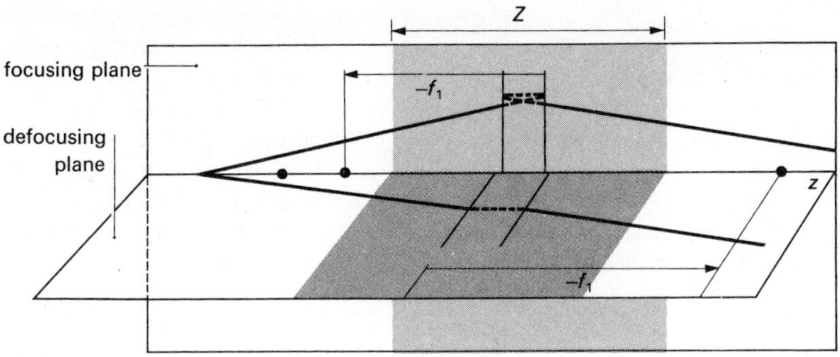

Figure 19

These formulae are in agreement with what is expected from the solution (equations **2.46**) of the equations of motion. In the plane where $k_j^2 < 0$, the lens is always defocusing because $Z/f_2 < 0$; in the plane where $k_j^2 > 0$, the lens is focusing provided $k_j Z_0 < \pi$ (Figure 19). We notice that if the lens is 'thin', that is, $|k_j| Z_0 \ll 1$, the focal length in the focusing and in the defocusing plane has the same magnitude $|Z_0/f| \simeq |k_j^2| Z_0^2$.

If two quadrupole lenses are set up with a field-free gap of length s between them, and the defocusing plane of one coincides with the focusing plane of the other, the matrix characterizing this combined system is given by the product $\mathrm{M}^+\mathrm{M}_0\mathrm{M}^-$ where M_0 is $\mathrm{M}(k_j = 0)$ (see equation **2.13**). The focal length of this system is given by the (1, 2) element of this matrix:

$$-\frac{1}{F_1} = \frac{1}{F_2} = -(\mathrm{M}^+\mathrm{M}_0\mathrm{M}^-)_{12} = -(m_{21}^+ m_{11}^- + m_{22}^+ m_{21}^- + s m_{21}^+ m_{21}^-).$$

In general the two quadrupole lenses can be different, that is, the value of the parameter $|k_j| Z_0$ is different for the two lenses. If the two quadrupoles are identical, the expression for the focal length becomes relatively simple. With the notation $|k_j| Z_0 = \alpha$

$$-\frac{1}{F_1} = \frac{1}{F_2} = |k_j|(\sin\alpha \cosh\alpha - \sinh\alpha \cos\alpha + |k_j| s \sin\alpha \sinh\alpha).$$

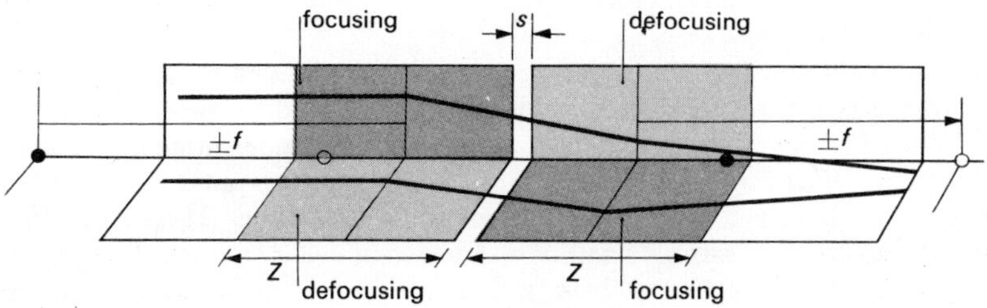

Figure 20

In general this system is astigmatic because the foci in the perpendicular symmetry planes are at different positions. They coincide, however, in these two planes if the component lenses are thin so that $\sin\alpha \cosh\alpha \simeq \sinh\alpha \cos\alpha$, that is, $\tan\alpha \simeq \tanh\alpha$. In this case $-1/F_1 = 1/F_2 > 0$, and we conclude that the system consisting of two identical thin quadrupole lenses with alternating focusing and defocusing planes always constitutes a focusing system (Figure 20). In section 2.8 we shall look at this problem in more general terms, and it will be seen that a sequence of any number of quadrupole lenses, arranged with alternate focusing and defocusing planes, can give an over-all focusing effect. Or putting it more appropriately, within such a system a beam can be confined to a finite cross-section over an arbitrarily long flight path. This property gives magnetic quadrupole lenses great importance in handling beams of high-energy charged particles.

2.6 Fields with one plane of mirror symmetry

Electric and magnetic fields characterized by having one plane of mirror symmetry are used in a great variety of charged-particle spectrometers. The best-known examples are the uniform magnetic field in which particles are collected after describing a semicircular orbit, various uniform magnetic 'sector'-fields and the electrostatic fields in a cylindrical condenser. The purpose of these devices is to select particles of a given energy or momentum out of a beam of particles with known charge-to-mass ratio.

Beta-ray spectrometers are typical devices of this class. A slightly different problem arises if the particles in a beam have approximately the same kinetic energy but different charge-to-mass ratios. In this case one may wish to select particles of a definite charge-to-mass ratio. This is achieved by the use of mass spectrometers.

The considerations underlying the design of the two kinds of spectrometer are somewhat different in detail but similar in general principles. In each case one wishes to separate particles characterized by a given value of one parameter, say a, and the performance of the device is judged by its 'resolution' which gives the spread δa of this parameter for the particles collected. In electron-optic terms, the task is to produce an image of the source, by means of particles for which the parameter a is in the range $a + \delta a$, and to separate this image from those produced by other particles. This separation occurs because of the chromatic aberration, which causes the images formed by particles of different parameter values a to have different positions.

It can be seen from the equations developed earlier that in fields of rotational symmetry, or in those with two mutually perpendicular planes of symmetry, the chromatic aberration arising from the spread of the momentum of the particles has two effects: (1) it shifts the image plane *along* the optical axis, and (2) it changes the linear magnification. These are second-order effects and therefore the resolution is relatively small, as was demonstrated for the solenoidal magnetic field spectrometer. In that case 'ring focusing' was introduced as a means of improvement. For fields with one plane of mirror symmetry, as will be seen, the optical axis is a curve whose position depends on the momentum of the particle. Consequently the chromatic aberration appears mainly as a first-order shift of the image in the image plane in a direction *perpendicular* to the optical axis.

Let us consider an electrostatic field described in terms of the Cartesian coordinates X_1, X_2, X_3, and let the plane of symmetry be the plane $X_3 = 0$. In this case symmetry conditions require that the potential is an even function of X_3:

$$\Phi(X_1, X_2, X_3) = \Phi(X_1, X_2, -X_3).$$

Consequently the field component parallel to X_3 is an odd function of X_3: $E_3(X_1, X_2, X_3) = -E_3(X_1, X_2, -X_3)$. Hence $E_3(X_1, X_2, 0) = 0$. Therefore, if at any instant the particle is moving in the $X_3 = 0$ plane it stays there all the time because there is no component of the force perpendicular to this plane.

A magnetic field, symmetrical with respect to the $X_3 = 0$ plane, can be derived from a vector potential $\mathbf{A}(X_1, X_2, X_3)$ whose components have the following properties:

$$A_3 = 0 \qquad A_j(X_1, X_2, X_3) = A_j(X_1, X_2, -X_3) \qquad (j = 1, 2).$$

Hence, for the magnetic field $\mathbf{B} = \nabla \times \mathbf{A}$:

$$B_j(X_1, X_2, X_3) = -B_j(X_1, X_2, -X_3) \qquad (j = 1, 2)$$

and $B_j(X_1, X_2, 0) = 0$. Thus the third component of the Lorentz force acting on a particle in the $X_3 = 0$ plane vanishes:

$$e(\mathbf{v} \times \mathbf{B})_3 = 0.$$

Therefore we conclude that a particle continues to move in the plane of mirror symmetry of an electromagnetic field, if it was initially moving in that plane.

If the field distribution satisfies certain further conditions besides being symmetrical with respect to the $X_3 = 0$ plane, it becomes possible that a particle launched with a particular initial momentum can move on a closed trajectory, usually called an *equilibrium orbit*. The problem of the general conditions of existence of closed orbits and the study of their properties is a difficult task of classical mechanics. We shall consider only the simple problem where the particles move in an electromagnetic field which is symmetrical with respect to the $X_3 = 0$ plane *and* is rotationally symmetric about the X_3 axis. Fields of this type and sectors of such fields play an important part in a great variety of beam devices, such as those mentioned above. The treatment of electron motion in such fields is relatively easy because the equilibrium orbits are circles in the symmetry plane centred on the X_3-axis.

The symmetry properties of the field suggest that the motion of the electron should be described in a cylindrical polar coordinate system (r, θ, z), where z measures the distance along the symmetry axis and $z = 0$ defines the symmetry plane. Thus the electric field is derived from a scalar potential, which is an even function of z,

$$\Phi(r, z) = \Phi(r, -z) \qquad \textbf{2.53a}$$

and independent of the azimuth angle θ. The magnetic field is derived from a vector potential $\mathbf{A}$, which has an azimuthal component only, with the same symmetry properties as the scalar potential, that is, it is independent of the azimuth angle:

$$A_\theta(r, z) = A_\theta(r, -z) \qquad A_r = A_z = 0. \tag{2.53b}$$

Note again that in the symmetry plane

$$\mathbf{E}(r, 0) = (\nabla\Phi)_{z=0} = \mathbf{e}_r E(r)$$

and $$\mathbf{B}(r, 0) = (\nabla \times \mathbf{A})_{z=0} = \mathbf{e}_z B(r),$$

that is, the electric field has only a radial component, and the magnetic field has only an axial component.

We can readily justify the claim that a particle with appropriately chosen initial momentum can move in the symmetry plane on a closed circular orbit. If the particle is moving in the plane, there is only a radial force acting upon it, and the equation of motion is

$$\ddot{r} - r\dot{\theta}^2 = \frac{e}{m}(E + r\dot{\theta}B).$$

This equation can be satisfied by a constant radius and constant angular velocity $r = R$ and $\dot{\theta} = \omega$, subject to the condition that

$$\omega^2 R = -\frac{e}{m}[E(R) + \omega R\, B(R)].$$

If an electron is launched at a radial position R with an initial momentum $\mathbf{p}$ tangential to the circle of radius R, that is, $\mathbf{p} = -p(\mathbf{e}_z \times \mathbf{e}_r)$, and we choose

$$p = m\omega R,$$

by substituting $\omega = p/mR$ into the above relation, we obtain

$$\frac{1}{R} = \frac{em}{p^2} E(R) + \frac{e}{p} B(R). \tag{2.54}$$

This equation gives the radius of the equilibrium orbit in terms of the field intensities and the momentum of the orbiting particle. Note that the particle is always moving perpendicularly to the fields and therefore there is no work done by them. Consequently equation **2.54** is relativistically correct. To express this fact, the total relativistic energy $W = mc^2 = (p^2c^2 + m_0^2 c)^{\frac{1}{2}}$ can be introduced and thus equation **2.54** takes the form

$$\frac{1}{R} = \frac{W}{p^2c^2}\, e\, E(R) + \frac{e}{p} B(R). \tag{2.54a}$$

If there is only an electric field or only a magnetic field present, equation **2.54** reduces to

$$\frac{1}{R_e} = \frac{em}{p^2} E(R) = \frac{W}{p^2c^2} e\, E(R) \qquad \textbf{2.55a}$$

or $$\frac{1}{R_m} = \frac{e}{p} B(R), \qquad \textbf{2.55b}$$

respectively, and in terms of these quantities,

$$R^{-1} = R_e^{-1} + R_m^{-1}. \qquad \textbf{2.54b}$$

If the electron is slightly displaced from its equilibrium orbit, where the electromagnetic forces and the centrifugal force acting on it are in balance, this balance is upset, and the electron will be subject to some net force. Considering only small displacements the forces can be determined by expanding the potential and the vector potential in a power series, in the neighbourhood of the equilibrium orbit, and taking first-order terms in the displacement into account. It is reasonable to expect that the resultant forces will be proportional to the displacement from the equilibrium orbit and, depending on the direction of the forces, the particle may oscillate about the equilibrium orbit. If the field distribution is sufficiently simple, more precise predictions can be made. Before undertaking more elaborate considerations, let us look at a few examples.

The simplest field which satisfies the symmetry conditions assumed is a uniform magnetic field parallel to the z-axis. A particle with any initial momentum perpendicular to this field will describe circular orbits. Thus there is a state of neutral equilibrium. The momentum of the particle defines the radius of the equilibrium orbit, and a small radial displacement shifts the particle to an adjacent equilibrium orbit of the same radius. Two particles emitted by a point source with momenta of the same magnitude, but of slightly different direction in the plane perpendicular to the magnetic field, describe circles of the same diameter which intersect at the source and – to a first-order approximation – at a point diametrically opposite. A displacement parallel to the z-axis is not counteracted by any restoring force.

The situation changes if the magnetic field has the required symmetry properties, but is not uniform. In this case, there is a uniquely defined symmetry plane and a uniquely defined symmetry axis, so there will be a uniquely defined equilibrium orbit, determined by equation **2.54**. For displacements in the plane of the equilibrium orbit, the rate of change of the field is the crucial quantity which determines whether or not the particle stays oscillating about the equilibrium orbit. Quantitative prediction is complicated by the fact that in a magnetic field angular momentum is not conserved, and therefore this question will be left open for the time being. However, we can show by qualitative arguments that, if the magnetic field is radially decreasing, particles displaced perpendicularly to the plane of the equilibrium orbit will be subject to a restoring force directed

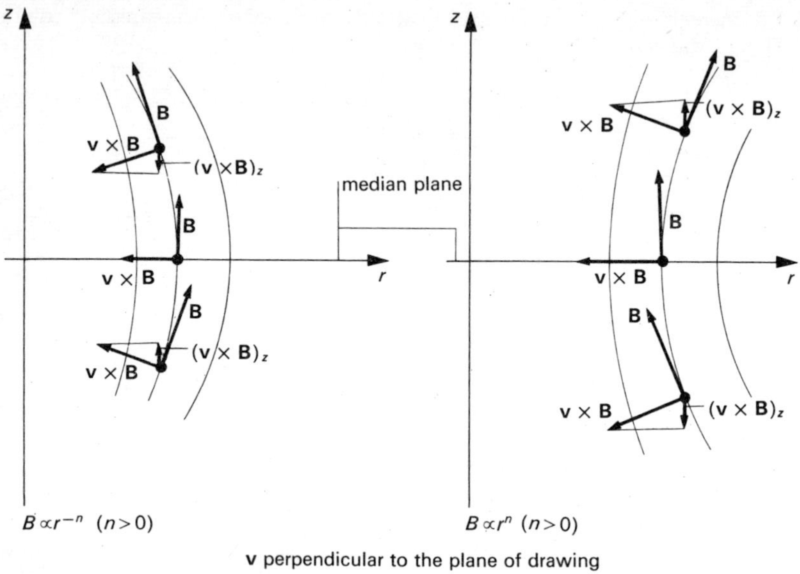

Figure 21

towards this plane. A glance at Figure 21 makes this point clear without any further explanation. A radially increasing magnetic field has the opposite effect, so far as motion perpendicular to the plane of the equilibrium orbit is concerned.

A simple electric field which satisfies the assumed symmetry conditions can be maintained between two coaxial cylindrical electrodes held at different potentials. The field distribution between the electrodes is given by $E(r) = E_0 R/r$, and the electrons are acted upon by a force $\mathbf{F} = -eE\mathbf{e}_r$. They will remain on an equilibrium orbit of radius $R = R_e$, if equation **2.55a** is satisfied by $E(R) = E_0$. The motion of electrons, slightly displaced radially in the plane of the equilibrium orbit (Figure 22), is described by defining the radial position of the particle with

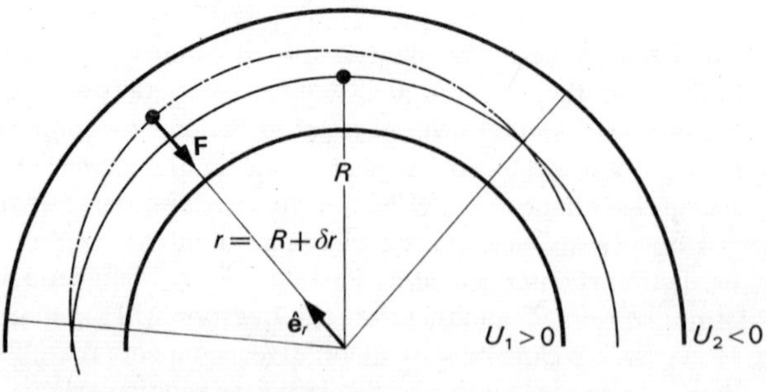

Figure 22

respect to the equilibrium position $r = R+\delta r$, and considering the solution of the equation of motion

$$\frac{d^2}{dt^2}(R+\delta r)-(R+\delta r)\left(\frac{d\theta}{dt}\right)^2 = -\frac{e}{m}E.$$

Since the motion takes place in an electrostatic field the angular momentum is conserved: $r^2\dot{\theta} = R^2\omega$. Close to the equilibrium orbit the field intensity can be approximated by $E(r) = E(R)+(\partial E/\partial r)_R\,\delta r$. Thus the equation of motion can be written in the form

$$\frac{d^2}{dt^2}\delta r-\left(1+\frac{\delta r}{R}\right)^{-3}R\omega^2 \simeq \frac{d^2}{dt^2}\delta r-\left(1-3\frac{\delta r}{R}\right)R\omega^2 = -\frac{e}{m}\left[E(R)+\left(\frac{\partial E}{\partial r}\right)_R\delta r\right].$$

We recognize that on the equilibrium orbit:

$$R\omega^2 = \frac{e}{m}E(R),$$

and therefore the displacement from the equilibrium orbit, δr, must satisfy the differential equation

$$\frac{d^2}{dt^2}\delta r = -\omega^2\left[3+\frac{(\partial E/\partial r)_R}{E_R/R}\right]\delta r,$$

where E_R represents $E(R)$. In the field between coaxial cylinders

$$\left(\frac{\partial E}{\partial r}\right)_R = -\frac{E_R}{R},$$

and therefore one obtains

$$\frac{d^2}{dt^2}\delta r = -2\omega^2\delta r.$$

This equation indicates that the electrons oscillate about the equilibrium orbit with $\sqrt{2}$ times the angular frequency of the orbital revolution on the equilibrium orbit. In other words, the particles cross the equilibrium orbit at points which are at an angular distance $\pi/\sqrt{2}$ apart. Since the field is everywhere perpendicular to the symmetry axis, a particle displaced perpendicularly to the symmetry plane is subject to no restoring force.

In order to describe quantitatively the electron motion in the neighbourhood of the equilibrium orbit, in the case of arbitrary electric and magnetic fields of the required symmetry properties, the approach followed in the previous example must be carried out in much more general terms. For this purpose the Hamiltonian formalism affords a powerful and elegant tool.

The arguments presented in the Appendix (A.6) show that the differential equations of the trajectories can be expressed in the canonical form, using *coordinates* which describe the position of the electrons *relative to the equilibrium orbit*. The electron's position is expressed in terms of orthogonal coordinates

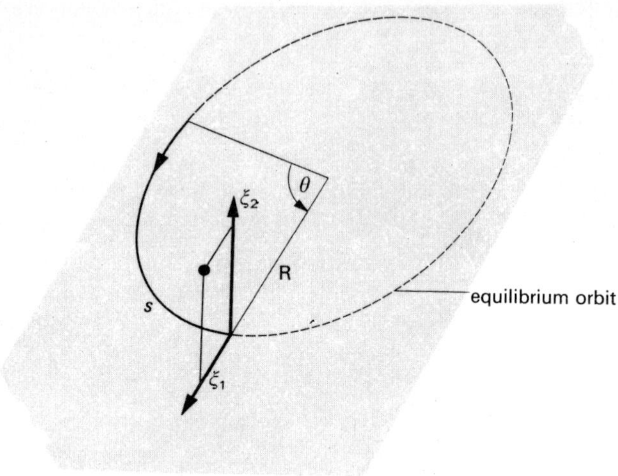

Figure 23

defined as follows: $\xi_1 = r-R$ is the radial distance of the particle from the equilibrium orbit, $\xi_2 = z$ is the normal distance from the plane of the equilibrium orbit and $s = R\theta$ is the length of arc measured along the equilibrium orbit from an arbitrary origin (Figure 23). In terms of these coordinates, the canonical momenta π_1, π_2, π_s and the total energy of the system W, we can determine a Hamiltonian-type function

$$G = G(\xi_1, \xi_2, s; \pi_1, \pi_2; W).$$

(Note: π_s does not occur among the independent variables.) With the aid of this function the canonical equations can be formed:

$$\frac{d\xi_1}{ds} = \frac{\partial G}{\partial \pi_1}, \qquad \frac{d\pi_1}{ds} = -\frac{\partial G}{\partial \xi_1},$$
$$\frac{d\xi_2}{ds} = \frac{\partial G}{\partial \pi_2}, \qquad \frac{d\pi_2}{ds} = -\frac{\partial G}{\partial \xi_2}. \qquad \textbf{2.56}$$

These equations are exact and, in principle, allow the determination of the functions $\xi_1(s)$ and $\xi_2(s)$, the components of the displacement of the electron from the equilibrium orbit, as a function of the distance measured along the equilibrium orbit. We wish to consider 'paraxial' trajectories only, that is, those adjacent to the equilibrium orbit. Therefore it is permissible to expand the Hamiltonian G in a power series, and to consider only terms not higher than second order in the coordinates and the canonical momenta. The expansion will contain terms of the following types. There will be a constant term, $G(0)$, which is the value of G on the equilibrium orbit, that is, when $\xi_1 = \xi_2 = \pi_1 = \pi_2 = 0$; and the coefficients of the first-order terms will be of the form $(\partial G/\partial \xi_j)_0$ and $(\partial G/\partial \pi_j)_0$ $(j = 1, 2)$, the subscript indicating that these quantities must be taken at

$\xi_j = \pi_j = 0$. This condition defines particles moving on the equilibrium orbit where the left-hand side of each of the equations in **2.56** must vanish. Thus the first-order terms of the expansion vanish. The coefficients of the second-order terms behave in the following manner:

$$\left(\frac{\partial^2 G}{\partial \xi_i \partial \pi_j}\right)_0 = 0 \qquad (i, j = 1, 2),$$

because they contain π_j linearly.

$$\left(\frac{\partial^2 G}{\partial \xi_i \partial \xi_j}\right)_0 = C_1 \delta_{ij}, \qquad \left(\frac{\partial^2 G}{\partial \pi_i \partial \pi_j}\right)_0 = C_2 \delta_{ij},$$

where C_1 and C_2 are constants, and δ_{ij} is the Kronecker delta. Thus the 'cross terms' vanish, because of the symmetry properties of the fields under consideration. Then the Hamiltonian takes the following approximate form:

$$G = G(0) + \frac{1}{2}\left\{\left(\frac{\partial^2 G}{\partial \xi_1^2}\right)_0 \Delta\xi_1^2 + \left(\frac{\partial^2 G}{\partial \xi_2^2}\right)_0 \Delta\xi_2^2 + \left(\frac{\partial^2 G}{\partial \pi_1^2}\right)_0 \Delta\pi_1^2 + \left(\frac{\partial^2 G}{\partial \pi_2^2}\right)_0 \Delta\pi_2^2\right\}, \qquad \textbf{2.57}$$

where $\Delta\xi_j$ and $\Delta\pi_j$ are displacements from the equilibrium orbit, and the corresponding canonical momenta are all small quantities.

For small displacements, the canonical equations, on the substitution of equation **2.57**, give

$$\frac{d}{ds}(\Delta\xi_j) = \left(\frac{\partial^2 G}{\partial \pi_j^2}\right)_0 \Delta\pi_j \qquad (j = 1, 2).$$

Differentiating once more with respect to s, and taking the second set of canonical equations into account, using the approximate Hamiltonian, one finally obtains

$$\frac{d^2}{ds^2}(\Delta\xi_j) = -\left(\frac{\partial^2 G}{\partial \pi_j^2}\right)_0 \left(\frac{\partial^2 G}{\partial \xi_j^2}\right)_0 \Delta\xi_j.$$

To obtain a more compact form the notation will be changed: $\Delta\xi_j \equiv x_j$ and

$$\left(\frac{\partial^2 G}{\partial \xi_j^2}\right)_0 \left(\frac{\partial^2 G}{\partial \pi_j^2}\right)_0 \equiv k_j^2.$$

k_j^2 is a real but not necessarily positive constant.

The results of these considerations can now be summarized. If the electromagnetic fields satisfy the symmetry conditions assumed, then the electron trajectories adjacent to the equilibrium orbit are determined by very simple differential equations:

$$\frac{d^2 x_j}{ds^2} = -k_j^2 x_j \qquad (j = 1, 2), \qquad \textbf{2.58}$$

where x_1 is a small displacement, in the radial direction, from the equilibrium orbit in its plane, and x_2 is a small displacement perpendicular to the plane of the equilibrium orbit; k_j^2 is a constant.

To determine k_j the explicit form of the Hamiltonian G must be known. From the Hamiltonian derived in the Appendix (A.6), one finds after somewhat lengthy but straightforward calculation that

$$k_2^2 = -R^{-2}\left[RR_e^{-1}\left\{1+\left(\frac{\partial E/\partial r}{E/R}\right)_R\right\}+RR_m^{-1}\left(\frac{\partial B/\partial r}{B/R}\right)_R\right] \quad \textbf{2.59a}$$

and $$k_1^2 = R^{-2}\left[1+\left(\frac{R}{R_e}\right)^2\right]-k_2^2, \quad \textbf{2.59b}$$

where R is the radius of the equilibrium orbit, R_e and R_m are the radii defined by equation **2.55** and E and B are the fields in the plane of the equilibrium orbit.

Thus the constants k_j are characteristic of the field distribution. It is this constant which determines the nature of the solution of equation **2.58**, and hence the properties of the paraxial trajectories. If $k_j^2 > 0$, the solution is given in terms of circular functions and the trajectories stay in the neighbourhood of the equilibrium orbit. If $k_j^2 < 0$, the solutions are given in terms of hyperbolic functions and the trajectories diverge from the equilibrium orbit.

As an illustration of the application of equation **2.59**, the reader should determine the constants k_j for fields which were discussed earlier, and see whether the qualitative predictions are in agreement with the quantitative results. Thus for a uniform magnetic field, one should take $E = 0$, $R_e \to \infty$, $R_m = R$, $B =$ constant. For the cylindrical electric field, $B = 0$, $R_m \to \infty$ and $R_e = R$.

The example below deals with a more complicated problem of considerable practical interest. Suppose that there is no electric field present ($\mathbf{E} = 0$), and hence $R_e = \infty$, $R_m = R$. In the plane of mirror symmetry, $z = 0$, the magnetic field is everywhere perpendicular to this plane, $\mathbf{B} = B\mathbf{e}_z$, and let $B = B_0(r/R)^n$ describe its distribution. In this case

$$\frac{(\partial B/\partial r)_R}{B_0/R} = n$$

and equations **2.59a** and **b** give

$$k_1^2 = R^{-2}-k_2^2 = (1+n)R^{-2}, \quad \textbf{2.60a}$$

$$k_2^2 = -nR^{-2}. \quad \textbf{2.60b}$$

The differential equation **2.58** shows that in order to obtain trajectories which are confined to a finite region about the equilibrium orbit one must have $k_j^2 > 0$. To satisfy this condition for *both* components one must choose

$$-1 < n < 0 \quad \textbf{2.61}$$

and in this case, both components of the displacement from the equilibrium orbit vary sinusoidally while the electron progresses in a direction parallel to the equilibrium orbit. The spatial period of the oscillation about the equilibrium orbit is given by $\lambda_j = 2\pi/k_j$,

i.e. $\lambda_1 = 2\pi R(1+n)^{-\frac{1}{2}}$ and $\lambda_2 = 2\pi R|n|^{-\frac{1}{2}}$.

The two periods are equal, $\lambda = 2^{\frac{1}{2}}\pi R$, if $n = -\frac{1}{2}$; if λ_1 is reduced, λ_2 increases and *vice versa*. Note that the condition **2.61** describes a field which is decreasing with the distance measured from the symmetry axis at a rate slower than linearly. This type of field is used as the guiding field in cyclic accelerators, which will be discussed later. Equation **2.58** shows that whatever the actual distribution of the electric and/or magnetic field is, provided the symmetry conditions are satisfied as requested at the outset, and k_j is determined by equations **2.59a** and **b**, then the paraxial trajectories and their slopes are described by the matrix relation

$$\begin{bmatrix} x_j(s) \\ x_j'(s) \end{bmatrix} = \mathrm{M}_j \begin{bmatrix} x_j(0) \\ x_j'(0) \end{bmatrix} \quad (j = 1, 2),$$

where the matrix M_j has the form given by equations **2.52a** and **2.52b**.

$$\mathrm{M} = \begin{cases} \mathrm{M}_j^+ = \begin{bmatrix} \cos k_j s & k_j^{-1} \sin k_j s \\ -k_j \sin k_j s & \cos k_j s \end{bmatrix} & \text{if } k_j^2 > 0, \\ \mathrm{M}_j^- = \begin{bmatrix} \cosh|k_j|s & |k_j|^{-1} \sinh|k_j|s \\ |k_j| \sinh|k_j|s & \cosh|k_j|s \end{bmatrix} & \text{if } k_j^2 < 0. \end{cases} \qquad \mathbf{2.62}$$

Note that the distance s measured along the equilibrium orbit plays the role of the distance measured along the optic axis.

In the previous considerations, the matrix formalism supplied a convenient method for tracing paraxial trajectories through adjoining regions of different field distributions, including field-free drift regions, provided the adjoining regions had a common optic axis. The same applies now, but the requirement of a common optic axis must be specified more precisely, since it is not quite as obvious as in the previous situations.

As already pointed out, the optic axis is represented by the equilibrium orbit which is defined for a particle of a given momentum. Two adjoining regions of different field distributions have a common optic axis for a particle of a specific momentum, if the equilibrium orbits of that particle in the two regions form a continuous curve with a common tangent at the boundary of the two regions. As a simple but important example, let us consider a field which is a 'sector' of a rotationally symmetric field over a range $0 \leqslant \theta \leqslant \theta_0$ of the azimuth angle, and is symmetrical with respect to the $z = 0$ plane:

$$\left.\begin{aligned} \Phi &= \Phi\left(\frac{\rho}{R}, z\right) = \Phi\left(\frac{\rho}{R}, -z\right) \\ A_z &= A\left(\frac{\rho}{R}, z\right) = A\left(\frac{\rho}{R}, -z\right) \end{aligned}\right\} \quad \text{if } 0 \leqslant \theta \leqslant \theta_0,$$

$$A_1 = A_2 = 0.$$

Outside the range specified Φ and A are constants, that is, neither an electric nor a magnetic field is present.

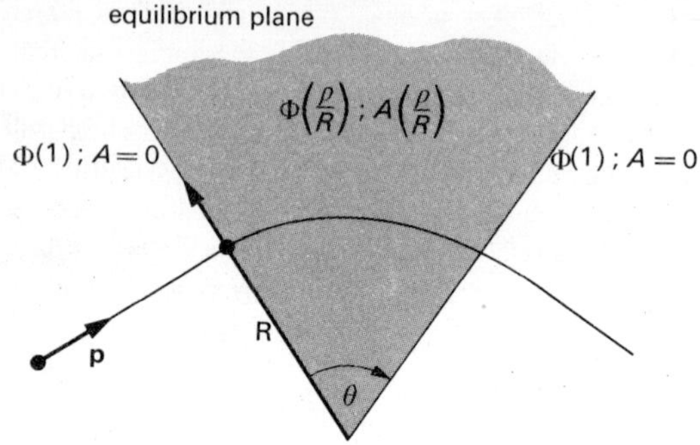

Figure 24

Let us assume that R is the radius of the equilibrium orbit of particles of momentum p in the field. Outside the field the particles move on straight lines (see Figure 24). If the constant potential outside the field is equal to that on the equilibrium orbit* the straight line which is perpendicular to the field boundaries, and adjoins the circular arc of radius R, constitutes the equilibrium orbit. This line is the optic axis for the particle of momentum p in the field-free regions adjoining the specified 'sector field'.

After these preliminary remarks, the paraxial imaging properties of such sector fields can readily be described in geometrical optic terms. For this purpose the field-free region of length d is characterized by the matrix

$$M_0(d) = \begin{bmatrix} 1 & d \\ 0 & 1 \end{bmatrix} \qquad \textbf{2.62a}$$

as established earlier. Let electrons of momentum p be emitted by a point $x_j(-d_1)$, at a distance $-d_1$ in front of the entrance boundary, with an initial slope $x'_j(-d_1)$. The length of the circular arc of the equilibrium orbit in the field is S. The position of the electron and the slope of its trajectory at a distance d_2 beyond the exit boundary is given by a relation analogous to equation **2.14**:

$$\begin{bmatrix} x_j(d_2) \\ x'_j(d_2) \end{bmatrix} = M_0(d_2)\, M_j(S)\, M_0(-d_1) \begin{bmatrix} x_j(-d_1) \\ x'_j(-d_1) \end{bmatrix} \qquad \textbf{2.63}$$

where $M_j(S)$ is given by equation **2.62**.

The cardinal elements can be obtained by the same arguments as before. One must remember, however, that trajectories *in the plane* of the equilibrium orbit, and those in a cylindrical surface containing the equilibrium orbit and *perpendicular* to its plane, are characterized by different parameters, k_1 and k_2 respectively. Consequently the two types of trajectory give different sets of cardinal elements. In other words, the field as a lens is astigmatic. With reference to

*Since on the equilibrium orbit $z = 0$, $\rho/R = 1$, we require that if $\theta < 0$ and $\theta > \theta_0$, then $\Phi = \Phi(0, 1)$. This condition is an essential one. If it were not satisfied, the kinetic energy and hence the momentum of the particle would change on crossing the boundary.

equations **2.19**, **2.20** and **2.21**, the distances of the foci d_{f1} and d_{f2} from the respective boundary planes are given by

$$\left(\frac{S}{d_{f2}}\right)_j = -\left(\frac{S}{d_{f1}}\right)_j = k_j S \tan k_j S \qquad \mathbf{2.64a}$$

and the distances of the principal planes, d_{p1} and d_{p2}, are given by

$$\left(\frac{d_{p1}}{S}\right)_j = -\left(\frac{d_{p2}}{S}\right)_j = \frac{1}{2}\,\frac{\tan \frac{1}{2}k_j S}{\frac{1}{2}k_j S}. \qquad \mathbf{2.64b}$$

For the focal length one finds

$$-\left(\frac{S}{f_1}\right)_j = \left(\frac{S}{f_2}\right)_j = k_j S \sin k_j S. \qquad \mathbf{2.64c}$$

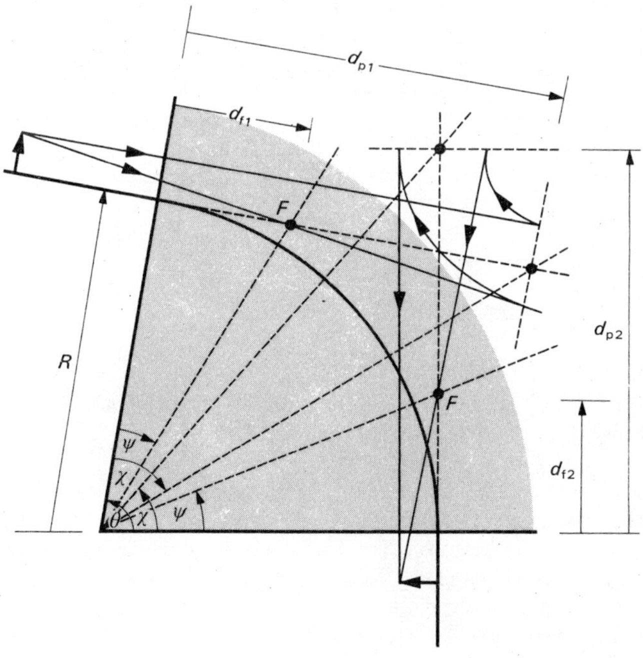

Figure 25

These formulae are directly applicable if $k_j^2 > 0$. If $k_j^2 < 0$, or $k_j = i|k_j|$, the required quantities are obtained in terms of hyperbolic functions, as expected.

The rules of the geometrical construction of the image of a source point must take into account the fact that the optic axis in the field is a circular arc (see Figure 25). To emphasize this let us reintroduce the angular aperture of the sector field $\theta = S/R$. The equations which define the cardinal elements of the lens are

$$d_{f2} = -d_{f1} = R(k_j R)^{-1} \cot (k_j R\theta) = R \tan \psi,$$
$$d_{p1} = -d_{p2} = R(k_j R)^{-1} \tan \tfrac{1}{2}(k_j R\theta) = R \tan \chi \qquad \mathbf{2.65}$$

where the angles ψ and χ are introduced by the definitions

$$\tan \psi = (k_j R)^{-1} \cot (k_j R\theta),$$
$$\tan \chi = (k_j R)^{-1} \tan \tfrac{1}{2}(k_j R\theta). \qquad \mathbf{2.65a}$$

The image construction based on these considerations is illustrated in Figure 25 for the case of a focusing sector field, where the numerical value $k_j \simeq 1{\cdot}5$ was chosen. The actual value of the parameter k_j depends on the distribution and nature of the field; a few examples representing the most frequently applied field distributions will be considered presently.

For the purpose of later reference it is useful to remember that if $x_j(-d_1)$ and $x_j(d_2)$ are coordinates of *conjugate* points, then the product matrix

$$\mathrm{M}_0(d_2)\, \mathrm{M}_j(S)\, \mathrm{M}_0(-d_1) = T_j,$$

which occurs in equation **2.63**, can be written explicitly in a very simple form. Denoting the elements of T_j by $t_{ik}^{(j)}$ one can write $t_{12}^{(j)} = 0$; from equation **2.21**, $t_{21}^{(j)} = -|f_j|^{-1}$; denoting the linear magnification by μ_j one finds $t_{11}^{(j)} = \mu_j$. The condition det $\mathrm{T} = 1$ reduces to $t_{11}^{(j)}\, t_{22}^{(j)} = 1$, and therefore

$$\mathrm{T} = \begin{bmatrix} \mu_j & 0 \\ |f_j|^{-1} & \mu_j^{-1} \end{bmatrix}. \qquad \mathbf{2.66}$$

Examples

Uniform magnetic deflecting field: $\mathbf{B} = B_0 \mathbf{e}_z$, $B_0 =$ constant. In the absence of an electric field $1/R_e = 0$ and $R_m = R$ is the radius of the equilibrium orbit for particles of momentum $p = eB_0 R$. From equation **2.59**

$$k_2 = 0 \qquad k_1 R = 1.$$

In other words, there is no restoring force and hence no focusing effect normal to the plane of the equilibrium orbit. The spatial period of the oscillation about the equilibrium orbit in its plane is $\lambda_1 = 2\pi R$. Therefore electrons emitted by a point source in a direction perpendicular to the field are focused after a half period, that is, after describing a semicircular trajectory in the field.

Electric field between coaxial *cylindrical* electrodes:

$$\mathbf{E} = E_0 \frac{R}{r} \mathbf{e}_r.$$

In the absence of a magnetic field, $1/R_m = 0$, $R_e = R$ is the radius of the equilibrium orbit for particles of momentum $p = (emE_0 R)^{\frac{1}{2}}$. From equation **2.59**

$$k_2 = 0 \qquad k_1 R = \sqrt{2}.$$

Again there is no restoring force normal to the equilibrium orbit. The spatial period of oscillations about the equilibrium orbit in its plane is $\lambda_1 = 2^{\frac{1}{2}}\pi R$, that is, in the plane of the equilibrium orbit, electrons emitted by a point source are focused after describing a circular arc corresponding to $\pi/\sqrt{2} \simeq 127°$ as seen earlier. If the field is confined to a region $0 \leqslant \theta \leqslant \theta_1$, the cardinal elements are given by equations **2.65** and **2.65a**.

Electric field between concentric *spherical* electrodes:

$$\mathbf{E} = E_0\left(\frac{R}{r}\right)^2 \mathbf{e}_r.$$

Any equatorial plane can be the plane of the equilibrium orbit. In the absence of a magnetic field $1/R_m = 0$, and $R_e = R$ is the radius of the equilibrium orbit for particles of momentum $p = (emE_0 R)^{\frac{1}{2}}$. From equation **2.59**

$$k_1 = k_2 = \frac{1}{R}.$$

There is a finite restoring force both normal to and in the plane of the equilibrium orbit. The spatial period of oscillation is the same in both planes:

$$\lambda_{1,2} = 2\pi R.$$

The positions of the cardinal elements are readily obtained from equation **2.65** if the field is confined to a given range of the azimuth angle, $0 \leqslant \theta \leqslant \theta_2$.

The 'Wien filter' is obtained by the superposition of crossed uniform electric and magnetic fields. In terms of Cartesian coordinates, the fields are described by, say,

$$\mathbf{B} = B_0 \mathbf{e}_z \qquad (B_0 = \text{constant}),$$
$$\mathbf{E} = E_0 \mathbf{e}_x \qquad (E_0 = \text{constant}).$$

The motion of the electron in such a field was briefly discussed in the Introduction, and it follows that if the electron has an initial velocity

$$\mathbf{v} = \frac{\mathbf{E}\times\mathbf{B}}{B_0^2} = -\frac{E_0}{B_0}\mathbf{e}_y,$$

it will move in a straight line parallel to the $-y$-axis, that is, in a direction perpendicular to both **E** and **B**.

In order to investigate the properties of the electron motion in electron-optic terms, one can argue as follows. The fields defined above are symmetric with respect to the plane $z = 0$, and they can be considered as rotationally symmetric with respect to an axis parallel to the z-axis and situated at $x = \infty$. The equilibrium orbit therefore will be in the $z = 0$ plane (or in any plane z = constant) and it will be a degenerate circle of zero curvature, $1/R = 0$, that is, a straight line parallel to the y-axis because the centre of the circle is at $x = \infty$. In other words the Wien filter has a straight optic axis. According to equation **2.54**, the condition

$1/R = 0$ implies that the electrons which can move on the equilibrium orbit have a momentum

$$p = -\frac{mE}{B}$$

directed parallel to the equilibrium orbit, in agreement with the previous result.

The quantities k_j, necessary for the description of the optical properties of the field, are obtained by taking the limits of the right hand sides of equation **2.59** as $R \to \infty$. Since both fields are uniform

$$k_2 = 0 \qquad k_1 = R_e^{-1} = meEp^{-2} = \frac{e}{m}\frac{B^2}{E}.$$

Thus there is no focusing effect normal to the plane $z = 0$ (that is, parallel to the magnetic field), and the spatial period of oscillation in the plane perpendicular to the magnetic field is

$$\lambda_1 = \frac{2\pi m}{e}\frac{E}{B^2}.$$

Thus the electrons emitted by a point source in a plane perpendicular to the magnetic field are focused after a flight path of length $\frac{1}{2}\lambda_1$. If the filter is of length S, the positions of the cardinal elements in the plane perpendicular to the magnetic field are obtained from equation **2.64** by substituting k_1; the length S is measured parallel to the y-axis, along which the potential is equal to the potential outside the filter.

2.7 Dispersion and resolving power

We wish to investigate the shift of the image of a source point resulting from a small change in the momentum of the electrons. Such a shift of the image has two causes. One is that the elements of the matrix M, characteristic of the imaging field, are functions of the parameter k_j which in turn depends on the particle momentum. The other reason is that the radius of the equilibrium orbit, that is, the optic axis, is defined in terms of the momentum. We can readily see that in a first-order approximation the second factor only is significant.

The matrix relating the coordinates and the slope of the trajectory of the electrons of momentum $p+\Delta p$ can be written, to a first-order approximation,

$$\mathrm{M}_j(p+\Delta p) = \mathrm{M}_j(p) + \frac{\partial \mathrm{M}_j}{\partial p}\Delta p \qquad (j = 1, 2).$$

If $y_j(s)$ and $y'_j(s)$ are the coordinates and the slope of the *paraxial* trajectory of an electron of momentum $p+\Delta p$, and these quantities are small, then to first-order approximation the equation

$$\begin{bmatrix} y_j(s) \\ y'_j(s) \end{bmatrix} = \mathrm{M}_j(p+\Delta p) \begin{bmatrix} y_j(0) \\ y'_j(0) \end{bmatrix}$$

reduces to $\begin{bmatrix} y_j(s) \\ y'_j(s) \end{bmatrix} = \mathrm{M}_j(p) \begin{bmatrix} y_j(0) \\ y'_j(0) \end{bmatrix}$.

But the quantities y_j and y'_j refer to the equilibrium orbit of the electron of momentum $p+\Delta p$ which is different from that of an electron of momentum p. It can be shown* that for an electron of momentum $p+\Delta p$ the equilibrium orbit remains in the symmetry plane, but its radius is $R+\Delta R$, where

$$\frac{\Delta R}{R} = \frac{1+RR_e^{-1}}{(k_1 R)^2} \frac{\Delta p}{p}. \qquad \textbf{2.67}$$

Therefore referring to the same optic axis (see Figure 26) one finds the relation

$$\begin{bmatrix} x_j(s) \\ x'_j(s) \end{bmatrix} = \begin{bmatrix} y_j(s)+\Delta R \\ y'_j(s) \end{bmatrix} = \begin{bmatrix} \Delta R \\ 0 \end{bmatrix} + \mathrm{M}_j \begin{bmatrix} y_j(0) \\ y'_j(0) \end{bmatrix}$$

$$= \begin{bmatrix} \Delta R \\ 0 \end{bmatrix} + \mathrm{M}_j \begin{bmatrix} x_j(0)-\Delta R \\ x'_j(0) \end{bmatrix}$$

$$= \mathrm{M}_j \begin{bmatrix} x_j(0) \\ x'_j(0) \end{bmatrix} + (\mathrm{I}-\mathrm{M}_j) \begin{bmatrix} \Delta R \\ 0 \end{bmatrix},$$

where I is the unit matrix. Using equation **2.67** to replace ΔR by Δp, the result above takes the form

$$\begin{bmatrix} x_j(s) \\ x'_j(s) \end{bmatrix} = \mathrm{M} \begin{bmatrix} x_j(0) \\ x'_j(0) \end{bmatrix} + aR(\mathrm{I}-\mathrm{M}_j) \begin{bmatrix} \Delta p/p \\ 0 \end{bmatrix}, \qquad \textbf{2.68}$$

where $a = (1+RR_e^{-1})(k_1 R)$.

*If the radius R of the equilibrium orbit of the electron emitted by a source at a potential $\Phi(R)$ with a momentum p satisfies equation **2.54**, then an electron emitted by the same source with a momentum $p+\Delta p$ has an equilibrium orbit of radius defined by the equation

$$(R+\Delta R)^{-1} = \frac{me\,E(R+\Delta R)}{P^2} + \frac{e\,B(R+\Delta R)}{P}, \qquad \textbf{i}$$

where P is determined by energy conservation:

$$\frac{1}{2m} P^2 + e\,\Phi(R+\Delta R) = \frac{1}{2m}(p+\Delta p)^2 + e\,\Phi(R).$$

To a first-order approximation

$$\Phi(R+\Delta R) = \Phi(R) + \frac{\partial \Phi}{\partial \rho}\Delta R = \Phi(R) - E(R)\Delta R.$$

$E(R+\Delta R)$ and $B(R+\Delta R)$ are given by similar expansions terminating with the first-order term in ΔR.

Taking into account the relation $R^{-1} = R_e^{-1} + R_m^{-1}$, and equations **2.59** for k_j, after somewhat lengthy but straightforward calculation, one finds that equation **2.67** follows from equation **i** above.

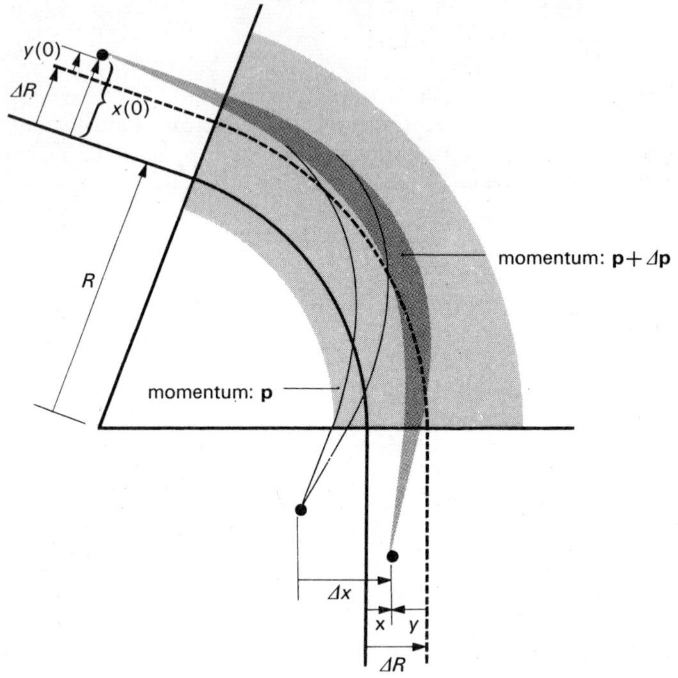

Figure 26

Equation **2.68** can be written more concisely in terms of a 3 × 3 matrix. If the elements of matrix M_j are denoted by m_{ik}, one can write

$$\begin{bmatrix} x_j(s) \\ x'_j(s) \\ \Delta p/p \end{bmatrix} = \begin{bmatrix} m_{11} & m_{12} & aR(1-m_{11}) \\ m_{21} & m_{22} & -aRm_{21} \\ 0 & 0 & 1 \end{bmatrix} \begin{bmatrix} x_j(0) \\ x'_j(0) \\ \Delta p/p \end{bmatrix} \qquad \textbf{2.69}$$

The benefit of replacing equation **2.67** by **2.69** is that in order to trace electrons through adjoining regions of different field distributions, having a common optic axis for electrons of momentum p, one can multiply the matrices characterizing the individual regions of the sequence. For a field-free drift region of length d, the relevant 3 × 3 matrix takes the form

$$\begin{bmatrix} 1 & d & 0 \\ 0 & 1 & 0 \\ 0 & 0 & 1 \end{bmatrix}$$

If $(x_j)_1$, $(x'_j)_1$, $(x_j)_2$ and $(x'_j)_2$ denote coordinates and trajectory slopes *at conjugate points*, for which the matrix T_j (equation **2.66**) is applicable, it is possible to write

$$\begin{bmatrix} (x_j)_2 \\ (x'_j)_2 \\ \Delta p/p \end{bmatrix} = \begin{bmatrix} \mu_j & 0 & aR(1-\mu_j) \\ -1/|f_j| & 1/\mu_j & aR/|f_j| \\ 0 & 0 & 1 \end{bmatrix} \begin{bmatrix} (x_j)_1 \\ (x'_j)_1 \\ \Delta p/p \end{bmatrix}. \qquad \mathbf{2.70}$$

From this equation, the distance between the coordinates of an image point, *in the plane of the equilibrium orbit*, produced by electrons of momentum $p+\Delta p$ and of p respectively, is given by

$$\Delta x \equiv (x_1)_2 - [(x_1)_2]_{\Delta p=0} = aR(1-\mu_1)\left(\frac{\Delta p}{p}\right). \qquad \mathbf{2.71}$$

Consider, for example, a uniform magnetic field in the absence of any electric field ($R_e^{-1} = 0$), and suppose that the source is set up relative to the field so that a real image is formed with unit magnification ($\mu = -1$). In this case, as seen earlier, $k_1 R = 1$ and hence $a = 1$, so that equation **2.71** reduces to

$$\Delta x = 2R(\Delta p/p).$$

The same result is found for a cylindrical electric sector field if the condition for the magnification is maintained.

Δx defined by equation **2.71** gives the shift of the image of a point source. If the source has a finite extension w, measured in the plane of the equilibrium orbit, in a direction perpendicular to the optic axis, its image will have an extension $\mu_1 w$ in the same direction. Electrons of momentum p and $p+\Delta p$ can be separated only if the image of the source is shifted by a distance greater than the width of the image: $\Delta x > \mu_1 w$,

and hence $$\frac{\Delta p}{p} \geqslant \frac{w}{aR}\frac{\mu_1}{1-\mu_1} \qquad \mathbf{2.72}$$

gives the condition which the fractional momentum deviation $\Delta p/p$ must satisfy to allow a complete separation. The equality sign in relation **2.72** defines the 'resolution' of the system as a momentum analyser. Under the conditions mentioned above (that is, $\mu_1 = -1$), the resolution given by a uniform magnetic or a cylindrical electric field is

$$\frac{\Delta p}{p} = \frac{w}{2R}.$$

It must be stressed that all the considerations above are valid in a first-order approximation only. The effects of higher-order aberrations were not taken into account.

The results so far obtained are directly applicable to problems where the beam under consideration consists of particles of the same e/m ratio, for example beta rays or a beam of protons etc. In mass spectrometers the situation is more complicated. Even if all the particles in the beam have the same charge, both factors, the mass m and velocity v of the particles, which determine the momentum $p = mv$ can vary independently of one another. If the radius of the equilibrium orbit of particles of mass m and velocity v is R, the radius of the equilibrium orbit of particles of mass $m+\Delta m$ and velocity $v+\Delta v$ is $R+\Delta R$ where*

$$\frac{\Delta R}{R} = a\frac{\Delta v}{v}+b\frac{\Delta m}{m},$$

$$\text{with} \quad a = (1+RR_e^{-1})(k_1 R)^{-2}, \qquad \textbf{2.73}$$

$$b = (k_1 R)^{-2}.$$

Since the purpose of a mass spectrometer is to select particles of well-defined mass irrespective of their velocities, conditions need to be found where the position of the image of a source is independent of the velocity spread Δv, and depends on the mass deviation Δm only.

The introduction of the momentum $p = mv$ into the expression of the radius of the equilibrium orbit shows that each term on the right-hand side of equation **2.54a** has the same mass dependence, but different velocity dependence. Hence it is possible to *find suitable combinations of electric and magnetic sector fields* (either by superposition or in sequence), *which compensate the effects of a velocity spread, but do not compensate the effect of the mass differences of the particles in the beam.*

The right-hand side of equation **2.73** can be considered as the scalar product of two vectors with orthogonal components (a, b) and $(\Delta v/v, \Delta m/m)$ respectively, that is, the equation can be rewritten as

$$\frac{\Delta R}{R} = [a \quad b]\begin{bmatrix}\Delta v/v \\ \Delta m/m\end{bmatrix}.$$

The same argument which led to equation **2.68** will give now

$$\begin{bmatrix}x_j(s) \\ x_j'(s)\end{bmatrix} = \mathrm{M}_j\begin{bmatrix}x_j(0) \\ x_j'(0)\end{bmatrix}+R(\mathrm{I}-\mathrm{M}_j)\begin{bmatrix}a & b \\ 0 & 0\end{bmatrix}\begin{bmatrix}\Delta v/v \\ \Delta m/m\end{bmatrix}$$

and this equation can be written in terms of a 4×4 matrix:

$$\begin{bmatrix}x_j(s) \\ x_j'(s) \\ \Delta v/v \\ \Delta m/m\end{bmatrix} = \begin{bmatrix}m_{11} & m_{12} & aR(1-m_{11}) & bR(1-m_{11}) \\ m_{21} & m_{22} & -aRm_{21} & -bRm_{21} \\ 0 & 0 & 1 & 0 \\ 0 & 0 & 0 & 1\end{bmatrix}\begin{bmatrix}x_j(0) \\ x_j'(0) \\ \Delta v/v \\ \Delta m/m\end{bmatrix}.$$

The relation between coordinates and trajectory slopes in *conjugate planes* can be written explicitly. For trajectories in the plane of the equilibrium orbit

$$\begin{bmatrix}(x_j)_2 \\ (x_j')_2 \\ \Delta v/v \\ \Delta m/m\end{bmatrix} = \mathrm{S}\begin{bmatrix}(x_j)_1 \\ (x_j')_1 \\ \Delta v/v \\ \Delta m/m\end{bmatrix},$$

*The argument leading to this result is essentially the same as that leading to equation **2.67**, but in the energy equation $p+\Delta p$ must be replaced by $(m+\Delta m)(v+\Delta v)$.

where, in terms of the focal length $|f|$ and magnification μ, the matrix S is defined as

$$S = \begin{bmatrix} \mu & 0 & aR(1-\mu) & bR(1-\mu) \\ -1/|f| & 1/\mu & aR/|f| & bR/|f| \\ 0 & 0 & 1 & 0 \\ 0 & 0 & 0 & 1 \end{bmatrix}. \qquad \mathbf{2.74}$$

Let us now consider a beam which traverses, in succession, two fields with a common optic axis for particles of momentum $p = mv$. The image formed by field (1) plays the part of a virtual source for field (2), and the relation between the final image and the original source is described by the product of two matrices S_2 and S_1, each of the type given by equation **2.74**. The position of the final image will be independent of the velocity spread $\Delta v/v$ if the matrix element

$$(S_2 S_1)_{13} = 0$$

or, explicitly, $\mu_2 a_1 R_1(1-\mu_1)+a_2 R_2(1-\mu_2) = 0.$ **2.75**

If this condition is satisfied, the displacement of the image in the plane of the equilibrium orbit in a direction perpendicular to the optic axis is given by

$$\Delta x = (S_2 S_1)_{14}\left(\frac{\Delta m}{m}\right).$$

Calculating the required matrix element and taking equation **2.75** into account, we find that

$$\Delta x = (1-\mu_1)\mu_2(a_2 b_1 - a_1 b_2)\frac{R_1}{a_2}\frac{\Delta m}{m}. \qquad \mathbf{2.76}$$

If the source is of width w, its image has a width $\mu_1 \mu_2 w$. Thus fractional mass differences $\Delta m/m$ can be resolved

if $\quad \Delta x \geqslant \mu_1 \mu_2 w$

or $\quad \dfrac{\Delta m}{m} \geqslant \dfrac{\mu_1}{1-\mu_1} \dfrac{a_2}{a_2 b_1 - a_1 b_2} \dfrac{w}{R_1}.$ **2.77**

For example, let us consider a system consisting of a cylindrical electric sector field followed by a uniform magnetic sector field, set up so that the radius of the equilibrium orbit of particles of momentum $p = mv$ is the same in both fields: $R_1 = R_2 = R$. Since the first field is electric, $R_{e1} = R$, and the second field magnetic, $1/R_{e2} = 0$. From the definitions of a and b (**2.73**), $a_1 = 1$, $a_2 = 1$, $b_1 = 2$ and $b_2 = 1$, and hence the resolution is $\Delta m/m = \mu_1(1-\mu_1)^{-1}wR^{-1}$. If the relative positions of the source and the fields are arranged to give real images and the first field gives unit magnification, $\mu_1 = -1$, the resolution is

$$\frac{\Delta m}{m} = \frac{w}{2R}.$$

Figure 27 shows such an arrangement.

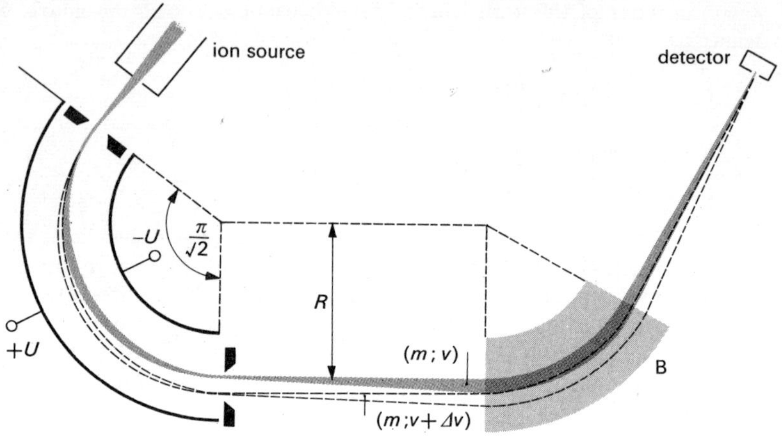

Figure 27

It may be noticed that although the shift of the image, corresponding to a fractional mass difference $\Delta m/m$, depends on the magnification of both fields, in the resolution only one occurs explicitly. The choice of $\mu_1 \simeq -1$ and $\mu_2 \simeq -1$ is advantageous because this gives the shortest total flight path between source and collector.

2.8 Beam focusing and beam confinement – general conditions

In the previous sections the phrase 'focusing' was used in the sense customary in geometrical optics, implying that such a lens is capable of producing a real image of a source, as opposed to a defocusing lens, which can produce a virtual image only. It was seen that the condition of focusing is that the particle is subject to a harmonic restoring force, making the particle follow a sinusoidal trajectory in the field. If a focusing lens is thin, its foci and the conjugate points are outside the field. This is not necessarily the case if the lens is thick or, rather, if the field is sufficiently long; this was the reason for the remark made at the end of section 2.2. For example, the particles in an axially symmetric magnetic field are always subject to an elastic-type restoring force and they may, depending on the length of the field, emerge in a divergent beam so that no real image can be produced outside the field.

There are electromagnetic fields whose action can be analysed profitably in electron-optic terms, although their function is not image production but the confinement of particles over a long flight path to a finite region in the vicinity of a prescribed trajectory. If this effect arises because of a linear restoring force resulting in the harmonic oscillation of the particle about a certain 'equilibrium orbit', the matrix characterizing the field will be of the type M^+ given by equation **2.62**. Defocusing fields are characterized by matrices of the type M^-, and they can easily be distinguished. If the matrix is of the type M^+ the field, if extended to any arbitrary length, will confine the particles to the neighbourhood of the equilibrium orbit throughout their flight paths whereas, for fields characterized by matrices of the type M^-, the beam becomes divergent.

Very often, however, particles pass through adjoining regions which are characterized by different matrices, and the resulting effect can be described only in terms of the product of these matrices. A situation of particular importance is one in which the field consists of a sequence of identical 'unit cells', each of which itself is a combination of a few sections of different properties. As an illustration one can think of a series of quadrupole lenses aligned with alternate focusing and defocusing planes, separated by gaps of a fixed length. How can one decide whether such a system is suitable for confining a beam, and if so, what are the conditions under which this effect can be achieved?

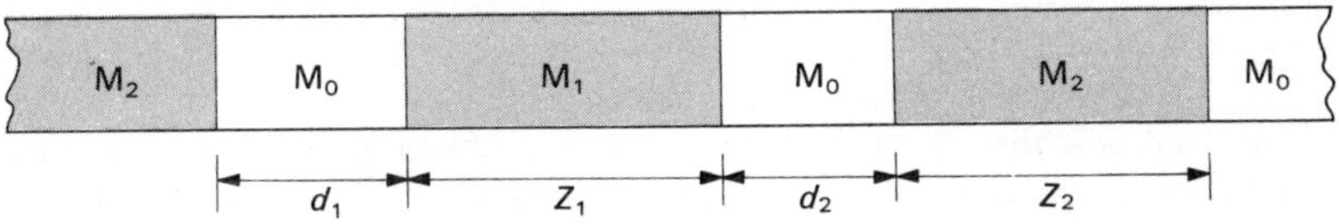

Figure 28

Figure 28 shows schematically such a system. In this example a unit cell consists of four sections; those of length d_1 and d_2 are field-free regions, while those of length z_1 and z_2 are characterized by matrices M_1 and M_2 alternately. Thus the unit cell is characterized by a matrix

$$M = M_2(z_2)\, M_0(d_2)\, M_1(z_1)\, M_0(d_1), \qquad \textbf{2.78}$$

where M_0 is of the type given by equation **2.62a**, matrices M_1 and M_2 are of the type given by equation **2.62**. The product matrix M is rather complicated. Moreover the total system is characterized by the matrix M^n, where n is the number of unit cells constituting the whole system. This may seem unmanageably complicated, but for guidance let us look at the matrices M^+ and M^- given by equation **2.62**. Both matrices satisfy the condition

$$\det M = 1, \qquad \textbf{2.79}$$

but there is an essential difference between them. Taking the sum of the diagonal elements, called the *trace* of the matrix, we find that in the case of the focusing field

$$|\mathrm{tr}\, M^+| = 2|\cos k_j s| \leqslant 2,$$

whereas in the case of the defocusing field

$$|\mathrm{tr}\, M^-| = 2|\cosh k_j s| \geqslant 2.$$

Now, if* the unit under consideration, like the one given in the example above, can be characterized by a quadratic matrix M, it will also satisfy condition **2.79**.

*The precise conditions under which the present considerations apply are as follows: (1) throughout the unit there exists a continuous equilibrium orbit; (2) denoting the displacement of the particles

We shall show that the condition of focusing is the same as in the previous case, that is, beam confinement can be achieved irrespective of the number of unit cells, if the matrix of the unit cell satisfies the condition

$$|\mathrm{tr}\,\mathrm{M}| < 2, \qquad \mathbf{2.80a}$$

whereas the beam is divergent if

$$|\mathrm{tr}\,\mathrm{M}| > 2. \qquad \mathbf{2.80b}$$

(i) First let us consider the implications of the condition **2.80a**. This condition is clearly satisfied if we choose as matrix elements:

$$m_{11} = \cos\mu - a\sin\mu \qquad m_{22} = \cos\mu + a\sin\mu$$

giving $\quad \mathrm{tr}\,\mathrm{M} = 2\cos\mu,$

where μ and a are arbitrary real numbers, neither of which is zero. Thus within the condition that $|\mathrm{tr}\,\mathrm{M}| < 2$, m_{11} and m_{22} are arbitrary. Also m_{12} can be arbitrarily chosen; it may be written in the form

$$m_{12} = b\sin\mu.$$

The matrix element m_{22} is no longer arbitrary, because condition **2.79** must be satisfied, and hence one finds

$$m_{21} = -\frac{1+a^2}{b}\sin\mu.$$

Thus we can write

$$\mathrm{M} = \begin{bmatrix} \cos\mu - a\sin\mu & b\sin\mu \\ -\dfrac{1+a^2}{b}\sin\mu & \cos\mu + a\sin\mu \end{bmatrix}$$

This matrix can be split into the sum of two matrices:

$$\mathrm{M} = \mathrm{I}\cos\mu + \mathrm{J}_1\sin\mu, \qquad \mathbf{2.81}$$

where I is the unit matrix, and

$$\mathrm{J}_1 = \begin{bmatrix} -a & b \\ -\dfrac{1+a^2}{b} & a \end{bmatrix}. \qquad \mathbf{2.82}$$

from the equilibrium orbit in its plane by x_1, and perpendicular to the plane by x_2, trajectories adjacent to the equilibrium orbit satisfy the differential equation

$$\frac{d^2x_j}{ds^2} = -k_j s,$$

where k_j is constant within a constituent section of the unit cell, but, in general, different for each one; s is the arc length measured along the equilibrium orbit.

Observe the properties of this matrix.

$$\det J_1 = 1,$$
$$\operatorname{tr} J_1 = 0,$$
$$J_1^2 = -I.$$

From this last property of the matrix J, it follows by direct calculation that

$$M^2 = I \cos 2\mu + J_1 \sin 2\mu$$

and, in general,

$$M^n = I \cos n\mu + J_1 \sin n\mu. \qquad \textbf{2.83}$$

If, therefore, the position of the particle and the slope of the trajectory at the entrance of the first unit cell are given by

$$\mathbf{X}_0 = \begin{bmatrix} x_0 \\ x'_0 \end{bmatrix},$$

then the position of the particle and the trajectory slope at the end of unit cell number n will be

$$\begin{bmatrix} x_n \\ x'_n \end{bmatrix} = M^n\mathbf{X}_0 = \mathbf{X}_0 \cos n\mu + J_1 \mathbf{X}_0 \sin n\mu. \qquad \textbf{2.84}$$

In this equation, $J_1 \mathbf{X}_0$ is a column vector which depends only on the matrix elements of a single unit cell, since the parameters μ, a and b can be readily expressed in terms of the matrix elements m_{jk}:

$$\cos \mu = \tfrac{1}{2}(m_{11}+m_{22}),$$
$$a = -\frac{1}{2}\frac{(m_{11}-m_{22})}{\sin \mu},$$
$$b = \frac{m_{12}}{\sin \mu}.$$

In successive addition of unit cells, only the factors $\cos n\mu$ and $\sin n\mu$ change, but these factors are limited to the range between $+1$ and -1. Therefore if $|\operatorname{tr} M| < 2$, the system can confine the beam irrespective of the total length of the flight path of the particles in it.

(ii) If $\quad |\operatorname{tr} M| > 2$

similar arguments show that M has the form

$$M = I \cosh \mu + J_2 \sinh \mu, \qquad \textbf{2.85}$$

where I is again the unit matrix, and

$$J_2 = \begin{bmatrix} -a & b \\ \dfrac{1-a^2}{b} & a \end{bmatrix},$$

where a, b and μ are characteristic of the unit cell, and are real numbers, different from zero. Observe that $\det J_2 = -1$, $\mathrm{tr}\, J_2 = 0$, and $J_2^2 = I$, and hence

$$M^n = I \cosh n\mu + J_2 \sinh n\mu. \tag{2.86}$$

Thus the relation between the positions and trajectory slopes at the entrance to the first unit and at the exit of the nth unit is given by the equation

$$\begin{bmatrix} x_n \\ x'_n \end{bmatrix} = \mathbf{X}_0 \cosh n\mu + J_2 \mathbf{X}_0 \sinh n\mu.$$

Thus, the beam is divergent, because both the displacement of the particle and the slope of its trajectory increases as additional units are traversed.

(iii) Finally the limiting case, $\mathrm{tr}\, M = 2$, has to be investigated. The matrix must take the form

$$M = \begin{bmatrix} 1-a & b \\ -\dfrac{a^2}{b} & 1+a \end{bmatrix} = I + aJ_3,$$

where a and b are real numbers.

$$J_3 = \begin{bmatrix} -1 & \dfrac{b}{a} \\ -\dfrac{a}{b} & 1 \end{bmatrix},$$

with the properties $\det J_3 = 0$, $\mathrm{tr}\, J_3 = 0$, $J_3^2 = 0$.

Hence $\quad M^n = I + naJ_3.$

Therefore the condition $\mathrm{tr}\, M = 2$ leads to a divergent system again.*

A field-free region of length d is characterized by the matrix

$$M_0 = \begin{bmatrix} 1 & d \\ 0 & 1 \end{bmatrix}$$

which is a special case of the previous one. One knows intuitively that in field-free space there is no focusing. The result of a formal calculation agrees with this, since

$$M_0^n = \begin{bmatrix} 1 & nd \\ 0 & 0 \end{bmatrix}$$

It is left as an exercise for the reader to determine the condition under which a pair of identical quadrupole lenses, with alternating focusing planes and a given

*In this case there is one particular input condition which allows particles to pass a unit cell and enter the subsequent one under the same condition. This is so because $\det J_3 = 0$ and hence it is possible that $J_3\mathbf{X}_0 = 0$ even if $\mathbf{X}_0 \neq 0$, namely if $x'_0 = (a/b)x_0$.

gap between them, will act as a focusing system, and compare the result with the conclusions of section 2.5.

Finally a few additional remarks must be made.

(a) In the above considerations, a unit cell was taken between two boundaries separating regions of different field distributions. If a section is considered between another pair of surfaces the same distance apart, the shift of the reference planes leaves the matrix M invariant.

Let the matrix A (with matrix elements a_{jk}) characterize the section between the original and the shifted reference planes, and let the matrix N characterize the overlapping section (det A = det N = 1; see Figure 29). In this case the output vector **X** at the shifted reference plane can be written in two alternative ways in terms of the input vector $\mathbf{X}_0$ given at the original reference plane:

$$\mathbf{X} = \mathrm{AM}\mathbf{X}_0 \qquad \text{or} \qquad \mathbf{X} = \mathrm{ANA}\mathbf{X}_0.$$

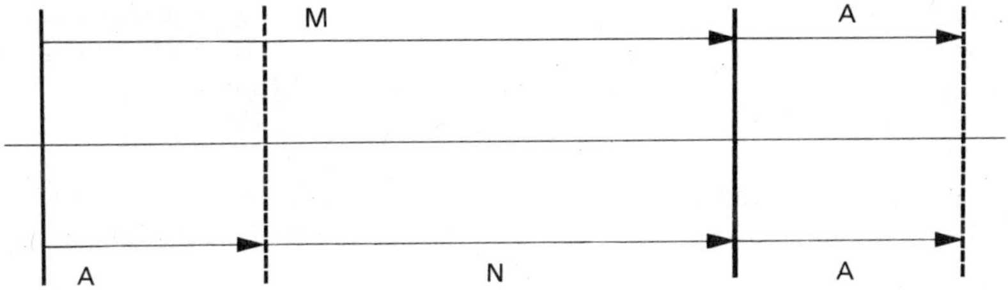

Figure 29

Hence the matrix representing the unit cell with the shifted reference plane is

$$\mathrm{M}' = \mathrm{AN} = \mathrm{AMA}^{-1}$$

where A^{-1} is the inverse matrix of A, that is, $\mathrm{AA}^{-1} = \mathrm{A}^{-1}\mathrm{A} = \mathrm{I}$, or explicitly:

$$\mathrm{A}^{-1} = \begin{bmatrix} a_{22} & -a_{12} \\ -a_{21} & a_{11} \end{bmatrix}.$$

By direct substitution one obtains

$$\mathrm{A}^{-1}\mathrm{J}_i\mathrm{A} = \mathrm{J}_i \qquad (i = 1, 2, 3)$$

and therefore $\mathrm{M}' = \mathrm{AMA}^{-1} = \mathrm{M}$. **2.87**

Thus, if M characterizes a unit cell, equation **2.87** is valid regardless of whether the cell is focusing or defocusing. This means that the matrix characteristic of the unit cell of a periodic electron-optic system is independent of the position of the reference plane.

(b) If the periodic electron-optic system is focusing (that is, $|\mathrm{tr\ M}| < 2$), the trajectories adjacent to the equilibrium orbit are periodic. It must be emphasized, however, that in general the trajectories do not have a sinusoidal shape. To obtain the trajectories it is necessary to investigate in detail the solution of the differential equation

$$\frac{d^2x_j}{dt^2} = -k_j^2(s)x_j, \qquad \mathbf{2.88}$$

where $k_j^2(s)$ varies periodically, and the period is equal to the length L of the unit cell, that is, $k_j(s+L) = k_j(s)$. Equation **2.88** is a Hill-type differential equation, and its analysis is simplified under the conditions we considered, namely, if $k_j(s)$ is sectionally constant. Under these conditions the analysis leads to the same stability conditions as those found above, and it gives the equations of the trajectories too. Differential equations of the type **2.88** with a periodic coefficient play an important part in the study of wave propagation in periodic structures, and the reader should consult the literature (for example, Brillouin, *Wave Propagation in Periodic Structures*, Chapter 8).

(c) In the preceding discussion it was assumed that the electrons were subject to externally applied fields only, and their mutual interaction could be neglected. This is equivalent to saying that the electric field acting on the electrons can be derived from a static potential Φ, which satisfies Laplace's equation $\nabla^2\Phi = 0$. The interaction between the electrons can be taken into account by considering the modification of the potential distribution due to the presence of electrons in the field, that is, by requiring that the potential satisfies Poisson's equation

$$\nabla^2\Phi = -\frac{\rho}{\varepsilon_0}, \qquad \mathbf{2.89}$$

where ρ is now the charge density, a function of the coordinates. The convection current density is, by definition, $\mathbf{j} = \rho\mathbf{v}$, and therefore, as a result of the modification of the potential distribution by the space charge, the trajectory equations will contain terms which depend on the beam current-density distribution. While one knows intuitively that the repulsive forces between electrons limit the charge which can be concentrated in unit volume, or the current density which can flow in a beam, the rigorous description of space-charge effects in quantitative terms presents very difficult mathematical problems.

The only problem we wish to discuss very briefly is the relationship between the electron beam current and the magnetic field confining it to a cylindrical region of constant diameter. The feasibility of such beam confinement by magnetic fields can be readily seen.

Let us consider the electron motion in a uniform magnetic field; the electrons are emitted by a source in the magnetic field and are accelerated by an electric field of rotational symmetry about an axis parallel to the magnetic field. Under these conditions the radial motion of the electron is determined by the differential equation

$$\frac{d^2r}{dt^2} - r\left(\frac{d\theta}{dt}\right)^2 = -\frac{e}{m}\frac{\partial\Phi}{dr} - \frac{e}{m}Br\frac{d\theta}{dt} \qquad \mathbf{2.90}$$

Because of the conservation of the canonical momentum (see equation **1.26a**), the motion is subject to the condition

$$\frac{d\theta}{dt} = \frac{1}{2}\frac{e}{m}B\left(1-\frac{r_0^2}{r^2}\right). \qquad \textbf{2.91}$$

where r_0 is the distance of the source point from the symmetry axis.

The electron will stay at a constant distance from the symmetry axis, if it has no radial velocity, and in this case $d^2r/dt^2 = 0$. Under this condition the combination of equations **2.90** and **2.91** gives

$$\frac{(r/r_0)^4-1}{(r/r_0)^3} = \frac{\partial\Phi/\partial r}{(e/4m)B^2 r_0}. \qquad \textbf{2.92}$$

For a given r_0 and $\partial\Phi/\partial r$, $r/r_0 \to 1$ as $B \to \infty$, that is, the constant radius r of the electron trajectory approaches the radius r_0 which defines the position of the source point, irrespective of the origin of the potential Φ.

Let us now consider a beam of electrons emitted by an extended source located in a uniform magnetic field. Suppose that after an initial acceleration parallel to the magnetic field, the electrons move in a region where there is no field in the absence of the electrons. In the presence of the electrons there will be a field determined by equation **2.89**, and the charge in a cylinder of radius r about the symmetry axis and of unit length is, by Gauss's theorem,

$$Q_1 = \int \rho\, dV = -\varepsilon_0 \int \nabla\Phi . d\mathbf{s} = -2\pi\varepsilon_0 r\frac{\partial\Phi}{\partial r}. \qquad \textbf{2.93}$$

If U_{eff} denotes an effective potential, which lies between the potential on the symmetry axis and that at radius r, an effective value for the drift velocity of the electrons can be introduced, that is, $v_{\text{eff}} = (2eU_{\text{eff}}/m)^{\frac{1}{2}}$, and in terms of these quantities the beam current is

$$I = Q_1\left(\frac{2e}{m}U_{\text{eff}}\right)^{\frac{1}{2}}. \qquad \textbf{2.94}$$

Thus equations **2.92**, **2.93** and **2.94** give

$$\left(\frac{r}{r_0}\right)^4 - 2a\left(\frac{r}{r_0}\right)^2 - 1 = 0, \qquad \textbf{2.95}$$

with $$\frac{1}{a} = 2^{\frac{1}{2}}\pi\varepsilon_0\left(\frac{e}{m}\right)^{\frac{3}{2}}\frac{B^2 U_{\text{eff}}^{\frac{1}{2}}}{I} r_0^2.$$

If r_0 is the radius of the cathode, equation **2.95** relates the radius of the confined beam to the beam current and the applied magnetic field.

If the cathode is in a region where there is no magnetic field, the conditions of beam confinement take a different form. Under these conditions the canonical momentum at the source vanishes if the initial electron velocity is zero (see section 1.2.2). Therefore equation **2.91** reduces to $d\theta/dt = \frac{1}{2}(eB/m)$, and hence equation **2.92** becomes

$$\frac{\partial \Phi}{\partial r} = \frac{eB^2}{4m} r,$$

which gives on integration

$$\Phi(r) = \Phi(0) + \frac{eB^2}{8m} r^2, \qquad \textbf{2.96}$$

where the first term gives the potential on the symmetry axis.

Assuming that the potential is independent of z, equation **2.89** takes the form

$$\frac{1}{r}\frac{\partial}{\partial r}\left(r \frac{\partial \Phi}{\partial r}\right) = -\frac{\rho}{\varepsilon_0},$$

and the potential distribution, given by equation **2.96**, requires a space-charge distribution

$$\rho = -\frac{\varepsilon_0}{2}\frac{e}{m}\Phi. \qquad \textbf{2.97}$$

As the field is conservative and there is no radial motion

$$\dot z^2 + (r\dot\theta)^2 = 2\frac{e}{m}\Phi$$

and hence $\quad \dot z = \left[2\frac{e}{m}\Phi(0)\right]^{\frac{1}{2}}.$

That is, the electron velocity in the present case is independent of r. Therefore the current carried by a beam of radius r is

$$I = \pi r^2 \rho \dot z = -\left(\frac{1}{2}\right)^{\frac{1}{2}} \varepsilon_0 \pi \left(\frac{e}{m}\right)^{\frac{3}{2}} B^2 \Phi_0^{\frac{1}{2}} r^2, \qquad \textbf{2.98}$$

where $\Phi_0 = \Phi(0)$ is the potential on the symmetry axis.

If a channel is formed by a conductor of radius r at the potential $\Phi_r = \Phi(r)$, the intensity of the beam filling this channel is given by substituting Φ_0 from **2.96** into **2.98**:

$$I = 8\pi\varepsilon_0 \left(\frac{e}{2m}\right)^{\frac{1}{2}} \frac{e}{8m} B^2 r^2 \left[\Phi_a - \frac{e}{8m} B^2 r^2\right]^{\frac{1}{2}}.$$

If Φ_r is kept constant, the intensity of the beam filling the channel of fixed radius r varies with the magnetic field B, and reaches a maximum if

$$\frac{e}{8m} B^2 r^2 = \frac{2}{3}\Phi_r. \qquad \textbf{2.99}$$

Hence the maximum current which can be accommodated in the channel of radius r if the boundary of the channel is at a potential Φ_r is

$$I_{max} = \frac{16}{3}\pi\varepsilon_0\left(\frac{e}{6m}\right)^{\frac{1}{2}}\Phi_r^{\frac{3}{2}}. \qquad \mathbf{2.100}$$

When an extended electron source is outside the uniform magnetic field and is adjusted to give maximum current in a given channel, as described above, the conditions of the 'Brillouin flow' are established. It is interesting to note that the relation between the beam intensity and the potential at the boundary of the beam is of the same form as the Childs–Langmuir law which will be discussed in section 5.2.2.

It must be emphasized that the conditions chosen in the above discussion are idealized beyond what can be achieved under practical circumstances. Already an extension of the paraxial ray equation to include the space-charge effect shows that, under the conditions of Brillouin flow, the beam envelope undergoes radial oscillations, with the result that the beam diameter shows periodic variations along the beam axis with a spatial period $2^{\frac{1}{2}}\pi(\dot{z}/\omega_c)$, where ω_c is the cyclotron frequency. When the beam is completely immersed in a very strong magnetic field resulting in a rectilinear flow, it is subject to instabilities and collapses if the intensity is increased beyond a threshold. For more information on these problems the reader should consult the literature on high-current electron beams.

2.9 Motion of neutral dipole particles in static fields

Atoms and molecules often have a magnetic moment $\boldsymbol{\mu}$ which is proportional to the intrinsic angular momentum $\mathbf{J}$ of the particle. If such a particle is situated in an electromagnetic field, its motion is affected by the interaction between the field and the dipole moment. An adequate analysis of this interaction requires quantum-mechanical considerations. However, many problems of practical significance can be described in classical terms if a few simple assumptions, justified by quantum mechanics, are introduced in addition to classical principles. The limits of the validity of this approach must be carefully observed and, in particular, care must be taken to avoid any conflict with the uncertainty principle. We shall begin with entirely classical considerations.

A magnetic moment $\boldsymbol{\mu}$ situated in a magnetic field of flux density $\mathbf{B}$ is subject to a torque

$$\mathbf{N} = \boldsymbol{\mu}\times\mathbf{B}, \qquad \mathbf{2.101}$$

which tends to align the magnetic dipole parallel to the field. If the angle between the dipole and the field is initially θ, and the dipole is rotated through an angle $d\boldsymbol{\theta} = \mathbf{e}_N\,d\theta$ about the direction of $\mathbf{N}$ ($\mathbf{e}_N$ is a dimensionless unit vector parallel to $\mathbf{N}$), the work done is

$$\begin{aligned} dW = \mathbf{N}.d\boldsymbol{\theta} &= (\boldsymbol{\mu}\times\mathbf{B}).d\boldsymbol{\theta} \\ &= \mu B \sin\theta\, d\theta \\ &= -\mu B\, d(\cos\theta) = -d(\boldsymbol{\mu}.\mathbf{B}). \end{aligned}$$

Hence the potential energy of the dipole $\boldsymbol{\mu}$, situated in the magnetic field of flux density $\mathbf{B}$, is

$$U = -\boldsymbol{\mu}.\mathbf{B} \qquad \mathbf{2.102}$$

and the force acting on the magnetic moment is

$$\mathbf{F} = -\nabla U = \nabla(\boldsymbol{\mu}.\mathbf{B}) = \boldsymbol{\mu}\times(\nabla\times\mathbf{B})+(\boldsymbol{\mu}.\nabla)\mathbf{B}. \qquad \mathbf{2.103}$$

This equation shows that in a uniform magnetic field, there is no force acting on the particle and it is subject to a torque only.

If the particle has an angular momentum parallel to its magnetic moment, $\boldsymbol{\mu} = \gamma\mathbf{J}$ with γ = constant, the torque produces a change in the angular momentum at a rate

$$\frac{d\mathbf{J}}{dt} = \mathbf{N}$$

and hence by equation **2.101**

$$\frac{d\mathbf{J}}{dt} = (\boldsymbol{\mu}\times\mathbf{B}) \qquad \mathbf{2.104}$$

or $$\frac{d\boldsymbol{\mu}}{dt} = \gamma(\boldsymbol{\mu}\times\mathbf{B}). \qquad \mathbf{2.104a}$$

Introducing $\quad \boldsymbol{\Omega} = -\gamma\mathbf{B} = -\frac{\mu}{J}\mathbf{B},$

equations **2.104** and **2.104a** give

$$\frac{d\mathbf{J}}{dt} = \boldsymbol{\Omega}\times\mathbf{J} \quad \text{and} \quad \frac{d\boldsymbol{\mu}}{dt} = \boldsymbol{\Omega}\times\boldsymbol{\mu}$$

respectively, indicating that the angular momentum and the magnetic moment rotate with an angular velocity $\boldsymbol{\Omega}$ about the direction of the magnetic field. This angular velocity depends on the magnetic field and the constant γ, the so-called magneto-gyric ratio.

How does this behaviour of a classical dipole moment compare with that of the magnetic moment of an atom? In classical physics any vector quantity, defined by its components, can be measured with precision. As a consequence of the uncertainty principle, however, the components of the angular momentum (or those of the magnetic moment) of an atom cannot be measured simultaneously. Simultaneous measurements can be performed with precision on one component of the angular momentum, say J_1, parallel to some specified reference direction, and on the length of the vector, without knowing the precise direction of the latter. In other words, one cannot determine both the magnitude of the vector and its orientation at the same time.

The component J_1 is always found to be either an odd or an even multiple of $\frac{1}{2}\hbar$(where $\hbar = h/2\pi$, h is Planck's constant) depending on the structure of the particular atom in question. In fact the 'angular momentum state' of the atom is characterized by the integral (or half-integral) number j, which gives the maximum angular momentum component $j\hbar$ parallel to the reference direction; the possible values of this component are $m\hbar$ with $m = -j, -j+1, \ldots, j-1, j$.

The length of the angular momentum vector is given by $[j(j+1)]^{\frac{1}{2}}\hbar$ instead of $j\hbar$ as one would expect classically. As a model of the quantum-mechanical results, one may visualize the angular momentum (and magnetic moment) vector as lying *somewhere* on a cone with the reference direction as symmetry axis, but without knowing its precise position.

The reference direction mentioned repeatedly can be specified as that of the direction of the magnetic field in which the particle is situated, but the model as described above still does not say anything about the precession of the magnetic moment about the magnetic field. To apply the concept of 'precession' to atomic magnetic moments, the classical result from equation **2.104** must be reinterpreted by means of quantum-mechanical considerations. This is beyond the scope of our discussion.

Of the classical formulae seen so far, equation **2.102** is in agreement with quantum theory if, in calculating the scalar product $\boldsymbol{\mu}.\mathbf{B} = \gamma\mathbf{J}.\mathbf{B} = \gamma J_1 B$, J_1 is taken as one of the allowed values. Hence one expects that the *translatory* motion of the dipole particle is determined by the equation of motion $M\ddot{\mathbf{r}} = \mathbf{F}$, where M is the mass of the particle. The force is

$$\mathbf{F} = \nabla(\gamma\mathbf{J}.\mathbf{B}) = \gamma J_1 \nabla B = \mu_1 \nabla B, \qquad \textbf{2.105}$$

where μ_1 is the value of the component of the magnetic moment parallel to the field $\mathbf{B}$, as allowed by quantum theory.

This assertion can be justified by quantum-mechanical arguments, and it is no more surprising than the fact that the motion of charged atomic particles in electromagnetic fields can be described by classical equations of motion. Equation **2.105** may be recognized as the basis of the description of the Stern–Gerlach experiment. The applicability of equation **2.105** to the description of the motion of *neutral* magnetic dipole particles in inhomogeneous magnetic fields naturally leads to the investigation of the possibility of developing 'atomic-beam optics'.

For simplicity let us restrict our further considerations to particles of 'spin $\frac{1}{2}$', for example, alkali atoms in ground state. This implies that the component of the angular momentum of the particle parallel to a reference direction has two possible values $J_1 = \pm\frac{1}{2}\hbar$. In this $\gamma = e/mc$, where e and m are the charge and mass of the *electron*, and c is the velocity of light. The quantity $\mu_1 = \frac{1}{2}(e\hbar/mc)$ is known as the Bohr magneton; the component of the magnetic moment parallel to the reference direction can be $\pm\mu_1$.

The equation of motion

$$\ddot{\mathbf{r}} = \pm\frac{\mu_1}{M}\nabla B,$$

for a dipole particle can be compared with that for a particle of mass M and charge $\pm q$ in an electric field $\mathbf{E}$:

$$\ddot{\mathbf{r}} = \pm\frac{q}{M}\mathbf{E}.$$

The magnetic moment μ_1 in the former case, and the charge q in the latter, obviously play the same role; the motion of the dipole particle is the same as that of the charged particle, provided the spatial distribution of the *gradient* of the modulus of the magnetic field in the first case is the same as the distribution of the electric field in the second case. In other words, all the methods of charged-particle beam optics are applicable to beams of neutral dipole particles, and identical effects are obtained if the magnetic field distribution is chosen appropriately. In the case of charged particles, the field is usually chosen to operate on particles of the same charge sign. Beams of dipole particles, on the other hand, usually consist of a mixture of particles in the opposite spin states, and in a given field these acquire the same acceleration, but in opposite directions. A very important practical consequence is the possibility of producing 'polarized' atomic beams, that is, beams which contain particles in one spin state only.

Before continuing it should be noted that devices of atomic beam optics are much less flexible than those of charged particle optics, because the dipole moments are very small and the fields available yield relatively small effects. Suppose, for example, that an atom of kinetic energy T_a passes through a field of uniform gradient of length s. The angular deflection is given by

$$\theta_a \simeq \frac{1}{2}\frac{\mu s \nabla B}{T_a}.$$

Compare this deflection with that suffered by an electron beam of kinetic energy T_e, in a uniform electric field of the same length:

$$\theta_e \simeq \frac{1}{2}\frac{eEs}{T_e}.$$

That is, the ratio of the deflections is

$$\frac{\theta_a}{\theta_e} \simeq \frac{\mu}{e}\frac{\nabla B}{E}\frac{T_e}{T_a}.$$

If one considers an atomic beam emerging from a source at a temperature higher than room temperature, the mean kinetic energy of the atoms in the beam is of the order of 0·1 eV. Since the Bohr magneton $\mu \simeq 10^{-23}$ A m^2, the deflection of such an atomic beam, in a magnetic field of uniform gradient of 1000 Wb m^{-3}, is smaller than the corresponding deflection of a beam of 1 keV electrons in a uniform electric field of 1 kV m^{-1}. Note that this electric field is quite small from a practical point of view, whereas the magnetic field gradient considered is in a range which represents the limit of technical feasibility.

The kind of magnetic field with uniform gradient mentioned above is used in a Stern–Gerlach experiment, and the main problem consists in realizing the uniformity of the gradient, either by shaping the pole-tips of the magnet or – if no iron is used – by a suitable arrangement of the current-carrying conductors. Without going into details, it must be mentioned that a magnetic field distribution, which can most readily be described in an analytical form, is that generated by the current flowing in opposite directions in two straight conductors parallel to one another. If the coordinate system is chosen as shown in Figure 30, and the two conductors can be considered infinitely long, the magnetic field can be derived from a vector potential with a z-component only, which is a function of x and y only; so $A_z = A(x, y)$, while $A_x = A_y = 0$.

$$A_z \propto \ln \frac{r_1^2}{r_2^2},$$

where $r_{1,2}^2 = (a \pm x)^2 + y^2$, and $2a$ is the distance between the conductors. It is left to the reader to show that in a small region such as that indicated in Figure 30, the field gradient can be considered constant. The 'two-wire field' can be generated in the gap between iron pole pieces, if they are shaped in such a way that their surfaces coincide with the (scalar) equipotential surfaces of the required field, and provided that the iron is sufficiently far from magnetic saturation.

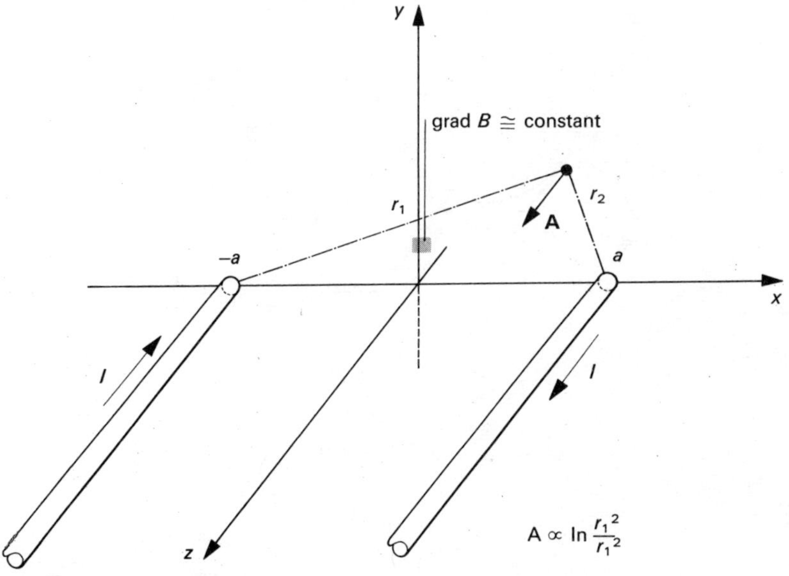

Figure 30

To consider the possibility of focusing a beam of dipole particles, a field must be found where the distribution of the gradient is reminiscent of the electric field distribution in an electron lens. This implies rotational symmetry, and the requirement of a quadratic increase with radius of the gradient of the modulus of the magnetic field.

Suppose $2n$ wedge-shaped magnetic pole pieces are arranged symmetrically with alternating polarity on a circular cylinder of radius R. Within the cylinder the flux density $\mathbf{B}$ can be derived from a scalar potential and, if the lengths of the pole pieces $L \gg R$ in the neighbourhood of the cylinder axis (that is, for radii $r \ll R$) the potential distribution is described to a good approximation by

$$U = \text{constant} \times r^n \cos(n\theta)$$

giving the flux density $\mathbf{B} = -\nabla U$. This field is, of course, not rotationally symmetrical, but one can readily see that its modulus,

$$B(r) = B_0\left(\frac{r}{R}\right)^{n-1} \qquad (B_0 \equiv B(R)),$$

has the required property. Taking six poles ($n = 3$; see Figure 31), one has the required field distribution.

$$B(r) = B_0\left(\frac{r}{R}\right)^2.$$

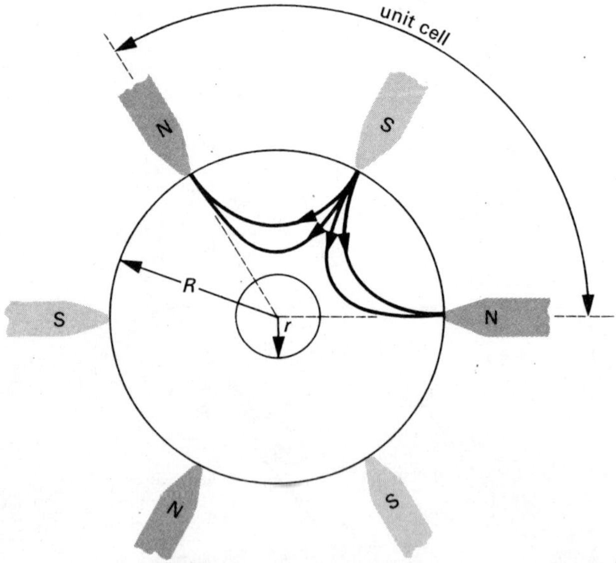

Figure 31

In such a field the motion of the particle with spin $\frac{1}{2}$ is described by the equation of motion

$$m\ddot{\mathbf{r}} = \pm\mu_1 \nabla B = \pm 2\mu_1 B_0 R^{-2}\mathbf{r},$$

that is, the particle is subject to a force proportional to its displacement from the cylinder axis. Hence – following the usual arguments – the differential equation of the trajectories adjacent to the cylinder axis (z-axis) can be written as

$$r'' = -K^2 r, \qquad \textbf{2.106}$$

$$\text{with} \quad K^2 = \pm R^{-2}\frac{\mu_1 B_0}{W_k}, \qquad \textbf{2.107}$$

where W_k denotes the kinetic energy of the particles.

It follows from equation **2.106** that the sextupole magnet acts as a lens for magnetic dipole particles and its optical properties can be described by the formalism established earlier (see sections 2.3 and 2.5). Particular attention must be paid to the fact that a given field acts as a focusing field on particles in one spin state, for which $K^2 > 0$, and at the same time it acts as a defocusing lens for particles in the other spin state, because for them $K^2 < 0$. This state of affairs is the basis of the use of sextupole fields for the polarization of atomic beams. The polarization of the beam emerging from the magnet can be characterized by the relative number of the particles in the two spin states. More precisely, if the numbers of particles emerging from the magnet in the two spin states are denoted by N^+ and N^- respectively, the polarization is defined as

$$P = \frac{N^+ - N^-}{N^+ + N^-}. \qquad \textbf{2.108}$$

This quantity can be estimated in the following way.

The solution of equation **2.106** can be written in the familiar form with matrices M of the type **2.62**. The maximum displacement of the particles from the axis within the field is limited by the structure of the magnet to a maximum value, say R, and this imposes a limitation on the range of the initial position r_0, and of the initial trajectory slope r'_0. If r_0 and r'_0 are plotted along the axes of a Cartesian coordinate system, a permissible pair of values (r_0, r'_0) is represented by a point, and all the points (r_0, r'_0) representing initial conditions for particles which can pass the field unobstructed, fill an area in the plane. This area can be used as a measure of the 'transmission' of the lens, if it can be assumed that all initial values are equally probable. This is the case if the lens is irradiated by an extended source. Under such circumstances the polarization can be defined in terms of the areas, A^+ and A^-, occupied by the representative points corresponding to particles in opposite spin states:

$$P = \frac{A^+ - A^-}{A^+ + A^-}. \qquad \textbf{2.109}$$

Thus our task is reduced to the determination of these areas.

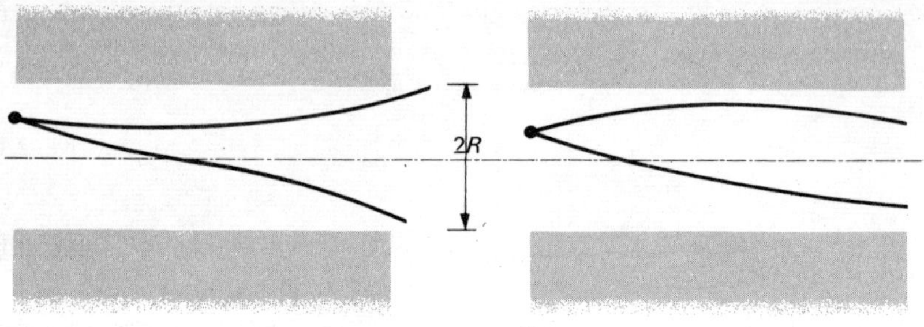

Figure 32

Consider first the particles which are *defocused* in the field. The coordinates and the trajectory slopes at the exit are related to the initial values by the equations

$$r = r_0 \cosh KL + \frac{r_0'}{K} \sinh KL,$$

$$r' = r_0 K \sinh KL + r_0' \cosh KL,$$

where L is the length of the field. The trajectories in the defocused beam are therefore always convex towards the optic axis (see Figure 32a) and particles can pass a channel of radius R provided $|r| \leqslant R$. Hence, using the first of the equations above, one obtains for the limiting values of the initial conditions

$$\frac{r_0}{R/(\cosh KL)} + \frac{r_0'}{KR/(\sinh KL)} = \pm 1.$$

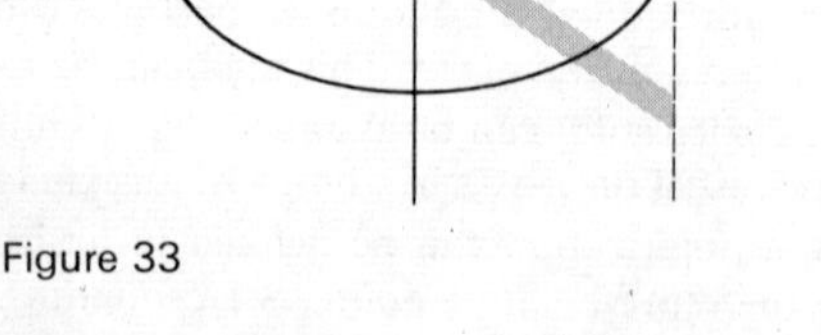

Figure 33

This equation defines two parallel straight lines on the (r_0, r_0') plane which, together with the condition $-R \leqslant r_0 \leqslant R$, determine the area

$$A^- = \frac{4KR^2}{\sinh KL} \qquad \textbf{2.110}$$

(see Figure 33).

For the *focused* beam, the coordinates and the trajectory slopes at the exit are related to the initial values by the equations

$$r = r_0 \cos KL + \frac{r'_0}{K} \sin KL,$$

$$r' = -r_0 K \sin KL + r'_0 \cos KL,$$

and the trajectories are always concave towards the optic axis (see Figure 32b).

To simplify the analysis of this case it will be assumed that $KL \geqslant \pi$. This condition can be met if the field is sufficiently high and extends over a sufficient length L. If this condition is satisfied, the particle must reach maximum displacement within the field, and where this occurs the slope of the trajectory is zero.

Taking the square of both equations above and adding, one obtains

$$\left(\frac{r_0}{R}\right)^2 + \left(\frac{r'_0}{KR}\right)^2 = \left(\frac{r}{R}\right)^2 + \left(\frac{r'}{KR}\right)^2.$$

Taking into account the conditions at maximum displacements, $r_{max} = R$ and $r'_{max} = 0$, the above equation defines the contour of the area within which the permissible values (r_0, r'_0) lie:

$$\left(\frac{r_0}{R}\right)^2 + \left(\frac{r'_0}{KR}\right)^2 = 1,$$

which is an ellipse of area

$$A^+ = KR^2\pi. \qquad \textbf{2.111}$$

Hence $$\frac{A^+}{A^-} = \frac{\pi}{4} \sinh KL$$

or from equation **2.109**:

$$P = \frac{\frac{1}{4}\pi \sinh KL - 1}{\frac{1}{4}\pi \sinh KL + 1}, \qquad \textbf{2.112}$$

where $KL \geqslant \pi$. For example, with $KL \simeq \pi$, equation **2.112** predicts that the beam emerging from the field is about 80 per cent polarized.

It has to be remembered that only spin $\frac{1}{2}$ particles have been considered. If the particles have higher angular momentum, the beam will consist of several pairs of focused and defocused components, each described by a different parameter K. One must also emphasize that throughout the considerations above, it was assumed that the particles have a permanent dipole moment, independent of the field in which they are situated. Furthermore the origin of the dipole moment effective in the interaction, which governs the motion in the field, was completely disregarded. The calculation performed with reference to spin $\frac{1}{2}$ particles tacitly assumed that the angular momentum and the accompanying magnetic moment are purely of electronic origin, that is, the effect of coupling between nuclear and electronic spin was disregarded. The validity of this assumption depends upon the actual experimental conditions, but its further investigation goes beyond the scope of this book.

Further reading

O. KLEMPERER, *Electron Optics*, 2nd edn, Cambridge University Press, 1953.

Monograph which gives a clear presentation of fundamental principles and of considerations essential to the design of electron-optic devices.

J. R. PIERCE, *Theory and Design of Electron Beams*, van Nostrand, 1954.

One of the basic reference books; chapters on space-charge flow and electron gun design are outstandingly valuable.

P. A. STURROCK, *Static and Dynamic Electron Optics*, Cambridge University Press, 1955.

A monograph for the specialist at an advanced level. It relates the theory of electron-optic instruments and particle accelerators to classical optics and particle dynamics. The treatment is based on the mathematical theory of the variational principles.

S. FLUGGE (ed.), *Encyclopedia ('Handbuch') of Physics*, Vol. 23: *Optics of Corpuscules*, Springer, Berlin, 1956.

Review papers at an advanced level on – among other topics – Electron and ion optics (by W. Glaser, in German); Mass spectrometers (by H. Ewald, in German); Beta-ray spectrographs (by T. R. Gerholm, in English).

K. SIEGBAHN (ed.), *Alpha-, Beta- and Gamma-Ray Spectroscopy*, Vol. 1, North Holland Publishing Co., 1965, Chapter 3.

Beta-ray spectrometer theory and design; magnetic alpha-ray spectroscopy; high-resolution spectroscopy. A review for the specialist.

E. D. COURANT and H. S. SNYDER, *Theory of the Alternating-Gradient Synchroton*, Annals of Physics, Vol. 3, 1958, pp. 1–48.

General theory of the confinement of beams of charged particles by means of periodic focusing–defocusing fields.

L. BRILLOUIN, *Wave Propagation in Periodic Structures*, McGraw-Hill, 1946, and Dover, 1953, Chapter 8.

'Mathieu's equation and related problems' is an excellent introduction to the theory and applications of second-order linear differential equations with periodic coefficients.

Chapter 3
Motion of Charged Particles in Time-Dependent Electromagnetic Fields

. . . a theory which explains things by making definite assumptions as to the mutual action of the parts of a system, must at the same time admit a transfer of energy.

H. A. Lorentz

3.1 Introduction

The preceding chapters dealt with the motion of charged particles in static electromagnetic fields, that is, subject to a force

$$\mathbf{F} = e(\mathbf{E}+\mathbf{v}\times\mathbf{B}), \qquad \textbf{3.1}$$

where the fields **E** and **B** were constant in time and functions of the coordinates only. In this chapter we shall study the interaction between charged particles and fields which vary in time. The force acting on a particle is again given by equation **3.1**, but now the fields depend explictly on time as well as on the coordinates. At first glance, one might think that the time dependence of the fields introduces only formal mathematical problems. There are, however, fundamental differences between the nature of static and time-dependent electromagnetic fields which lead to phenomena not encountered previously. A few general observations may, therefore, give useful guidance before embarking on the investigation of any particular problem.

3.1.1 In all practical circumstances, the electromagnetic field in which the motion of particles is studied is confined to a region of finite dimensions, and the particles spend a finite length of time τ in the field. The time dependence of the fields has significant consequences only if the fields change appreciably during the transit time of the particles, that is, if the change in field intensity during this time is comparable with its mean value,

$$\mathbf{E}(t+\tfrac{1}{2}\tau)-\mathbf{E}(t-\tfrac{1}{2}\tau) \simeq \mathbf{E}(t), \qquad \mathbf{B}(t+\tfrac{1}{2}\tau)-\mathbf{B}(t-\tfrac{1}{2}\tau) \simeq \mathbf{B}(t).$$

If τ is so small that a first-order series expansion is appropriate, these conditions reduce to relations of the form

$$\frac{1}{\alpha}\frac{\partial\alpha}{\partial\tau}\tau \simeq 1,$$

where α represents the components of **E** and **B**.

If the fields vary periodically $\mathbf{E}(t+T) = \mathbf{E}(t)$, $\mathbf{B}(t+T) = \mathbf{B}(t)$, the effects arising from their time dependence will be noticeable, provided the transit time of the particles is comparable with the periodic time, $\tau \simeq T$. To illustrate this effect in a simple example, let us consider the deflection of an electron in a transverse electric field, which is uniform in space and varies sinusoidally in time with an angular frequency ω. This is the usual condition of operation of the electric deflector of a cathode ray tube.

If the initial momentum of the electrons is parallel to the z-axis of a Cartesian coordinate system, $\mathbf{p} = p\mathbf{e}_z$, and the field is $\mathbf{E} = \mathbf{e}_x E_0 \sin \omega t$, parallel to the x-axis, the electrons leave the field at an angle θ with respect to the z-axis.

$$\tan \delta = \frac{p_x}{p_z} = \frac{eE_0}{p} \int_{t-\frac{1}{2}\tau}^{t+\frac{1}{2}\tau} \sin \omega t' \, dt' = \frac{e\,E(t)\tau}{p} \frac{\sin \frac{1}{2}\omega\tau}{\frac{1}{2}\omega\tau},$$

where t is the instant at which the electron passes the middle of the deflecting field, and τ is the transit time. Had the field been static ($\omega = 0$) with an intensity equal to that of the alternating field at the instant t, the angle of deflection would have been given by

$$\tan \delta_s = \frac{e\,E(t)}{p} \tau.$$

Thus the effect of the time dependence of the field appears through the factor $(\sin \frac{1}{2}\omega\tau)/(\frac{1}{2}\omega\tau)$.

One might expect that all the results obtained for static fields would remain valid with the kind of modification obtained in the example above. This is often the case. But if the orders of magnitude of electron energies and frequencies make the factor $(\sin \frac{1}{2}\omega\tau)/(\frac{1}{2}\omega\tau)$ significantly different from unity, this general conclusion is not valid.

If the path length of the electrons in the field along the direction of the initial velocity, $v = p/m$, is l, then $\tau = l/v$, and, introducing the free-space wavelength of the alternating electromagnetic field, $\lambda = 2\pi c/\omega$, one finds

$$\theta \equiv \tfrac{1}{2}\omega\tau = \pi \frac{c}{v} \frac{l}{\lambda}.$$

Thus $(\sin \theta)/\theta$ is significantly different from unity if $l/\lambda \simeq v/c$.

If the velocity of the electron is obtained by accelerating it through a potential difference U, then

$$\frac{v}{c} = \left[1 - \left(1 + \frac{eU}{m_0 c^2}\right)^{-2}\right]^{\frac{1}{2}}$$

where $m_0 c^2 \simeq 500$ keV is the rest energy of the electron.

Therefore, unless $eU \ll m_0 c^2$, transit-time effects can be expected only if the path length of the particles is comparable with the free-space wavelength of the alternating field. One must, therefore, investigate whether the *physical* assumptions made in the course of the treatment of electron motion in static fields remain valid if the fields under consideration are confined to regions whose linear dimensions, and the path lengths of electrons in them, are comparable with the wavelength of the alternating fields.

3.1.2 The spatial distribution of electrostatic and magnetostatic fields are independent of one another. This was assumed throughout the considerations in the preceding chapters, and the electron motion was described in fields arising from the superposition of arbitrarily chosen electric and magnetic fields, when often one or the other was absent. If, however, fields are time-dependent, there is an inseparable relation between them, which manifests itself, for example, in the phenomenon of electromagnetic induction. The relationship between electric and magnetic fields and their sources is given by Maxwell's electromagnetic field equations, which read

$$\nabla . \mathbf{D} = \rho, \qquad \mathbf{3.2a}$$

$$\nabla . \mathbf{B} = 0, \qquad \mathbf{3.2b}$$

$$\nabla \times \mathbf{E} = -\frac{\partial \mathbf{B}}{\partial t}, \qquad \mathbf{3.2c}$$

$$\nabla \times \mathbf{H} = \mathbf{j} + \frac{\partial \mathbf{D}}{\partial t}, \qquad \mathbf{3.2d}$$

where ρ and $\mathbf{j}$ are the true charge and current densities, $\mathbf{E}$ and $\mathbf{D}$ are the electric field and displacement, and $\mathbf{B}$ and $\mathbf{H}$ are the magnetic flux density and field respectively. To describe the fields $\mathbf{E}$ and $\mathbf{B}$ as functions of coordinates and time, the set of partial differential equations **3.2a–d** have to be solved, taking into account the boundary conditions appropriate to the physical situation. In order to make such a solution possible, relationships of **D** to **E**, **H** to **B**, and **j** to **E** and **B** are required, in addition to the above equations. These relationships are characteristic of the medium carrying the electromagnetic field; *in vacuo*

$$\mathbf{D} = \varepsilon_0 \mathbf{E} \qquad \varepsilon_0 = \frac{10^{-9}}{36\pi} \text{ A s V}^{-1}\text{ m}^{-1},$$

$$\mathbf{B} = \mu_0 \mathbf{H} \qquad \mu_0 = 4\pi + 10^{-7} \text{ V s A}^{-1}\text{ m}^{-1}$$

($\varepsilon_0 \mu_0 = c^{-2}$; c = velocity of light). The convection-current density $\mathbf{j}$ arises as a consequence of the motion of charges:

$$\mathbf{j} = \rho \mathbf{v}$$

and by virtue of charge conservation it satisfies the continuity relation

$$\nabla . \mathbf{j} + \frac{\partial \rho}{\partial t} = 0. \qquad \mathbf{3.3}$$

The relationship between this convective current and the fields is implicitly contained in its definition, because $\mathbf{v}$ is related to the force given by equation **3.1** through the equation of motion

$$\frac{d}{dt}(m\mathbf{v}) = e(\mathbf{E} + \mathbf{v} \times \mathbf{B}). \qquad \mathbf{3.4}$$

Even in the absence of charges ($\rho = 0$) the electric field and the magnetic flux density distributions can be independent of one another only if both of them are constant in time, and in this case, they are conservative. Earlier we have frequently made use of the validity of this statement. If however the fields are time dependent, $\partial \mathbf{E}/\partial t \neq 0$, $\partial \mathbf{B}/\partial t \neq 0$, then $\nabla \times \mathbf{E} \neq 0$ and $\nabla \times \mathbf{B} \neq 0$. In this case neither **E** nor **B** can be derived from a scalar potential, and neither of the fields is conservative.

If, however, a vector potential, **A**, is introduced, such that

$$\mathbf{B} = \nabla \times \mathbf{A}, \qquad \textbf{3.5a}$$

then from equation **3.2c**, one finds

$$\nabla \times \left(\mathbf{E} + \frac{\partial \mathbf{A}}{\partial t}\right) = 0$$

Therefore the vector sum in the brackets can be derived from a scalar potential $\Phi(\mathbf{r}, t)$.

$$\mathbf{E} + \frac{\partial \mathbf{A}}{\partial t} = -\nabla \Phi,$$

and hence $\mathbf{E} = -\nabla \Phi - \frac{\partial \mathbf{A}}{\partial t}$. **3.5b**

Closer examination shows that there is some ambiguity in both Φ and **A**. If Φ and **A** are defined as above, the same field is described by the electromagnetic potentials

$$\Phi' = \Phi + \frac{\partial u}{\partial t},$$

$$\mathbf{A}' = \mathbf{A} + \nabla u,$$

where $u(\mathbf{r}, t)$ is an arbitrary scalar function. The equations above define a 'gauge transformation' of the electromagnetic potentials; it leaves the electromagnetic fields invariant. This freedom in the choice of the electromagnetic potentials led Lorentz to impose the condition

$$\nabla . \mathbf{A} + \frac{1}{c^2} \frac{\partial \Phi}{\partial t} = 0, \qquad \textbf{3.5c}$$

where c is the velocity of light *in vacuo*. Alternatively the condition $\nabla . \mathbf{A} = 0$ can be imposed ('Coulomb gauge'). For a discussion of the deeper implications of the choice of the apparently arbitrary conditions, the reader must consult some advanced textbook on the classical theory of electromagnetic fields.

It should be remembered that, whereas the general properties of electromagnetic fields are described by the Maxwell equations, their actual distribution becomes uniquely determined only by additional boundary conditions. Considering the boundary surface between two media (1) and (2), and denoting the unit vector normal to the surface by $\hat{\mathbf{n}}$, the boundary conditions are

$$\hat{\mathbf{n}} \times (\mathbf{E}_1 - \mathbf{E}_2) = 0 \qquad \hat{\mathbf{n}} . (\mathbf{B}_1 - \mathbf{B}_2) = 0,$$

$$\hat{\mathbf{n}} . (\mathbf{D}_1 - \mathbf{D}_2) = \sigma \qquad \hat{\mathbf{n}} \times (\mathbf{H}_1 - \mathbf{H}_2) = \mathbf{j}_s, \qquad \mathbf{3.6}$$

where σ and $\mathbf{j}_s$ denote the surface charge density and the surface current density respectively. The first pair of these relations expresses the continuity of the tangential component of **E** and that of the normal component of **B** on any boundary. According to the second pair of these equations, the charge density and the current density carried by the boundary surface are related to the discontinuity of the normal component of **D** and the discontinuity of the tangential component of **H** respectively. In subsequent sections there will be problems relating to the motion of free electrons in fields excited in wave guides or cavities, the walls of which will be assumed to be perfectly conducting. Since the fields vanish inside such conductors, the tangential component of **E** and the normal component of **B** must vanish in the immediate vicinity of such walls. The tangential component of **H** and the normal component of **D**, however, need not vanish in the vicinity of the walls, since they carry surface charges and surface currents.

After this brief summary of the general properties of electromagnetic fields, it must be stressed that the inherent coupling between time-dependent electric and magnetic fields does not mean that there cannot be regions in which the magnetic field is negligible in comparison with the electric field, and *vice versa*. As mentioned above, the actual field distributions depend on the boundary conditions. Considering, for example, a 're-entrant cavity' (see Figure 34), one finds that between the faces of the two protruding posts, there is practically no magnetic field but an electric field alone.

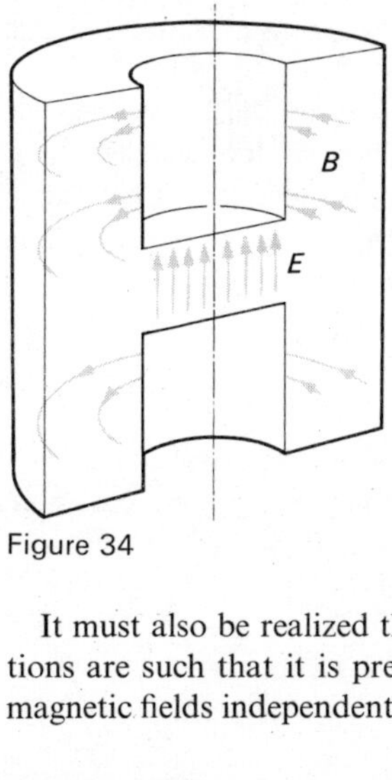

Figure 34

It must also be realized that there are situations in which the physical conditions are such that it is preferable to consider the time-dependent electric and magnetic fields independent from one another, as in the case of static fields. This is clearly the case if (i) at any given instant the surfaces of the electrodes designed to maintain the electric field are equipotential surfaces, that is, the line integral of the electric field between two electrodes is independent of the points and the path of integration chosen, and (ii) the instantaneous current in the conductors, designed to produce the magnetic field, is the same at any cross-section of the conductor. Whether these requirements are satisfied or not depends on the rate at which the fields change. In the case of periodically varying fields, the conditions are met if the gap between the electrodes and the distance between any two points on the conductors is negligible, in comparison with the free-space wavelength of the alternating field. It is useful to remember that the above requirements must also be met if one wishes to characterize the electric properties of the electrodes and of the current-carrying conductors in terms of a 'capacitance' and an 'inductance' respectively. Therefore briefly one can say that the fields behave in a 'quasi-static' manner if the electrodes and magnetizing coils form parts of a network which is adequately described in terms of 'lumped parameters'. For later reference it should be mentioned that in such networks the electromagnetic energy is stored in reactive elements (capacitance, inductance) and is dissipated in the resistive elements, but none can escape in the form of electromagnetic radiation.

3.1.3 The most significant features distinguishing the interaction of charged particles and time-dependent fields from that between charged particles and static fields arise from the fact that time-dependent electromagnetic fields are not conservative. Consider, for example, an alternating field confined to a finite region, such as a cavity resonator; outside this region the potential is everywhere the same. An electron emitted by a source in the field-free region, after passing the cavity, can have a kinetic energy different from the initial one, although it returned to a region where the potential is the same as that at its source. Energy is conserved only if the electromagnetic field, and the charges moving in it, are considered together as constituting a closed system.

With the aid of the Maxwell equations we can readily establish a relation which gives the energy balance between the field and the charges. If one takes the scalar product of both sides of equation **3.2d** with **E** and of equation **3.2c** with $-\mathbf{H}$, adding the equations thus obtained, one finds, using the vector identity

$$\nabla.(\mathbf{E}\times\mathbf{H}) = \mathbf{H}.\nabla\times\mathbf{E}-\mathbf{E}.\nabla\times\mathbf{H},$$

$$-\nabla.(\mathbf{E}\times\mathbf{H}) = \mathbf{j}.\mathbf{E}+\mathbf{E}.\frac{\partial\mathbf{D}}{\partial t}+\mathbf{H}.\frac{\partial\mathbf{B}}{\partial t}.$$

If the volume integral of each term is formed, the left-hand side can be turned into a surface integral by means of Gauss's theorem, to give the familiar equation:

$$-\int_{\mathrm{S}}(\mathbf{E}\times\mathbf{H}).d\mathbf{S} = \int_{\mathrm{V}}\mathbf{j}.\mathbf{E}\,dV+\frac{\partial}{\partial t}\int_{\mathrm{V}}(\tfrac{1}{2}\mathbf{E}.\mathbf{D}+\tfrac{1}{2}\mathbf{B}.\mathbf{H})\,dV \qquad 3.7$$

where S refers to the closed surface which forms the boundary of the volume V; $d\mathbf{S} = \hat{\mathbf{n}}\,dS$, and $\hat{\mathbf{n}}$ is the outward unit vector normal to dS.

Note that, dimensionally, each term represents the rate of change of some 'energy': the integrand on the left-hand side $\mathbf{E}\times\mathbf{H} = \mathbf{N}$ is the Poynting vector representing the electromagnetic energy flow per unit area per unit time across the boundary surface S; the integrand of the second term on the right-hand side is the electromagnetic energy stored in unit volume.

To interpret the first term on the right-hand side, write $\mathbf{j} = \rho\mathbf{v}$ and solve the equation of motion (**3.4**) for $\mathbf{E}$. Thus one finds

$$\begin{aligned}\mathbf{j}.\mathbf{E} &= \frac{m}{e}\rho\mathbf{v}.\frac{d\mathbf{v}}{dt} = \frac{1}{2}\frac{m}{e}\rho\frac{d}{dt}v^2\\ &= \frac{1}{2}\frac{m}{e}\rho\left(\frac{\partial v^2}{\partial t}+\mathbf{v}.\nabla v^2\right)\\ &= \frac{1}{2}\frac{m}{e}\left(\rho\frac{\partial v^2}{\partial t}+\mathbf{j}.\nabla v^2\right).\end{aligned}$$

To bring this equation to a more convenient form, the second term in the brackets can be rewritten using the vector identity $\nabla.(v^2\mathbf{j}) = \mathbf{j}.\nabla v^2+v^2\nabla.\mathbf{j}$, and taking account of the equation of continuity (**3.3**).

Thus $$\mathbf{j}.\nabla v^2 = \nabla.(v^2\mathbf{j})+v^2\frac{\partial\rho}{\partial t}$$

and hence $$\mathbf{j}.\mathbf{E} = \frac{1}{2}\frac{\partial}{\partial t}\left(\frac{\rho}{e}mv^2\right)+\frac{1}{2}\nabla.\left(\frac{m}{e}\rho v^2\frac{\mathbf{j}}{\rho}\right).$$

Since ρ/e is the volume density of the electrons, $\frac{1}{2}(\rho/e)mv^2 \equiv w$ is the total kinetic energy of the electrons present in unit volume, and by definition $\mathbf{j}/\rho = \mathbf{v}$. Therefore

$$\mathbf{j}.\mathbf{E} = \frac{\partial w}{\partial t}+\nabla.(w\mathbf{v}).$$

The next step is to form the volume integral of both sides of this equation. The second term on the right-hand side can be transformed into a surface integral by means of Gauss's theorem, and therefore

$$\int_V \mathbf{j}.\mathbf{E}\,dV = \frac{\partial}{\partial t}\int_V w\,dV+\int_S w\mathbf{v}.d\mathbf{S}.$$

The two terms on the right-hand side are readily interpreted. The first is the rate of change of the energy *stored*, as kinetic energy of the electrons, in the volume under consideration; the second term is the *net* energy *carried away* by the electrons leaving the volume in unit time. With the aid of the above equation, equation **3.7** gives the energy balance of the system which consists of the field and the charged particles, in the following form:

$$-\frac{\partial}{\partial t}\int_V w\,dV-\int_S w\mathbf{v}.d\mathbf{S} = \frac{\partial}{\partial t}\int_V \tfrac{1}{2}(\mathbf{E}.\mathbf{D}+\mathbf{B}.\mathbf{H})\,dV+\int_S(\mathbf{E}\times\mathbf{H}).d\mathbf{S}, \qquad \mathbf{3.8}$$

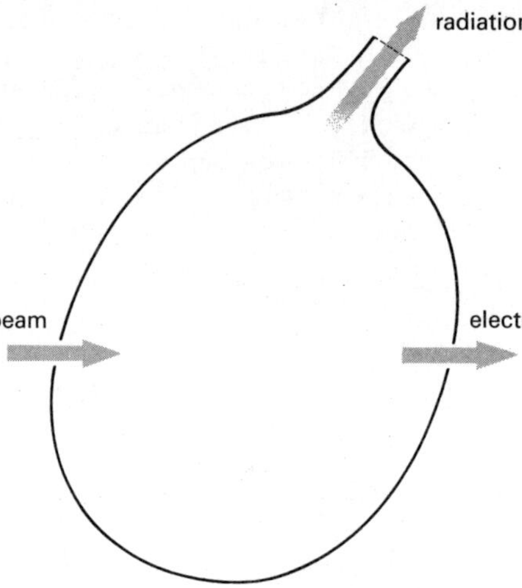

Figure 35

Let us now consider a cavity enclosed by perfectly conducting walls except for a small port through which electromagnetic radiation can flow; let electrons also flow through the cavity (see Figure 35). Suppose that both the field in the cavity and the electrons are in a stationary state of oscillation: in any volume element both the field and the kinetic energy of the particles vary periodically about a mean value. In this case the average energy *stored* in a volume element dV, calculated over a full cycle, vanishes, because

$$\frac{1}{T}\int_0^T \left(\frac{\partial w}{\partial t}\,dV\right) dt = 0$$

and $$\frac{1}{T}\int_0^T \left[\frac{1}{2}\frac{\partial}{\partial t}(\mathbf{E}.\mathbf{D}+\mathbf{B}.\mathbf{H})\,dV\right] dt = 0.$$

In other words, there is *no change in the average* energy stored in the cavity. There is, however, an energy *transfer process*, the net result of which is not zero. Taking the time average over a full cycle of each term of equation **3.8**, only those which involve the volume integrals vanish, and this leaves

$$-\frac{1}{T}\int_0^T \left(\int_S w\mathbf{v}.d\mathbf{S}\right) dt = \frac{1}{T}\int_0^T \left(\int_S (\mathbf{E}\times\mathbf{H}).d\mathbf{S}\right) dt. \qquad \mathbf{3.9}$$

The left-hand side of this equation is the difference between the kinetic energy of the electrons entering and those leaving the cavity, averaged over a cycle; the right-hand side represents the mean energy leaving the cavity in the form of electromagnetic radiation during a cycle. In other words, at the expense of the mean kinetic energy of the electrons, electromagnetic radiation can be generated, and conversely, at the expense of the radiation fed into the cavity, the electrons can gain a net kinetic energy.

If the fields are quasi-stationary in the sense discussed in the previous section, no energy can leave the system as electromagnetic radiation. In this case, the mean difference between the kinetic energy of the electrons entering and those leaving the field appears in some external network coupled to the system as a.c. power (gain or loss) – as will be discussed in some detail later.

The energy conversion processes described above form the basis of the operation of all electronic *amplifiers* and *generators*, in which *a.c. power is gained* at the expense of the kinetic energy of the electrons moving in an alternating electromagnetic field. The magnetron, in which the potential energy of the electrons is directly converted into electromagnetic energy, will be the only exception described. The reverse process provides the mechanism for the operation of resonance type *accelerators*, in which electrons or other charged particles *gain* kinetic energy at the expense of alternating electromagnetic fields.

Before analysing the principles of these devices further, we shall discuss a few problems in which the net energy transfer between fields and charged particles is negligible. We shall see phenomena the end results of which are reminiscent of those discussed in the preceding chapters.

3.2 Charged particles in time-dependent transverse fields

If the electromagnetic field is everywhere perpendicular to the direction of motion of charged particles, no work is done by the fields, and the energy of the particles is conserved during their passage through the fields. Where there are small deviations from this condition, the change in particle energy is still negligible compared with the initial energy. This condition is applicable to all the problems we shall investigate in this section.

A simple example of charged particle motion in a time-dependent electric field has already been given in the previous section. It was shown that if particles enter a sinusoidally varying uniform electric field in a direction perpendicular to the field, they leave the field deflected at an angle θ, which depends on the transit time τ. For small deflections

$$\theta = \frac{e\,E(t)\,\tau}{p}\,\frac{\sin\frac{1}{2}\omega\tau}{\frac{1}{2}\omega\tau}, \qquad \textbf{3.10}$$

where $E(t)$ is the instantaneous field intensity when the particle passes the mid-plane of the field, p is the initial momentum perpendicular to the field and ω is the angular frequency of the field. If l is the length of the field measured along the initial direction of the momentum, and the free-space wavelength is introduced, we have

$$\tfrac{1}{2}\omega\tau = \pi\frac{l}{\lambda}\frac{c}{v},$$

where $v = p/m$ is the particle's velocity.

There are two extreme cases of particular interest:

(i) $(l/\lambda)(c/v) \ll 1$, or $(\sin \frac{1}{2}\omega\tau)/(\frac{1}{2}\omega\tau) \simeq 1$, and therefore the deflection corresponds to the instantaneous field seen by the particle when it is half-way in the field. If $E(t)$ varies sinusoidally, particles which cross the mid-plane at $E(t) = 0$ are undeflected. If two such fields are separated by a field-free 'drift space' of length L (see Figure 36), particles crossing the first field at $t = 0$ pass the second one at $t = L/v$. If there is a phase difference δ between the two fields

$$E_1 = E_0 \sin \omega t \qquad \text{and} \qquad E_2 = E_0 \sin(\omega t - \delta),$$

the particle crossing the first field at $t = n\pi/\omega$ (n is an integer) will pass both fields undeflected provided that

$$\frac{\omega L}{v} - \delta = k\pi \qquad (k = 1, 2, \ldots).$$

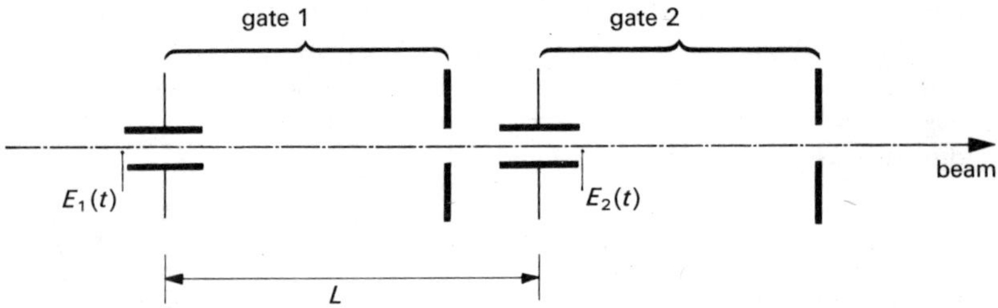

Figure 36

Thus two such fields with suitable apertures along the axis defined by the incident beam can select particles of velocity

$$\frac{v}{c} = \frac{2L}{\lambda}\left(k + \frac{\delta}{\pi}\right)^{-1}.$$

One can readily see that the spread in the velocity of the particles selected in this way is necessarily finite, and depends on the details of construction and conditions of operation. In any case the aperture diameters are finite, and so is the rate at which the deflecting field can change. Consequently there is a finite interval $\pm\Delta t_1$ about the instant $t = 0$, during which particles can cross field number 1 and pass aperture number 2. Similarly field number 2 and aperture number 3 form a 'gate' which is open for a finite interval $\pm\Delta t_2$. Plotting the position of the particle between apertures 1 and 2 against time, the slope of the straight lines give the velocities (see Figure 37), and the reader is left to draw his own conclusions from the graph.

(ii) In the other case of interest, the transit time of the particles in the field is an integral number of periods. In this case $(\sin \frac{1}{2}\omega\tau)/(\frac{1}{2}\omega\tau) = 0$ and, as seen from

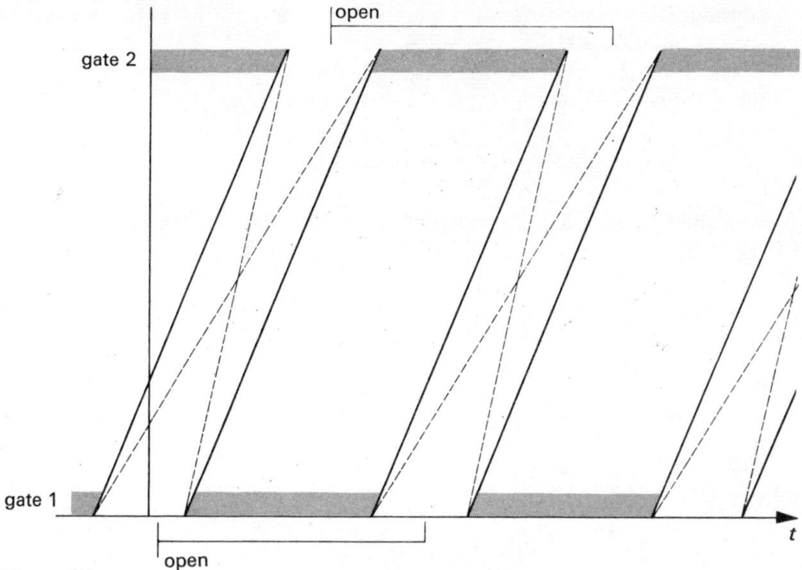

Figure 37

equation **3.10**, the particles can again pass the field without deflection. The relevant condition in this case has the form

$$\frac{l}{\lambda}\frac{c}{v} = n \qquad (n = 1, 2, \ldots). \tag{3.11}$$

Since a time-dependent but uniform electric field, with negligible magnetic field coupled to it, can be realized only if $l \ll \lambda$, the condition **3.11** requires that $v \ll c$, that is, the particles must be non-relativistic. Putting $v = (2eU/m)^{\frac{1}{2}}$, where U is an accelerating voltage, condition **3.11** shows that in the case of *ions*, a selection of the undeflected particles means a selection with respect to the specific charge:

$$\frac{e}{m} = \frac{1}{n^2}\left(\frac{cl}{\lambda}\right)^2 \frac{1}{U} \qquad (n = 1, 2, 3, \ldots).$$

Thus the process of deflection of charged particles in alternating electric fields offers a principle on which a 'radio-frequency mass spectrometer' can be designed. One of the characteristics of such a device is that, because of the factor $1/n^2$, it does not select particles with a uniquely defined value of e/m. In practice, one would select the undeflected particles by a small aperture centred on the axis of the incident beam in a plane downstream from the deflecting field. This method, however, produces a very inefficient system. As we shall see presently, particles which leave the field without an angular deflection, $\theta = 0$, emerge displaced from the axis defined by the initial momentum. Therefore the output beam has a larger cross-section than that of the incident beam, and a small aperture either causes considerable loss in transmitted intensity or accepts particles even if they do not satisfy condition **3.11**.

The displacement of the particles at the end of the field is obtained by solving the equation of motion

$$m\frac{d^2x}{dt^2} = e\,E(t) \quad \text{where} \qquad E(t) = E_0 \sin \omega t,$$

with the following initial conditions: at $t = t_0 - \frac{1}{2}\tau$, $x = 0$ and $dx/dt = 0$. The result is readily found:

$$x(t_0, \omega\tau) = \frac{e}{2m}E(t_0)\tau^2 \left\{\frac{\sin \frac{1}{2}\omega\tau}{\frac{1}{2}\omega\tau} - \cos \omega t_0 \frac{\sin \frac{1}{2}\omega\tau - \frac{1}{2}\omega\tau \cos \frac{1}{2}\omega\tau}{(\frac{1}{2}\omega\tau)^2}\right\}. \qquad \textbf{3.12}$$

If $\quad \frac{1}{2}\omega\tau = n\pi \qquad (n = 1, 2, \ldots),$

the above expression reduces to

$$x(t_0, n) = (-1)^n \frac{e}{m}\frac{E_0}{n\pi} \cos \omega t_0, \qquad \textbf{3.13}$$

where t_0 is the time at which the particle crosses the middle plane of the deflecting field.

Considerable improvement can be achieved by using two sequential and parallel deflecting fields of the same frequency and same length. In this case, the field-free drift space between the two fields can be chosen such that the displacement is cancelled for particles which satisfy the condition $\frac{1}{2}\omega\tau = n\pi$ in both fields. If the distance between the mid-planes of the two fields is L, and the particle velocity v satisfies **3.11**, the resulting displacement vanishes for all values of t_0 if

$$x(t_0, n) + x\left(t_0 + \frac{L}{v}, n\right) = 0.$$

Equation **3.13** shows that this condition is satisfied if

$$\frac{\omega L}{v} = (2k+1)\pi \qquad (k = 1, 2, 3, \ldots)$$

$$\text{or} \qquad \frac{L}{\lambda}\frac{c}{v} = k + \frac{1}{2} \qquad (k = 1, 2, 3, \ldots). \qquad \textbf{3.14}$$

Particles whose velocity satisfies simultaneously condition **3.11** and condition **3.14** form a beam which is neither deflected nor shifted relative to the incident beam. Particles of all other velocities can, in principle, be filtered out by means of suitably situated apertures. It must be pointed out, however, that the practical significance of such a device is limited by the fact that the function $(\sin x)/x$ varies slowly in the neighbourhood of its zeros, and this imposes a severe limitation on the resolving power. In addition, the requirement that two conditions must be satisfied simultaneously requires extremely high precision in construction.

3.2.1 *Alternating quadrupole fields*

Far superior performance can be obtained with time-dependent quadrupole fields. We shall only consider electric quadrupole fields in which the field is quasi-static, and its time dependence sinusoidal.

It follows from the previous considerations that if the transit time of the particles is negligible compared with the period of the alternating field, the effect of the quadrupole is the same as described in section 2.5, except that the parameter k_j, which determines the optical properties, varies with time. It is still true that particles passing through the field at a given time will be focused in one of the symmetry planes, and those moving in the other plane will be defocused. For a fixed position of the source, however, the image plane will sweep back and forth along the optic axis.

If the particles are moving so slowly that their passage through the field takes several periods, then the quadrupole can have a focusing action in both symmetry planes. In this case particles moving in either of the two symmetry planes will be subject to a focusing field during one half-period and a defocusing field in the alternate half-period. The resulting effect on particles spending an integral number of periods would be expected to be a focusing one, because – as was shown in the static case – pairs of identical quadrupole fields with alternating focusing and defocusing planes always give an over-all focusing effect.

The situation becomes more complex if the quadrupole field consists of a time-independent component upon which a time-dependent one is superimposed. To give a quantitative account of the effect of such a quadrupole lens on particles with long transit times, the properties of the equations of motion and their solutions must be investigated in detail.

The potential distribution in any plane perpendicular to the z-axis of a Cartesian coordinate system is given by

$$\Phi = \Phi_0(x^2 - y^2)r_0^{-2},$$

as in the static case, but here Φ_0 is time dependent. Let

$$\Phi_0 = U + V \cos \omega t$$

be the potential at a point, $x = r_0$, $y = 0$, at the time t (see Figure 38). The equations of motion take the form:

$$\left.\begin{aligned} m\frac{d^2x}{dt^2} &= -2e(U+V\cos\omega t)r_0^{-2}x, \\ m\frac{d^2y}{dt^2} &= 2e(U+V\cos\omega t)r_0^{-2}y, \\ m\frac{d^2z}{dt^2} &= 0. \end{aligned}\right\} \qquad 3.15$$

The third of these equations expresses the conservation of the z-component of the momentum. The first two belong to a type of differential equations, called *Mathieu's equations*, which are usually written in the standard form,

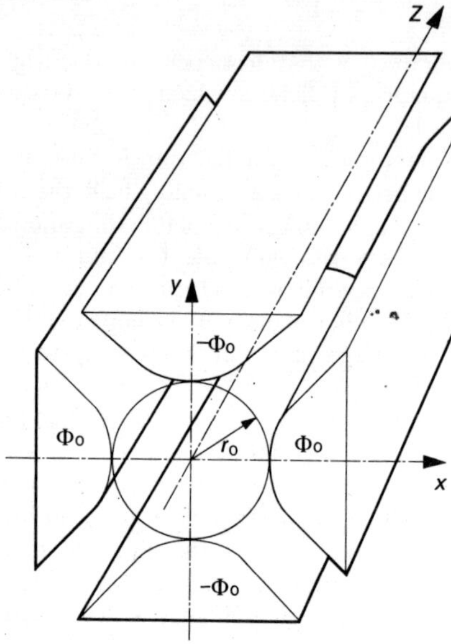

Figure 38

$$\frac{d^2w}{du^2}+(a+2q\cos 2u)w = 0, \qquad 3.16$$

and play an important part in the theory of non-linear oscillations, as well as in the theory of wave propagation in periodic structures. The first of the equations **3.15** takes this form if we write

$$w = x, \qquad u = \tfrac{1}{2}\omega t,$$
$$a = \frac{8eU}{mr_0^2\omega^2} \equiv a_x, \qquad q = \frac{4eV}{mr_0^2\omega^2} \equiv q_x \qquad 3.17$$

and, to obtain the second of the equations **3.15**, one puts

$$w = y, \qquad a = -a_x \equiv a_y, \qquad q = -q_x \equiv q_y.$$

The general solution of equation **3.16**, according to Floquet's theorem, is of the form (see the Appendix; A.6),

$$w = A_1 \exp(\mu u) f(u) + A_2 \exp(-\mu u) f(-u), \qquad 3.18$$

where A_1 and A_2 are arbitrary constants, $f(u)$ is a periodic function with period π and $\mu = \mu(q, a)$ is a function of the constant parameters q and a. According to the values of these parameters, μ can be a real, imaginary or complex number.

If $q = 0$, equation **3.16** reduces to the familiar equation of a linear oscillator $d^2w/du^2 = -aw$, the solution of which is

$$w = B_1 \exp(\mathrm{i}a^{\frac{1}{2}}u) + B_2 \exp(-\mathrm{i}a^{\frac{1}{2}}u). \qquad \textbf{3.18a}$$

This solution agrees with equation **3.18** if $f(u) =$ constant and $\mu = \mathrm{i}a^{\frac{1}{2}}$. Note that $\mathrm{i}a^{\frac{1}{2}}$ is imaginary if $a > 0$ and real if $a < 0$; in the first case, **3.18a** is a periodic function with a finite amplitude, whereas in the second case it grows to infinity with u.

Similarly, the solution **3.18** of Mathieu's equation changes its character depending on whether μ is imaginary, real or complex. Writing $\mu = \alpha + \mathrm{i}\beta$, where α and β

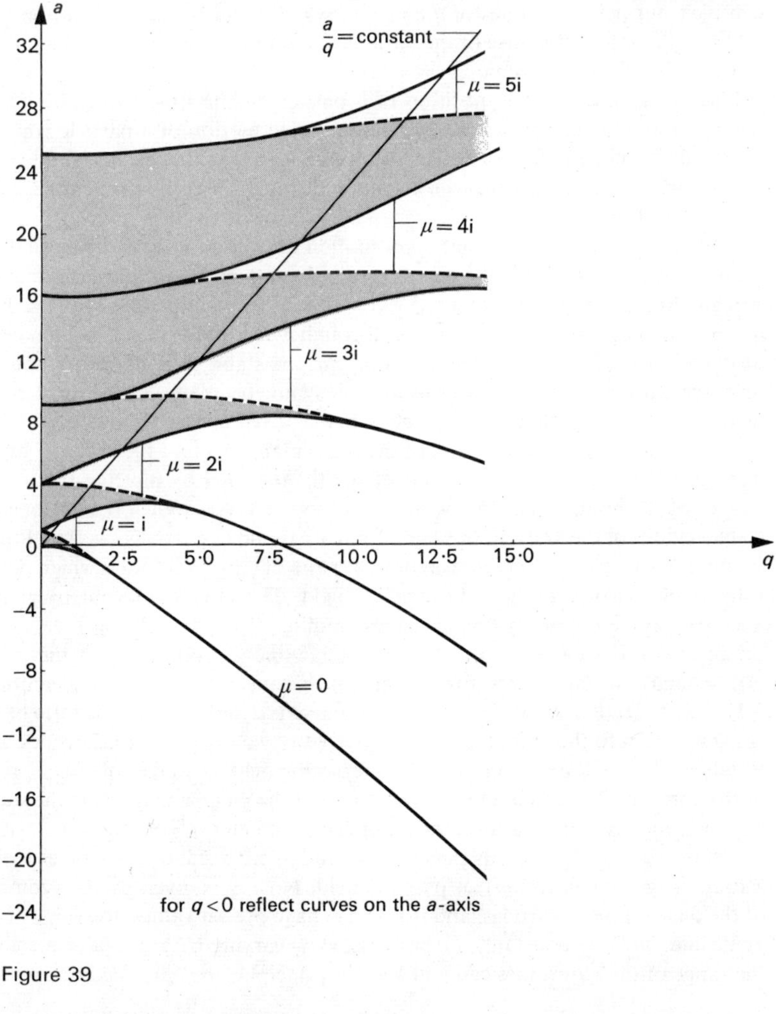

Figure 39

are real numbers, equation **3.18** takes the form

$$w = A_1 \exp(\alpha u) \exp(i\beta u) f(u) + A_2 \exp(-\alpha u) \exp(-i\beta u) f(-u).$$

This function remains *finite* for all values of u if and only if $\alpha = 0$, and in this case $w(u)$ is *periodic*.

Measuring the values of the real parameters q and a along the two axes of a Cartesian coordinate system, one can plot curves along which (q, a) is constant. Closer examination shows that the curves along which $(q, a) = ik$ ($k = 0, 1, 2, \ldots$) branch out from points of the a-axis at $a = k^2$, and divide the (q, a) plane into regions within which μ is alternately imaginary (shaded regions in Figure 39) and complex or real (blank regions). From this plot therefore, one can pick out pairs of values of q and a to which there belongs a solution $w(u)$ which is finite for all values of the variable u, and ones for which $w(u)$ increases indefinitely with increasing u.

Since $w(u)$ describes the position of a particle at the time $t = 2u/\omega$, if $w(u)$ remains finite for any value of u it represents the motion of a particle which is confined to a finite region of space. Such particles are said to be in a stable state of motion. Solutions of $w(u)$ which grow indefinitely with u represent unstable states of motion.

If the parameters of the equation of motion are varied in such a way that the ratio a/q is kept constant, the representative points in the (q, a) plane define a straight line passing through the origin. If $q > 0$, and the slope of the line, a/q = constant, is large, one passes through wide stable bands and narrow unstable ones; with decreasing values of a/q, the stable regions shrink and the unstable ones expand. The motions in both symmetry planes of the quadrupole field are stable simultaneously, if both pairs of parameter values (q_x, a_x) and $(-q_x, -a_x)$ are in a stable region of the (q, a) plane. To find the regions, superimpose the first and third quadrant of the (q, a) plane, by plotting the mirror images of the boundary curves with respect to the q-axis. When this is done the stable regions of the first and third quadrant are found to partially overlap. Points within these overlapping regions represent pairs of $(|q|, |a|)$ values which permit fully stable particle motion through the field. The only significant overlap is enclosed by the boundary curves corresponding to $\mu = 0$ and $\mu = i$, as shown on an enlarged scale in Figure 40. The 'guide line' passing through the origin is the locus of the points for which a/q = constant. From the definitions **3.17**, $a/q = 2U/V$, that is, the slope of this curve is determined by the ratio of the constant 'bias' to the amplitude of the alternating voltage on the electrodes, and is independent of the frequency of the alternating field, as well as of the e/m ratio of the particles. Note that only a finite length of the guide line is inside the stable region, and this defines a finite range of the parameter $q = 4eV/mr_0^2\omega^2$ within the stable region. This means that, if V, r_0 and ω are fixed, only particles within a fixed range of e/m ratios can pass the field. Now r_0 is given by the geometry of the quadrupole electrodes, and if V and ω have pre-set values, the slope of the guide line, $2U/V$ = constant, can be adjusted by varying U. As its value increases the range of the e/m values can – at least in principle – be narrowed down to an

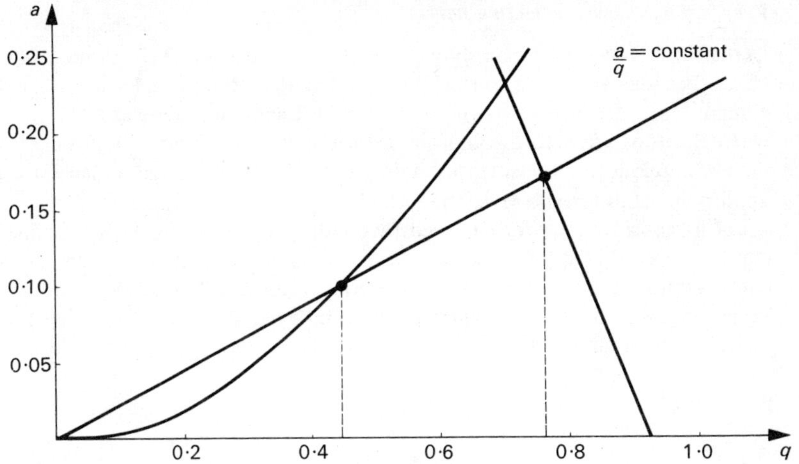

Figure 40

infinitesimally small value corresponding to a specific $|q|$ = constant. In this limiting case, any e/m value can be selected by varying the frequency of the field, and keeping all the other parameters constant, since the particles which can pass through the field under these circumstances must have a specific charge e/m = constant $\times\, \omega^2$. The reader should show that with $r_0 \simeq 1$ cm, the voltages and frequencies required for handling ions ($e/m \simeq 10^6$–10^8 C kg^{-1}) can be readily produced without violating the condition that the alternating field behaves in a quasi-stationary manner.

So far as the stability of the particle motion is concerned, only quantities which occur in the definition of the parameters q and a (see equation **3.17**) have a part to play. The initial conditions of entry determine the constants A_1 and A_2 in the solution given by equation **3.18**; the maximum excursion of the particle from the z-axis depends on the constants A_1 and A_2, as well as on the stability parameters q and a. Thus if the maximum excursion of the particle from the z-axis is limited, as is the case in all practical situations, the solid angle subtended by the beam at the source is also limited, and for a fixed set of experimental parameters the accepted solid angle depends on the specific charge e/m of the particles.

It is important to realize that the resolving power of the quadrupole mass selector has an inherent limitation associated with the transit time of the particles. The particles are subject to an alternating field only during their transit time τ. In other words, the particles are acted upon by a *pulsed* high-frequency field of carrier frequency ω and duration τ. Such a pulse has a spectrum spread over a frequency range $\omega \pm 2\pi/\tau$. Thus even if other conditions permit the selection of a sharply defined specific charge e/m = constant $\times\, \omega^2$, the finite transit time limits the fractional resolution to

$$\frac{\Delta(e/m)}{(e/m)} \simeq \frac{4\pi}{\omega\tau}.$$

3.2.2 *Very high-frequency deflecting fields*

Earlier in this section it was shown that for fast particles ($v \simeq c$) time of flight effects become noticeable only if the path length of the particles in the field is comparable with the wavelength of the field. Such conditions are usually met when electrons are fired through the field of a resonant cavity. In this case, the field has no longer a 'quasi-static' character and our previous arguments are not applicable without some revision.

Let us consider a cavity enclosed by conducting walls of arbitrary shape. It can be shown that if the condition $\nabla.\mathbf{A} = 0$ is imposed and there are no free charges within the cavity the scalar potential vanishes, $\Phi = 0$. Therefore under these conditions the electromagnetic field can be derived from a time-dependent vector potential alone, $\mathbf{A} = \mathbf{A}(x_1, x_2, x_3; t)$:

$$\mathbf{B} = \nabla \times \mathbf{A}, \qquad \mathbf{E} = -\frac{\partial \mathbf{A}}{\partial t}.$$

From the Maxwell equations it follows that the vector potential satisfies the wave equation

$$\nabla^2 \mathbf{A} - \frac{1}{c^2}\frac{\partial^2 \mathbf{A}}{\partial t^2} = 0 \qquad \mathbf{3.19}$$

and the solution of this equation is subject to the boundary conditions imposed by the walls of the cavity. Thus the determination of the field distribution in the cavity boils down to the solution of a 'boundary value problem', the complexity of which depends essentially on the shape of the cavity.

Earlier, the boundary conditions at the surface of perfect conductors were stated for the *fields*. If the tangential component of the vector potential **A** vanishes at the surface, the boundary conditions described by **3.6** are automatically satisfied.

Since the field is completely determined by the vector potential, it will be useful to express the force acting on a particle in terms of this quantity. In section 1.2 the Lagrangian of a charged particle moving in an electromagnetic field was given as

$$L = L_0 + L_f$$

with*

$$L_0 = -m_0 c^2 \left(1 - \frac{v^2}{c^2}\right)^{\frac{1}{2}}$$

and in the present case

$$L_f = e\mathbf{v}.\mathbf{A}.$$

From the equation of motion

$$\frac{d}{dt}\frac{\partial L}{\partial \dot{x}_j} - \frac{\partial L}{\partial x_j} = 0 \qquad (j = 1, 2, 3)$$

*Since our present considerations were taken up with reference to high-velocity particles, the relativistic form must be used.

the force acting on the particle is given by

$$\frac{dp_j}{dt} = -e\left[\frac{dA_j}{dt} - \frac{\partial}{\partial x_j}(\mathbf{v}.\mathbf{A})\right]. \qquad \mathbf{3.20}$$

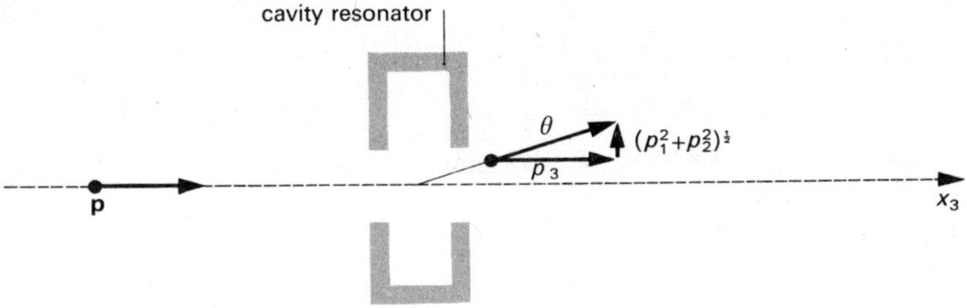

Figure 41

If the electron enters the cavity with an initial momentum $\mathbf{p} = p\hat{\mathbf{e}}_3$ parallel to the x_3 axis (see Figure 41), and leaves the cavity at time $t = \tau$, the angular deflection at the exit is given by

$$\tan\theta(\tau) = \frac{[p_1^2(\tau)+p_2^2(\tau)]^{\frac{1}{2}}}{p_3(\tau)}.$$

To determine the momentum components at the exit, it is necessary to integrate equation **3.20**. To simplify the task we make two restrictive assumptions, both easily satisfied in practical situations.

(i) During its passage through the field, the particle's trajectory is not significantly altered, that is, if the particle enters the field at $x_1(0) = x_{10}$, $x_2(0) = x_{20}$, $x_3(0) = 0$, it leaves the cavity at $x_1(\tau) = x_{10}$, $x_2(\tau) = x_{20}$, $x_3(\tau) = d$, just as if there were no field present. Clearly this condition is met if the field in the cavity is sufficiently weak.

(ii) The shape and position of the cavity is such that its end walls, through which the particle enters and leaves, are perpendicular to the x_3-axis. In this case the boundary conditions require that, at the points of entry and exit of the electron, the vector potential components perpendicular to the x_3-axis always vanish

$$A_j(x_1, x_2, 0; t) = A_j(x_1, x_2, d; t) = 0 \qquad (j = 1, 2).$$

To determine the *transverse* momentum components, p_1 and p_2, equation **3.20** must be integrated with respect to time over the time of flight.
The first term on the right-hand side vanishes by virtue of the boundary conditions, and only the contribution of the second term remains:

$$p_j(\tau) = e\int_0^\tau \frac{\partial}{\partial x_j}(\mathbf{v}.\mathbf{A})\,dt$$

$$p_j^{(\tau)} = e \int_0^\tau \left(v_1 \frac{\partial A_1}{\partial x_j} + v_2 \frac{\partial A_2}{\partial x_j} + v_3 \frac{\partial A_3}{\partial x_j} \right) dt \quad (j = 1, 2).$$

As a result of our assumption (i) we take $v_1\, dt = dx_1 = 0$, $v_2\, dt = dx_2 = 0$, and hence

$$p_j(d) = e \int_0^d \frac{\partial A_3}{\partial x_j}\, dx_3 \qquad (j = 1, 2).$$

Introducing the two-dimensional vector operator

$$\nabla_2 = \mathbf{e}_1 \frac{\partial}{\partial x_1} + \mathbf{e}_2 \frac{\partial}{\partial x_2},$$

the transverse momentum gained by the electron during its passage through the cavity can be written in the concise form

$$\begin{aligned} p_{\text{tr}}(d) &\equiv p_1 \mathbf{e}_1 + p_2 \mathbf{e}_2 \\ &= e \int_0^d (\nabla_2 A_3)\, dx_3 \end{aligned} \qquad \mathbf{3.21}$$

and the longitudinal momentum *gain* becomes, from equation **3.20**,

$$\Delta p_3 = e[A_3(x_{10}, x_{20}, d, \tau) - A_3(x_{10}, x_{20}, 0, 0)]. \qquad \mathbf{3.22}$$

These equations show that the deflection of the particle, $\theta = p_{\text{tr}}/p_3$, is completely determined by the properties of the vector potential component A_3 parallel to the initial direction of electron motion. To interpret this result in more familiar terms we must recall some of the most important features of the field distribution in resonant cavities of simple shapes.

For a *cylindrical* cavity it is indicated that the direction parallel to the cylinder axis has a preferential role. Indeed such a cavity can be excited in two distinctly different modes of oscillation.

(a) A cavity is said to be excited in a 'transverse electric mode' (TE-mode), if the electric field in the cavity is everywhere perpendicular to the cylinder axis. If the cylinder axis is parallel with the x_3-axis of an orthogonal coordinate system, this condition can be expressed by the requirement that in the cavity $E_3 = 0$ everywhere and at any time. In terms of the vector potential this amounts to the condition that $A_3 = 0$.

(b) A cavity is said to be excited in a 'transverse magnetic mode' (TM-mode), if the magnetic field in the cavity is everywhere perpendicular to the cylinder axis, or in other words: $B_3 = 0$ everywhere and at any time. This condition is met if $(\nabla \times \mathbf{A})_3 = 0$; but A_3 does not vanish identically.

A rectangular box with conducting walls can be considered as a degenerate cylindrical cavity. It is degenerate in the sense that none of the directions parallel to the edges of the box plays a preferential part. Using a Cartesian coordinate system, the axes of which are parallel with the edges of the rectangular cavity, and again taking the x_3-axis for reference, the cavity is excited in a TE-mode if $E_3 = 0$, $A_3 = 0$, and in a TM-mode if $B_3 = (\nabla \times \mathbf{A})_3 = 0$.

Figure 42a shows the simplest example of a rectangular cavity excited in a TE-mode, with the distributions of the electric and magnetic fields indicated schematically. The electric field is everywhere parallel to the x_2-axis, and varies sinusoidally along the other two axes, thereby satisfying the boundary conditions. The magnetic field lines form closed loops in planes perpendicular to the x_2-axis, so its direction with respect to the x_3-axis varies.

Figure 42b shows a rectangular cavity excited in a TM-mode. The magnetic field lines form closed loops, which are everywhere perpendicular to the x_3-axis. The electric field lines meet the walls at right angles, but inside the cavity there are regions where they run perpendicularly, and other regions where they run parallel to the x_3-axis.

The figures indicate that if the point of entry is appropriately chosen, electrons passing through the cavity parallel to the x_3-axis 'see' alternating electric and magnetic fields, both perpendicular to the direction of motion. Qualitatively, therefore, one would not expect any fundamental difference between the deflect-

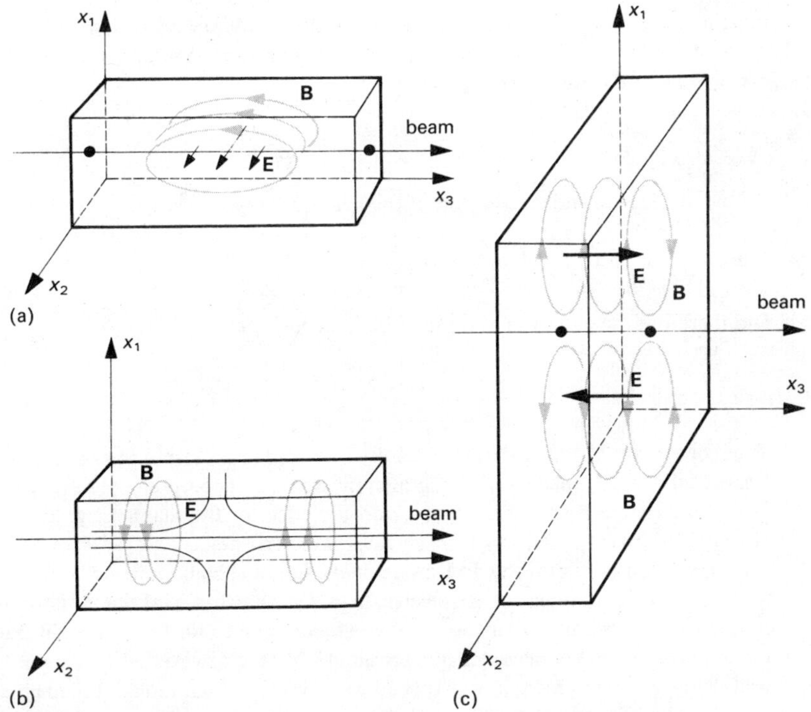

Figure 42

ing effect of the fields in a TE- or a TM-mode. Equation **3.21**, however, leads to another conclusion.

If the cavity is excited in a TE-mode, $A_3 \equiv 0$, and the transverse momentum component of the emerging electrons is $p_{tr} = 0$. In other words, no deflection can be produced by a cavity in this mode of excitation. One can say that, in this case, the electric and magnetic deflection of the electrons is mutually cancelled, regardless of the time of entry and the transit time.

If the cavity is excited in a TM-mode, $A_3 \neq 0$, and the integral in equation **3.21** can yield $p_{tr} \neq 0$ and predict the possibility of a beam deflection. To consider this case in further detail, let us resolve the electric field into a longitudinal component, $E_3 \mathbf{e}_3$, and a transverse one, $\mathbf{E}_{tr} = E_1 \mathbf{e}_1 + E_2 \mathbf{e}_2$. If the field varies sinusoidally in time, with an angular frequency ω,

then $E_3(x_1, x_2, x_3) = E_{30}(x_1, x_2) \cos k_3 x_3,$

where $k_3 = n\pi/d$ for a cavity of length d ($n = 0, 1, 2, \ldots$), and k_3 is subject to the condition that the quantity

$$\left(\frac{\omega}{c}\right)^2 - k_3^2 \equiv k_c^2$$

must assume a specific value determined by the boundary conditions set by the walls of the cavity. The transverse and the longitudinal components of the field are related by the equation

$$\mathbf{E}_{tr} = -\frac{ik_3}{k_c^2} \nabla_2 E_3.$$

Due to the sinusoidal variation of the field quantities in time,

$$\mathbf{E} = -\frac{\partial \mathbf{A}}{\partial t} = i\omega \mathbf{A},$$

and therefore the equation above can be re-arranged as

$$\nabla_2 A_3 = \frac{k_c^2}{\omega k_3} \mathbf{E}_{tr}. \qquad \mathbf{3.23a}$$

Thus equation **3.21** shows that if the cavity is excited in a TM-mode, the deflection is essentially determined by the integral of the transverse component of the electric field along the path of the electrons, and the effect of the magnetic field is not expected to cancel the effect of the transverse electric field.

It may be noted that if the TM-mode of excitation is characterized by $k_3 = 0$, the transverse component of the electric field, $\mathbf{E}_{tr}$, vanishes, and the longitudinal component E_3 becomes independent of x_3. Under these conditions, the right-hand side of equation **3.23a** becomes indeterminate. $\nabla_2 A_3$, however, still represents a well-defined vector which, in this special case, has the magnitude of the magnetic flux density. To show this it must be recognized that, in the present case, $A_1 = A_2 = 0$ and $A_3 = A_3(x_1, x_2)$. Therefore,

$$\mathbf{B} = \frac{\partial A_3}{\partial x_2}\mathbf{e}_1 - \frac{\partial A_3}{\partial x_1}\mathbf{e}_2$$

and using the relation $\mathbf{e}_1 = \mathbf{e}_2 \times \mathbf{e}_3$ etc. cyclically one obtains

$$\nabla_2 A_3 = -\mathbf{B} \times \mathbf{e}_3. \qquad \textbf{3.23b}$$

In this particular case, therefore, the deflection is entirely of magnetic origin.

Taking account of the assumption that, in the cavity, the electrons move essentially parallel to the x_3 axis at a constant velocity v, one can write

$$\nabla_2 A_3 = \mathbf{f}(x_1, x_2, x_3; t) \simeq \mathbf{f}\left(x_{10}, x_{20}, t_0 + \frac{x_3}{v}\right),$$

where $\mathbf{f}$ is a vector perpendicular to the x_3-axis, and the subscripts 0 refer to the coordinates and time of entry. In general

$$\mathbf{p}_{\mathrm{tr}} = \int_0^d \mathbf{f}\, dx_3 \neq 0.$$

However, the result of the integration will be a function of the time of entry, t_0, and of the transit time $\tau = d/v$, so that in this respect the deflection produced by a cavity excited in a TM-mode is similar to that produced by a quasi-stationary deflecting field. Once more, fields in a TE-mode never produce any deflection.

3.3 Velocity modulation and phase focusing

In Chapters 1 and 2 the description of the motion of charged particles, in or after their passage through static fields, was given in terms of time-independent trajectories. This cannot be done if the fields themselves are time dependent. In that case, the motion of the particle and its behaviour after its passage through the field depend both on the time of entry and the transit time. This is illustrated by the results of the previous sections, where the effects of time-dependent transverse fields were considered. If one investigates the kinematic effects of time-dependent longitudinal fields, the same general state of affairs is found, but, in addition, certain novel effects can be discovered which are peculiar to time-dependent fields and do not occur in static fields at all. Such a phenomenon is 'phase focusing'.

Let us consider the motion of a particle *after* its passage through a time-dependent electric field. To be quite specific we shall make the following assumptions.

(i) The field is parallel to the direction of motion of the particle. Choosing this direction as that of the x_3-axis of an orthogonal coordinate system: $\mathbf{E} = \mathbf{e}_3 E(x_1, x_2, x_3; t)$.

(ii) The field is confined to a short section of length d along the x_3-axis, that is, irrespective of the time, $E(x_1, x_2, x_3) = 0$, if $x_3 \leqslant 0$ and $x_3 \geqslant d$.

(iii) The rate of change of the field is sufficiently small for the field to be considered constant during the transit time τ of the particle: $(\partial \mathbf{E}/\partial t)\tau \ll \mathbf{E}(t)$.

By virtue of condition (i) a particle incident along the x_3-axis will suffer no deflection. This condition, together with (ii), is obviously satisfied close to the x_3-axis provided the field is maintained between two parallel electrodes each having an aperture centred on the x_3-axis. We shall be particularly interested in the effect of sinusoidally varying fields, which, for moderate frequencies, can be maintained by connecting the electrodes to a source of alternating potential difference. In the case of very high frequencies, for reasons already explained, the electrodes form parts of the wall of a resonant cavity.

Assumption (iii) is certainly valid, if (a) the velocity of the incident electrons is so high that the transit time τ_0 in the absence of the field satisfies the relation $(\partial \mathbf{E}/\partial t)\tau_0 \ll \mathbf{E}(t)$, and (b) the energy change which would take place if the field intensity had the instantaneous value $\mathbf{E}(t)$ throughout the passage of the particle is small compared with the kinetic energy of the incident particle, that is,

$$e \int E(x_3, t)\, dx_3 \ll \tfrac{1}{2}mu^2$$

at any time t; $u \ll c$ is the incident velocity.

Note that the integral of the instantaneous field taken along the x_3-axis between the points of entrance and exit,

$$\int_0^d E(x_3, t)\, dx_3 = V(t), \qquad \textbf{3.24}$$

defines a quantity which has the dimension of potential. If the field varies so slowly that it can be considered quasi-stationary, $V(t)$ is the instantaneous potential difference between the electrodes. If the electric field cannot be derived from a scalar potential alone, the term 'potential difference' between two electrodes has no meaning. However the integral in equation **3.24** does exist, and we shall refer to the quantity $V(t)$, defined above, as the 'instantaneous voltage difference' between the exit and entrance points of the particle.

Consider now particles which have the same initial velocity, but cross the field at different instants $t_1, t_2, \ldots$. Since the kinetic energies acquired in the field are different, $e\,V(t_1)$, $e\,V(t_2)$, ..., after crossing the field the particles have different velocities. If the field is increasing, each particle will acquire a larger velocity than the one preceding it, and somewhere downstream one can catch up with the other. Figure 43 shows this process pictorially. Time is plotted on the horizontal axis of a coordinate system and the distance covered by the particles is plotted along the vertical axis. Thus the velocity of each particle is represented by the slope of a straight line. The lines are parallel at $x_3 \ll 0$, indicating that the incident particles have the same velocity. The diagram shows that there is a distance X_3 past the field where the particles are 'bunched', and that further downstream the bunch separates again. If we now consider the whole ensemble of particles, the number of particles passing the field *in unit time* is found to be constant, whereas the rate at which they cross the plane at $x_3 = X_3$ is not. Let us put these qualitative observations into quantitative terms.

If the energy of the incident particles is

$$\tfrac{1}{2}mu^2 = eU_0,$$

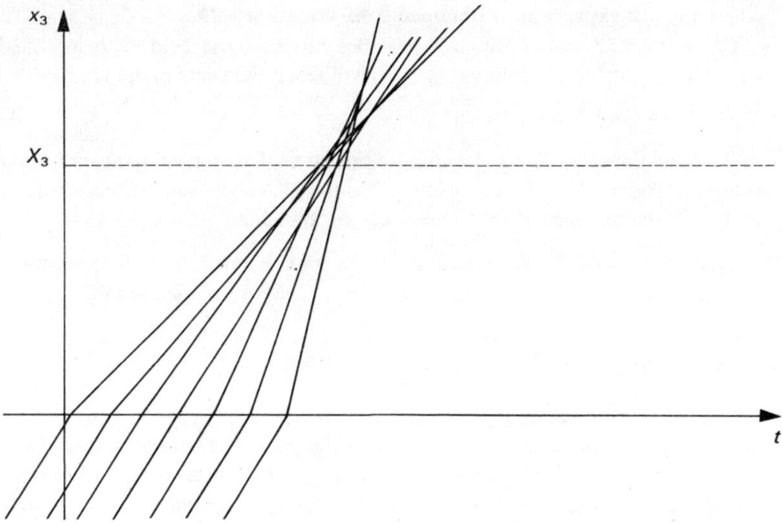

Figure 43

a particle crossing the field at t_0 leaves the time-dependent field with an energy

$$\tfrac{1}{2}mv^2 = e[U_0+V(t_0)]$$

and hence with a velocity

$$v(t_0) = u\left[1+\frac{V(t_0)}{U_0}\right]^{\frac{1}{2}}. \qquad 3.25$$

Thus the particles acquire velocities which depend on the instant t_0 when they cross the bunching field. The process is called 'velocity modulation'.

A particle which crosses the bunching field at t_0, at a later instant t will have covered a distance

$$x_3(t_0, t) = v(t_0)(t-t_0). \qquad 3.26$$

Another particle crossing the field at a time $t_0+\Delta t$ will cover, by the *same* time t, a distance

$$x_3(t_0+\Delta t, t) = x_3(t_0, t)+\frac{\partial x_3}{\partial t_0}\Delta t+(\text{terms of higher order in } \Delta t). \qquad 3.27$$

To a first-order approximation, the distance covered by the time $t = t_0+T$ is independent of the delay Δt, if $\partial x_3/\partial t_0 = 0$. According to equation **3.27** this condition is satisfied if

$$\frac{1}{T} \equiv \frac{1}{t-t_0} = \frac{\partial v/\partial t_0}{v(t_0)} = \frac{1}{2}\frac{\partial V/\partial t_0}{U_0+V(t_0)}, \qquad 3.28$$

where the last expression is obtained from equation **3.25**.

Equation **3.27** shows that all particles crossing the field with an incident energy eU_0, during the interval $t_0 \pm \Delta t$, will reach the same plane

$$x_3 \equiv X_3 = v(t_0)\, T(t_0) \qquad \textbf{3.27a}$$

at the same time $t = t_0 + T$. Because of this effect the alternating field considered is often called a 'phase-focusing lens'; T is the phase-focusing or 'bunching' time and X_3 is the distance of the phase focus or 'bunching' distance.

Note that T and X_3 are functions of the time of entry, t_0. If, for example, the field E varies sinusoidally, $V(t_0) = V_0 \cos \omega t_0$, equation **3.28** gives

$$\frac{1}{T(t_0)} = -\frac{\omega V_0}{2U_0} \frac{\sin \omega t_0}{1+(V_0/U_0)\cos \omega t_0}. \qquad \textbf{3.28a}$$

Next, the *rate* at which particles reach a plane, fixed at $x_3 = x_{30}$, can be determined. If the number of particles crossing the bunching field in unit time is N_0, those crossing during the interval between t_0 and $t_0 + dt_0$ is $N_0\, dt_0$. Assuming that no particle is lost, *all* these particles will arrive at the plane x_{30} in an interval between t and $t + dt$ where

$$t = t_0 + \frac{x_{30}}{v(t_0)}. \qquad \textbf{3.29}$$

If the number of particles which arrive at x_{30} in unit time is denoted by N, the conservation of particles demands that $N\, dt = N_0\, dt_0$, and hence

$$N = \frac{N_0}{dt/dt_0}. \qquad \textbf{3.30}$$

From equation **3.29** one finds

$$\frac{dt}{dt_0} = 1 - \frac{x_{30}}{v^2(t_0)} \frac{\partial v}{\partial t_0} = 1 - \frac{x_{30}}{X_3(t_0)},$$

where equations **3.27a** and **3.28a** have been taken into account. Hence,

$$N = N_0 \left[1 - \frac{x_{30}}{X_3(t_0)}\right]^{-1}. \qquad \textbf{3.31}$$

With the electron charge e, $eN_0 = I_0$ and $eN = I$ give the (convection) current in the planes $x_3 = 0$ and $x_3 = x_{30}$ respectively. Note that I is a function of time, even if I_0 is constant, and the time dependence of I arises from the displacement of the position of the phase-focus $X_3(t_0)$ relative to the fixed reference plane x_{30}, at which the current is measured. The current I seems to become infinite when the phase focus passes through the reference plane, $X_3 \to x_{30}$. This divergence is a consequence of the approximation used, which permits perfect phase focusing. This amounts to saying that the bunches can be compressed without any limit. In reality the interaction between the particles precludes this possibility and the current is limited to a finite maximum.

If the bunching field varies sinusoidally, the time dependence of the current I is readily found by calculating $X(t_0) = v(t_0)\, T(t_0)$ from equations **3.26** and **3.28a**, and substituting the result into equation **3.31**. One should recall that assumptions (i–iii), made earlier, require among other conditions that $V_0 \ll U_0$, and therefore it is sufficient to retain terms which are first order in V_0/U_0. Thus one obtains

$$I = I_0\left(1+\frac{1}{2}\frac{x_{30}}{u}\frac{V_0}{U_0}\,\omega \sin \omega t_0\right)^{-1} = I_0(1+\xi \sin \omega t_0)^{-1} \qquad 3.32$$

The coefficient of the sine term,

$$\frac{1}{2}\frac{x_{30}}{u}\frac{V_0}{U_0}\,\omega \equiv \xi, \qquad 3.33$$

is usually called the 'bunching parameter'.

The current I is clearly a periodic function of t_0 with period $2\pi/\omega$. For small values of the bunching parameter, $\xi \ll 1$, I varies sinusoidally:

$$I \simeq I_0(1-\xi \sin \omega t_0),$$

but for ξ comparable with unity it has sharp peaks. If $\xi \geqslant 1$ the peaks go to infinity for reasons already explained. It is very important to note that the amplitude of the current I depends on the bunching parameter only. This means that at a given frequency ω, and at any arbitrarily small but finite bunching voltage V_0, one can obtain any predetermined current amplitude, provided the coordinate of the reference plane, x_{30}, is chosen sufficiently large. Thus *velocity modulation* can provide a very efficient method for the production of an *intensity-modulated* electron beam, a prerequisite for the development of electronic amplifiers and oscillators, as we shall see in Chapter 5.

Equation **3.32** has a somewhat impractical feature. The time dependence of I is given in terms of the time t_0, when the particles cross the bunching field, and not in terms of the time t, when they cross the reference plane x_{30} where the current is observed. The relation between the two time-scales is given by equation **3.29**. For the sinusoidally varying bunching field considered before, **3.29** can be written in the form

$$\omega\left(t-\frac{x_{30}}{u}\right) = \omega t_0 - \xi \cos \omega t_0. \qquad 3.34$$

There is no simple way of solving this equation for ωt_0, in terms of ωt. It is clear, however, that since the current I is periodic in $\phi \equiv \omega t_0$ with period 2π, it must be a periodic function of the variable $\omega(t-x_{30}/u)$ also. Therefore one can expand I in a Fourier series.

$$I = I_1 + \sum_1^\infty a_n \cos n\omega\left(t-\frac{x_{30}}{u}\right) + \sum_1^\infty b_n \sin n\omega\left(t-\frac{x_{30}}{u}\right), \qquad 3.35$$

where $$I = \frac{\omega}{2\pi}\int_0^{2\pi/\omega} I(t)\,dt,$$

$$a_n = \frac{\omega}{\pi}\int_0^{2\pi/\omega} I(t)\cos n\omega\left(t-\frac{x_{30}}{u}\right)dt,$$

$$b_n = \frac{\omega}{\pi}\int_0^{2\pi/\omega} I(t)\sin n\omega\left(t-\frac{x_{30}}{u}\right)dt.$$

Note that apart from the constant I_1 each term in equation **3.35** represents a travelling wave.

To evaluate I_1 we recall that, by virtue of the conservation of the total number of particles, $I\,dt = I_0\,dt_0$, and since the periodicity in t and in t_0 is the same, one obtains

$$I_1 = I_0.$$

To evaluate the Fourier coefficients a_n and b_n, one has to substitute **3.34**, and taking into account the relation $I\,dt = I_0\,dt_0$ again, one finds

$$a_n = \frac{I_0}{\pi}\int_0^{2\pi}\cos n(\phi-\xi\cos\phi)\,d\phi = \begin{cases}(-1)^{\frac{1}{2}n} I_0 2J_n(n\xi) & \text{if } n \text{ is even,}\\ 0 & \text{if } n \text{ is odd;}\end{cases}$$

$$b_n = \frac{I_0}{\pi}\int_0^{2\pi}\sin n(\phi-\xi\cos\phi)\,d\phi = \begin{cases}(-1)^{\frac{1}{2}(n-1)} I_0 2J_n(n\xi) & \text{if } n \text{ is odd,}\\ 0 & \text{if } n \text{ is even;}\end{cases}$$

where $\phi = \omega t_0$ and the functions J_n are the Bessel functions of the first kind. Thus equation **3.35** can be written in the form

$$\frac{I}{I_0} = 1+2J_1(\xi)\sin\omega\left(t-\frac{x_{30}}{u}\right)+2\sum_{n=1}^{\infty}(-1)^n J_{2n}(2n\xi)\cos\left[2n\omega\left(t-\frac{x_{30}}{u}\right)\right]+$$

$$+2\sum_{n=1}^{\infty}(-1)^n J_{2n+1}[(2n+1)\xi]\sin\left[(2n+1)\omega\left(t-\frac{x_{30}}{u}\right)\right]. \qquad \textbf{3.36}$$

To appreciate this expression, it must be realized that the shapes of the Bessel functions J_n (see Figure 44) are reminiscent of the wave form of damped oscillations, but that the zeros of J_n are not equidistant from one another. Hence, although in general the frequency spectrum of the current I is very rich in harmonics, by choosing the value of the bunching parameter ξ, one or other of the harmonics can be enhanced or suppressed. This will be shown to be of importance from the point of view of the amplification and generation of harmonics by klystrons.

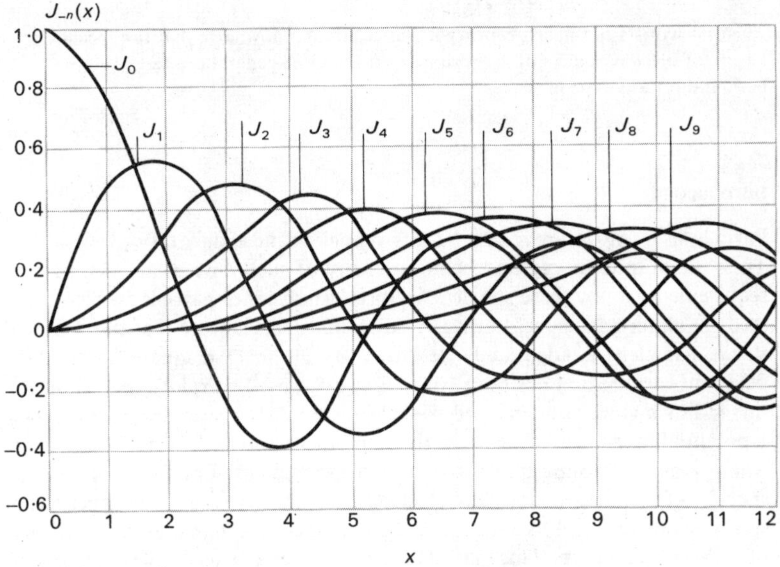

Figure 44

Further reading

W. K. H. PANOFSKY and M. PHILLIPS, *Classical Electricity and Magnetism*, 2nd edn, Addison-Wesley, 1962.

J. D. JACKSON, *Classical Electrodynamics*, Wiley, 1962.

Both are textbooks written at an advanced level. They introduce the basic concepts of the theory of electromagnetic fields, and of their interaction with charged particles. In both books the subject is presented with great attention to aspects dictated by further developments in modern physics. The selection of, and the emphasis on, particular topics is different in the two books.

W. PAUL, H. P. REINHARD and H. V. ZAHN, 'The electric mass-filter as mass spectrometer and isotope separator', *Zeitschrift für Physik*, Vol. 152, 1958, pp. 143–182.

Original paper – in German – on the theory and design of the radio-frequency-driven electric-quadrupole lens for mass-spectroscopic purposes.

L. BRILLOUIN, *Wave Propagation in Periodic Structures* (see Further reading, Chapter 2).

J. STOKER, *Non-linear Vibrations*, Interscience, 1950.

Chapter 6 gives a rigorous treatment of Floquet's theorem, and of the stability problem for Hill's and Mathieu's equations.

Chapter 4
Particle Accelerators

. . . for scientists of the last century it was common knowledge that the special nature of the waves did not matter and that the same general properties could be found for any type of waves.

L. Brillouin

4.1 Introduction

Particle accelerators are research tools of nuclear and elementary-particle physics. Thus the output energy and beam intensity of such devices are determined by requirements set by these subjects. Before the advent of particle accelerators, the study of alpha-particle scattering by heavy nuclei had shown that the electric field about a nucleus is adequately described by the inverse square law (Coulomb field) outside a region of about 10^{-15} metre radius. To explore the region within this radius by means of charged particles, one needs projectiles which can climb a potential barrier which requires them to have a kinetic energy $W_k = Ze/4\pi\varepsilon_0 r$, where Z is the atomic number of the target nucleus. For light atoms, kinetic energies of the order of a million electron volts are required. High-precision mass spectroscopy revealed that the mass of a nucleus is less than the sum of the masses of its constituent parts. This mass defect is interpreted in terms of the mass–energy equivalence as the binding energy of the nucleus. In order to release a particle bound in a nucleus one requires, again, a projectile whose kinetic energy is a few million electron volts. The picture presented grossly over-simplifies the situation, there being nuclear reactions which can be studied with incident particles whose energy is below a million electron volts, whereas others require incident energies an order of magnitude higher than this.

The requirements in elementary-particle physics are best illustrated by the energies needed to produce pairs of elementary particles. Some of these have masses a few hundred times that of an electron; the proton and the neutron are roughly two thousand times heavier than the electron. Thus, for the creation of a proton–antiproton pair, about 2 GeV* is required in the reference frame in which the centre of mass of the pair created is at rest.†

*1 MeV = 10^6 eV (electron volts),
1 GeV = 10^9 eV,
1 TeV = 10^{12} eV.

† It should be noted that in the 'centre-of-mass system', the energy available for producing a reaction is always less than the kinetic energy of the particle as measured in the laboratory system, if the target is at rest there. If the target nucleus is at rest (mass M_0) in the laboratory system, and the projectile of rest mass m_0 has a kinetic energy ε, then relativistic considerations show that in the proper frame of the centre of mass, the system has a total energy W, given by the relation (see **A.28** in the Appendix, **A.7**).

$$W = [(E_p+E_t)^2+2E_t\varepsilon]^{\frac{1}{2}},$$

The most obvious way of accelerating charged particles is used in low-energy devices: the particles are allowed to fall through a constant potential difference of the desired value. When this method is extended to higher and higher energies, serious difficulties arise in the production of high potential differences and in applying these potential differences to some system of electrodes without electric breakdown. Thus the problem of the design of such accelerators is primarily one of high-voltage engineering. The guiding of the particles from the source to the target is based on the principles of rotationally symmetric electric electron lenses. Accelerators of this type are the Cockroft–Walton generator and the Van de Graaf generator, both limited to a few million electron volts output energy, corresponding to practical limitations in the production of high d.c. voltages. These generators will not be discussed in the subsequent sections.

An alternative approach which has proved extremely fruitful, and is applied up to the highest energies obtainable at present, is the basis of all the *resonance* accelerators. The idea is the same as that applied when a swing is boosted to higher and higher amplitudes, by imparting to it relatively small impulses repeatedly at appropriate intervals. Accelerators of this kind can be classified roughly as linear accelerators, in which the particle trajectory is essentially a straight line, and cyclic accelerators. In the latter case a magnetic field guides the particles on a spiral or a closed orbit.

It goes without saying that an accelerator becomes a useful research tool only if, at the desired energy, it delivers particles in a sufficiently high-intensitv beam. Thus beyond finding a suitable mechanism for accelerating particles to high energies, one must ensure that particles are not lost during the acceleration process. In other words the acceleration of charged particles to high energies presents two problems: (1) to transfer energy to the particles in such a manner that the effect is cumulative; (2) to ensure that the particles move in the neighbourhood of a stable equilibrium orbit.

The solution of problem (1) is, at least in principle, not too difficult and the relevant techniques were developed relatively early. The main development in the design of accelerators yielding higher and higher particle energies is characterized by the ever-increasing sophistication in handling problem (2).

4.2 Linear accelerators

Figure 45 shows four different schemes for the acceleration of particles. In scheme (a), a *constant* potential difference U_0 is applied to produce a uniform electric field, and particles falling through this field gain a kinetic energy eU_0. In scheme (b), the same potential difference is divided into smaller steps and the particles

where $E_p = m_0c^2$ and $E_t = M_0c^2$ are the rest energy of the projectile and of the target respectively. Thus the energy available for producing a reaction is $\Delta W = W-(E_p+E_t)$. In other words to produce a reaction which requires an energy ΔW, the kinetic energy of the projectile in the laboratory system must be

$$\varepsilon \geqslant \Delta W\left(1+\frac{E_p}{E_t}+\frac{\Delta W}{2E_t}\right).$$

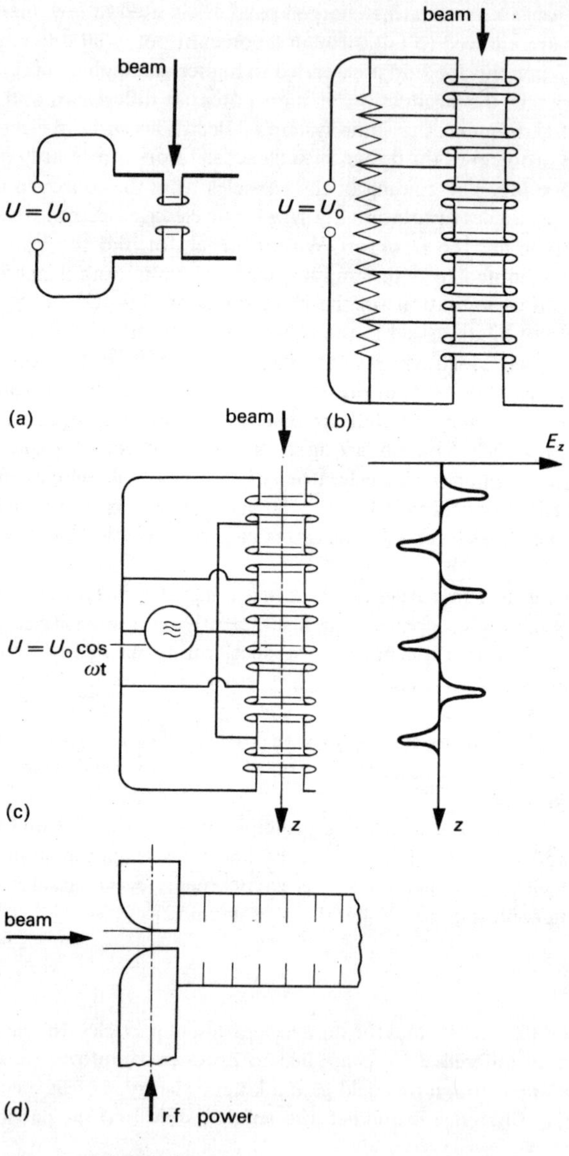

Figure 45

gain kinetic energy when they cross the gaps between cylindrical electrodes; in the electrodes themselves there are field-free drift spaces. The subdivision of the total potential difference means that a good electron-optic design can be used and this helps in overcoming high-voltage difficulties, but the ultimate energy is determined by the total applied potential difference U_0.

In scheme (c), alternate electrodes are connected and there is the same *time-dependent* potential difference $U_0 \sin \omega t$ between each pair of adjacent electrodes. Thus at any instant the field is oppositely directed in alternate gaps. Inside the cylindrical electrodes, however, there is no field, and if the transit time of a particle in such a drift region is half the periodic time of the field, that is, $\tau = \pi/\omega$, the particle will see in each gap a field of the same direction and of the same intensity. The particle crosses the successive gaps at $t_0 + n\tau/\omega$ ($n = 0, 1, \ldots$), and after passing n gaps it will have gained a kinetic energy $neU_0 \sin \omega t_0$, provided that, during the passage of the particle through a gap, the field can be considered constant. The condition, that the time of flight in each drift section is a half-period of the alternating field, defines the lengths l_n of the 'drift tubes':

$$l_{n+1} = \frac{\pi}{\omega} v_n = \frac{1}{2} \frac{v_n}{c} \lambda, \qquad \textbf{4.1}$$

where v_n is the velocity of the particle when it leaves gap n, and λ is the free-space wavelength of the field.

In scheme (d), the accelerating field is an electromagnetic wave travelling in a wave-guide, excited in a TM-mode, that is, in such a manner that the electric field on the axis is parallel to the axis, and is given by

$$E = E_0 \sin \omega \left(t - \frac{z}{v} \right). \qquad \textbf{4.2}$$

Thus a particle travelling with the velocity v is subject to a constant field intensity, and the kinetic energy gained is

$$\int_0^\tau E(t_0) v \, dt \qquad (t_0 = \text{time of entry}, \ \tau = \text{time of flight}),$$

determined by the time of flight in the accelerating tube, and can be increased at will by extending the length of the tube. Next we shall consider qualitatively some further features of scheme (c), and then scheme (d) will be discussed quantitatively in the context of acceleration of relativistic particles.

The lengths l_n of the drift tubes determined by equation **4.1**, are correct for the acceleration of particles with a sharply defined entrance phase ωt_0, since in each gap the particle acquires a kinetic energy ε which, expressed in units of the rest energy, is given by

$$\frac{\varepsilon}{m_0 c^2} = \frac{eU_0}{m_0 c^2} \sin \omega t_0$$

and hence the velocities in subsequent drift tubes are

$$\frac{v_n}{c} = \left\{ 1 - \left(1 + n \frac{eU_0}{m_0 c^2} \sin \omega t_0 \right)^{-2} \right\}^{\frac{1}{2}}.$$

In fact if the drift tubes are suitable for the acceleration of particles which enter with a phase angle $0 \leqslant \omega_0 t \leqslant \frac{1}{2}\pi$, particles with entrance phase $\pi - \omega t_0$ will be accelerated as well. If, however, one compares the behaviour of particles with entrance phase angles slightly different from ωt_0 and $\pi - \omega t_0$ respectively, an important difference is found. It can readily be seen that particles crossing a gap with the same velocity in the neighbourhood of a phase angle in the range $0 \leqslant \omega t_0 \leqslant \frac{1}{2}\pi$ are subject to phase focusing, whereas particles in the neighbourhood of the phase angle $\pi - \omega t_0$ are phase defocused. We leave it to the reader to show by qualitative arguments that if a particle, entering a gap with a velocity v_0 and phase angle $0 \leqslant \omega t_0 \leqslant \frac{1}{2}\pi$, passes the subsequent gaps with phase angles $\omega t_0 + n\pi$ ($n = 1, 2, 3, \ldots$), then particles which enter the first gap with velocity v_0 and phase angle $\omega t_0 + \delta\phi$, carry out 'phase oscillations', that is, they cross the successive gaps with phase angles $\omega t_0 + n\pi + \Delta\phi$, where $\Delta\phi$ varies periodically with some small amplitude. For this reason, the particle which crosses the gaps with phase $\omega t_0 + n\pi$ is said to be in a state of *stable* phase equilibrium.

The fact that there exists a state of stable phase equilibrium has two implications of utmost importance. (i) A particle entering the first accelerating gap with velocity v_0 and phase angle ωt_0 will continue to be accelerated, even if a small perturbation occurs causing the phase angle and/or the entrance velocity at the successive gaps to become slightly different from the nominally required one. (ii) Particles which cross the first gap with small deviations from the nominal initial velocity and the nominal phase angle can still be accelerated. The range of acceptable initial velocities and phase angles is of vital importance from the point of view of the output beam intensity.

Equation **4.1** shows that once the particle velocities become very nearly equal to the velocity of light ($v_n \simeq c$) the accelerating gaps have to be equidistant. At this stage of the process one can no longer speak of 'acceleration' in the usual sense of the word. All the same, the kinetic energy of the particle keeps increasing and the energy increment manifests itself in the increase of the relativistic mass.

Let us consider a section of the accelerator in which the gaps are equidistant. The instantaneous field plotted against the distance along the accelerator axis has the shape shown in Figure 45c, and $E(t)$ varies sinusoidally. Such a field distribution can be analysed into a Fourier series of *standing* waves, and each component of the standing wave can be expressed as the superposition of two waves *travelling* in opposite directions. The waves will have the same frequency as the alternating field and wavelengths $\lambda_0, \frac{1}{2}\lambda_0, \frac{1}{3}\lambda_0, \ldots$. Thus the Fourier expansion will contain sine and cosine terms of the argument

$$\omega t \pm \frac{2\pi n}{\lambda} z = \omega\left(t \pm \frac{nz}{v_0}\right),$$

where $$v_0 = \frac{\omega\lambda_0}{2\pi}$$

is the phase velocity of the component of wavelength λ_0.

If this field is viewed from a coordinate system moving with an electron of velocity v_0, the component of wavelength λ_0 travelling in the same direction appears stationary, and all the other components appear time dependent varying sinusoidally with different frequencies. Thus one expects that the component of wavelength λ_0 will have a cumulative effect on the electron motion, whereas the other components cause some periodic perturbation. The situation is obviously simpler if one considers the effect of one travelling wave only. This is, however, not a simplification of theoretical interest only, but the actual situation in scheme (d) described earlier.

The problem we propose to investigate is this. There is a wave-guide in which the wave propagates along its axis with a phase velocity v_0. The electric field on the axis is parallel to the axis and is given by equation **4.2**. Let us determine how the momentum (or energy) of a particle moving along the axis varies with its position, as a result of the interaction with the field.

Note that the electromagnetic field in a travelling wave is not conservative, for, disregarding space charge, the electric field in the wave-guide can be derived from a vector potential **A** such that $\mathbf{E} = -\partial \mathbf{A}/\partial t$. Thus, if the field is given by equation **4.2**, the vector potential is

$$\mathbf{A} = A(z, t)\mathbf{e}_z = \mathbf{e}_z \frac{E_0}{\omega} \cos \omega \left(t - \frac{z}{v}\right) \qquad \mathbf{4.3}$$

and the Hamiltonian

$$H = \{m_0^2 c^4 + c^2(\boldsymbol{\pi} - e\mathbf{A})^2\}^{\frac{1}{2}}$$

is *an explicit function of time*; $\boldsymbol{\pi}$ is the canonical momentum related to the linear momentum $\mathbf{p}$ by the equation $\boldsymbol{\pi} = \mathbf{p} - e\mathbf{A}$.

In order to set up the Hamiltonian of a system, one is not compelled to use the coordinates x_i, which are measured with respect to axes at rest in the laboratory. One can, for example, use any other set of quantities X_i suitable for the unique determination of the position of the particle, and momentum components P_i which are canonically conjugate to X_i. It can be shown that if the position of the particle is defined with respect to a coordinate system which is moving with the phase velocity v_0 of the travelling wave, the Hamiltonian becomes independent of time. One expects instinctively that this should happen, since the field appears independent of time for an observer who is travelling with the field. In other words, when described with reference to such a frame, the field appears static and hence conservative. However we do not intend to discuss the electron motion as it appears to an observer stationed in a coordinate system moving with the phase velocity of the travelling wave. Instead, as observers fixed in the laboratory, we only wish to use a moving coordinate system in which the position of the electron moving along the x_3-axis is given by

$$X_1 = x_1 = 0 \qquad X_2 = x_2 = 0 \qquad X_3 = z - v_0 t \equiv Z \qquad \mathbf{4.4}$$

and to set up the appropriate Hamiltonian. Calculations (see the Appendix, A.8) show that the Hamiltonian can be written in the form

$$H = [(m_0c^2)^2 + p^2c^2]^{\frac{1}{2}} - pv_0 - ev_0A(Z), \tag{4.5}$$

where p, the *linear momentum* of the particle *in the laboratory frame*, is canonically conjugate to Z, and

$$A(Z) = \frac{E_0}{\omega}\cos\left(\frac{\omega}{v_0}Z\right) \tag{4.6}$$

is independent of time. Consequently the Hamiltonian **4.5**, characteristic of the system with reference to the chosen frame, is a constant of motion (although it is not the total energy). Thus the variation of the particle's momentum as a function of Z can be represented by curves, H = constant, in the phase plane (Z, p), similarly to the way discussed in the context of a simple pendulum (see section 1.2.3). It is advantageous to use dimensionless scales; therefore p/m_0c is plotted against $\omega Z/v_0 \equiv (2\pi/\lambda_0)Z$ as related by the equation

$$\left[1+\left(\frac{p}{m_0c}\right)^2\right]^{\frac{1}{2}} - \frac{p}{m_0c}\frac{v_0}{c} - \frac{eE_0}{m_0c^2}\frac{\lambda_0}{2\pi}\cos\left(2\pi\frac{Z}{\lambda_0}\right) = \frac{H}{m_0c^2} = \text{constant}. \tag{4.7}$$

where $\lambda_0 = 2\pi v_0/\omega$ is the wavelength of the wave travelling in the wave-guide.

If the frequency is fixed, there are two parameters which can be chosen in an arbitrary manner. Firstly, the amplitude E_0 of the electric field, and secondly, the phase velocity v_0. The first is essentially determined by the power applied to excite the field in the wave-guide, and the second is determined by the structure of the wave-guide.* As we shall see later two cases have to be investigated separately: (i) where $v_0 < c$ and (ii) where $v_0 \simeq c$. It is worth noting that, in the equation above, the cofficient of the cosine term has a simple meaning: it gives the energy (in units of the rest energy) acquired by the particle in $1/2\pi$ wavelengths if it is subject to the maximum accelerating field.

Figure 46 shows contour lines H/m_0c^2 = constant in the case $v_0 < c$. The ordinate scale on the right-hand side gives p/m_0c, that on the left-hand side gives the kinetic energy in units of the rest energy. The curves show obvious similarity to those obtained for a simple pendulum (Figure 4) and they fall into two categories. First there are closed curves all enclosing the point $Z = 0$,

$$\frac{p}{m_0c} = \frac{\gamma_0 v_0}{c} \equiv \frac{p_0}{m_0c},$$

where $\gamma_0 = (1 - v_0^2/c^2)^{-\frac{1}{2}}$. This point represents a particle moving with the

*The wave propagation in a wave-guide is similar to that in a dispersive medium, that is, the phase velocity is a function of the frequency. In a simple wave-guide, such as a pipe of uniform cross-section, the phase velocity is always larger than the velocity of light. If the wave-guide is divided by regularly spaced apertures into sections of equal length, the wave propagation shows the characteristic features of propagation in periodic structures. The phase velocity becomes a function of the frequency, and varies over a range between zero and a maximum, determined by the spacing and the diameter of the apertures.

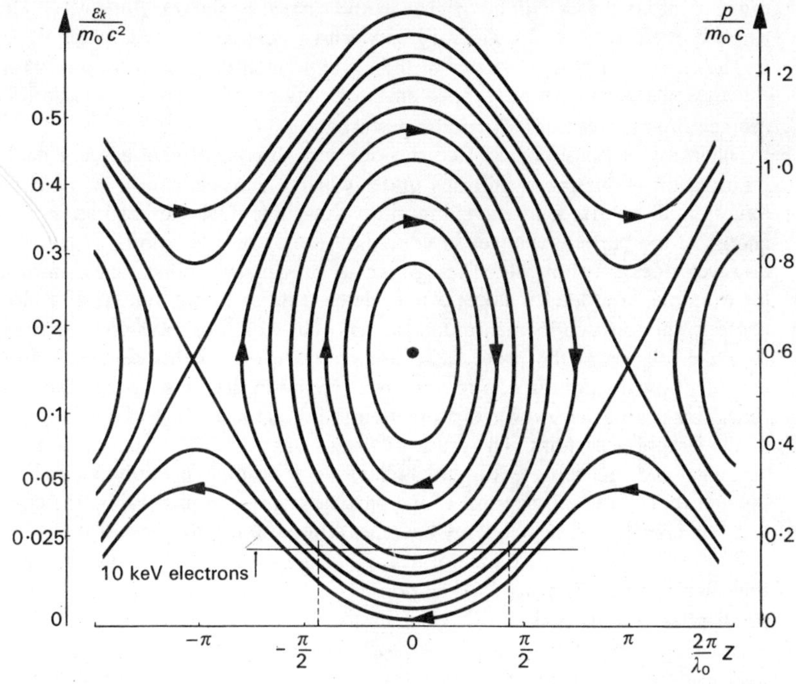

Figure 46

phase velocity v_0 of the wave and is locked at a phase where

$$-\sin\left(\frac{2\pi Z}{\lambda_0}\right) = \sin\omega\left(t-\frac{z}{v_0}\right) = 0.$$

Note that at this phase position the force on the particle is zero, and one can see why it represents a state of stable equilibrium. If the phase of the particle is advanced ($Z > 0$) it is subject to a retarding force, and if the particle is lagging behind ($Z < 0$) it is subject to an accelerating force. Particles of the same momentum are also in equilibrium at $Z = \pm\pi$, but these positions are unstable since the conditions are the opposite to those just described. In fact the closed and the oscillating contour lines are separated by those which go through the points $(\pm\pi, p_0/m_0 c)$.

A particle represented by a point on a closed contour line has a periodically varying momentum, and oscillates within a finite range of phases about the position of stable phase equilibrium, $Z = 0$. If a particle of momentum p_1 enters the field at $z = 0$, $t = t_1$, its initial phase is $Z_1 = v_0 t_1$, and if $(Z_1, p_1/m_0 c)$ is a point on a closed contour curve, the particle will move in such a manner that its representative point stays on this contour curve. Figure 46 indicates that particles which have a momentum higher than a threshold value, and enter within a finite range of initial phases, will be *trapped* in and carried by the travelling wave. Once trapped, their momentum will vary between two extreme values, which depend on the entrance phase. Particles of the same momentum, but having a 'wrong' entrance phase, will not be trapped and experience a relatively small variation in momentum as a result of a periodic perturbation.

Diagrams in phase space, like the one just discussed, give a great deal of information about the conditions under which a particle can be trapped in a longitudinal electric field travelling in a wave-guide. One can also see how the energy of the particle changes in course of the phase oscillations. This picture, however, does not show how these processes take place in time. One cannot tell, for example, how long it takes a particle starting at the minimum of a closed contour line to reach its maximum, that is, what length of wave-guide is needed to make full use of the process. To answer questions of this kind, one has to consider how the motion of a representative point in the phase space develops in time. This problem, however, is more involved than the analogous one in the case of the simple pendulum. The main difference between the phase diagram just discussed and that representing the behaviour of a simple pendulum arises from the fact that, if the momentum of the particle changes considerably, the corresponding change in kinetic energy becomes comparable with the rest energy, and therefore the mass variation is not negligible. This is the reason why, in general, the calculation of the period of the phase oscillations is a complicated task. For small phase oscillations, however, a first-order approximation can be readily found.

Considering a particle moving in the vicinity of the stable phase, $Z = 0$, one can apply the same kind of arguments as those used in section 2.6, for the study of electron trajectories adjacent to an equilibrium orbit. Accordingly, one expands the Hamiltonian **4.5** about the equilibrium point $(0, p_0)$ up to terms of second order in the small quantities ΔZ and $\Delta p = p - p_0$. The first-order terms and the 'cross'-terms vanish because the expansion is carried out in the neighbourhood of the equilibrium position, and therefore

$$H = H(0, p_0) + \frac{1}{2}\left[\frac{\partial^2 H}{\partial Z^2}(\Delta Z)^2 + \frac{\partial^2 H}{\partial p^2}(\Delta p)^2\right], \qquad \textbf{4.8}$$

where ΔZ and Δp are canonically conjugate. Using this Hamiltonian and the canonical equations, one obtains

$$\frac{d}{dt}(\Delta Z) = \frac{\partial H}{\partial(\Delta p)} = \frac{\partial^2 H}{\partial p^2}\Delta p,$$

and hence $$\frac{d^2}{dt^2}(\Delta Z) = \frac{\partial^2 H}{\partial p^2}\frac{d}{dt}(\Delta p) = -\frac{\partial^2 H}{\partial p^2}\frac{\partial^2 H}{\partial Z^2}\Delta Z,$$

or simply $$\frac{d^2}{dt^2}(\Delta Z) = -\Omega^2 \Delta Z, \qquad \textbf{4.9}$$

with $\Omega^2 = \frac{\partial^2 H}{\partial p^2}\frac{\partial^2 H}{\partial Z^2}$. **4.10**

Equation **4.9** clearly shows that the particle carries out harmonic oscillations about its equilibrium position if $\Omega^2 > 0$, and in this case Ω gives the frequency of the phase oscillation. Using the Hamiltonian **4.5** with equation **4.6**, one finds

$$\frac{\partial^2 H}{\partial Z^2} = \frac{\omega}{v_0} eE_0 \quad \text{and} \quad \frac{\partial^2 H}{\partial p^2} = (m_0 \gamma_0^3)^{-1}. \qquad \textbf{4.11}$$

Substitution of these expressions in **4.10** gives

$$\Omega^2 = \frac{\omega}{v_0}\frac{eE_0}{m_0 \gamma_0^3}. \qquad \textbf{4.12}$$

(The quantity $m_0 \gamma_0^3$, a function of the velocity v_0, is often called the 'longitudinal mass'.)

Finally, let us estimate the extreme values of the kinetic energy which a particle carried by the travelling wave can have. We observe that since the Hamiltonian H = constant and, of course, $H(0, p_0)$ = constant, the bracketed term in equation **4.8** must be constant. Therefore the instantaneous deviation ΔZ from the stable phase and the instantaneous deviation Δp from the equilibrium momentum satisfy the relation

$$\frac{2\pi eE_0}{\lambda_0}(\Delta Z)^2 + (m_0 \gamma_0^3)^{-1}(\Delta p)^2 = C^2 = \text{constant}, \qquad \textbf{4.13}$$

where the coefficients are those given by equations **4.11**. Equation **4.13** is the equation of an ellipse; the amplitudes of the phase oscillation, and of the momentum variation, are given by the extreme values of ΔZ and Δp respectively:

$$(\Delta Z)_{\text{extr}} = \pm C\left(\frac{2\pi eE_0}{\lambda_0}\right)^{-\frac{1}{2}},$$

$$(\Delta p)_{\text{extr}} = \pm C(m_0 \gamma_0^3).$$

Now, the kinetic energy of the particle of momentum $p_0 + \Delta p$ is by definition

$$W_k = [m_0^2 c^4 + (p_0 + \Delta p)^2 c^2]^{\frac{1}{2}} - m_0 c^2.$$

Expanding this expression to the first order in Δp, and denoting the kinetic energy of the particle in the equilibrium phase by

$$(m_0^2 c^4 + p_0^2 c^2)^{\frac{1}{2}} - m_0 c^2 = W_{k_0}$$

one obtains $\qquad W_k' = W_{k_0} + v_0 \Delta p,$

or the extreme values of the kinetic energy

$$(W_k)_{\text{extr}} = W_{k_0} + v_0 (\Delta p)_{\text{extr}}.$$

Eliminating the constant C by forming $(\Delta p)_{\text{extr}}/(\Delta Z)_{\text{extr}}$, one finds in terms of the extreme phase deviation,

$$(W_k)_{\text{extr}} = W_{k_0} + v_0(\tfrac{1}{2}\pi\lambda_0 eE_0 m_0 \gamma_0^3)^{\frac{1}{2}} \frac{(\Delta Z)_{\text{extr}}}{\frac{1}{2}\lambda_0},$$

or in units of the rest energy,

$$\frac{(W_k)_{\text{extr}}}{m_0 c^2} = \frac{W_{k_0}}{m_0 c^2} + \pi \frac{v_0}{c} \gamma_0^{\frac{3}{2}} \left(\frac{\lambda_0 eE_0}{2\pi m_0 c^2}\right)^{\frac{1}{2}} \frac{(\Delta Z)_{\text{extr}}}{\frac{1}{2}\lambda_0}. \qquad \textbf{4.14}$$

It must be stressed that the approximation used is valid only if $(\Delta Z)_{\text{extr}} \ll \frac{1}{2}\lambda_0$. Therefore if one takes the value $(\Delta Z)_{\text{extr}} \simeq \frac{1}{2}\lambda_0$, which limits the stable phase oscillations, one obtains *overestimated* limits for $(W_k)_{\text{extr}}$; the term which appears in addition to W_{k_0} turns out to be too large by a factor $\frac{1}{2}\pi$.

As mentioned before, if the frequency of the electromagnetic field is given, there are two parameters which affect the processes studied. Keeping the phase velocity, v_0, fixed, and changing the amplitude of the field, the contour lines H = constant will change their shape. Large E_0 will make the closed contour lines more elongated in the vertical direction, that is, the extreme values of the kinetic energy of a particle oscillating about the stable phase will grow, and the frequency of the phase oscillations increases also. It goes without saying that the changes described refer to the effect on a particle of a given charge and rest mass. If the rest mass is increased, replacing, for example, electrons by protons, the result is the same as if the amplitude of the field had been reduced by the same factor.

It is important to realize that if the amplitude of the field is finite – as it must be – and the phase velocity is less than the velocity of light, the maximum kinetic energy obtainable cannot be increased beyond that given by equation **4.14** by extending the interaction time, that is, the flight path in the field.

If the phase velocity of the wave is changed, all other parameters being fixed, the pattern of the contour lines changes drastically. For one thing, the momentum (or kinetic energy) of the particle in equilibrium phase changes and it grows to infinity as $v_0 \to c$. Furthermore, the range over which the kinetic energy of a particle can vary in the neighbourhood of the stable phase becomes infinite and 'phase oscillations' cease to have a meaning because the oscillation frequency vanishes ($\Omega \to 0$). In fact, as $v_0 \to c$, the validity of the approximations based on the power series expansion of the Hamiltonian (see equation **4.8**) breaks down, and meaningful conclusions can be drawn only by considering the rigorously valid equation **4.7**. In the limiting case $v_0 \to c$, one obtains from equation **4.7**

$$\frac{p}{m_0 c} = \frac{1 - \left(\dfrac{H}{m_0 c^2} + \dfrac{\lambda_0}{2\pi}\dfrac{eE_0}{m_0 c^2}\cos\dfrac{2\pi}{\lambda_0}Z\right)^2}{2\left(\dfrac{H}{m_0 c^2} + \dfrac{\lambda_0}{2\pi}\dfrac{eE_0}{m_0 c^2}\cos\dfrac{2\pi}{\lambda_0}Z\right)^2}.$$

If the constant

$$\frac{H}{m_0 c^2} > \frac{\lambda_0 e E_0}{2\pi m_0 c^2},$$

the contour lines are periodic (but not sinusoidal) curves with minima at $Z = n\lambda_0$, and maxima at $Z = (2n+1)\frac{1}{2}\lambda_0$ ($n = 0, 1, 2, \ldots$). If, on the other hand,

$$\frac{H}{m_0 c^2} < \frac{\lambda_0 e E_0}{2\pi m_0 c^2},$$

$$\frac{p}{m_0 c} \to \infty \quad \text{at phases defined by} \quad \cos\left(2\pi\frac{Z}{\lambda_0}\right) = -\frac{2\pi}{\lambda_0}\frac{H}{eE_0}.$$

A family of contour lines $H/m_0 c^2$ = constant is shown in Figure 47. We see that if a particle's initial momentum is larger than a threshold value, and the phase of entry is within a finite range about $Z = 0$, its representative point in phase space moves on a curve which runs to infinity. Such a particle is trapped in the travelling wave, and as the representative point approaches asymptotically a well-defined phase position $-\frac{1}{2}\lambda_0 < Z_0 < 0$, the particle reaches a state of motion in which it is subject to an essentially constant accelerating field. Although the particle will never reach the phase velocity of the field, that is, the velocity of light, its momentum and its energy increase monotonically and manifest themselves in monotonically increasing mass. Without carrying out any calculations, one may expect, by analogy with the previously discussed case, that the particle settles fairly quickly near the asymptotic phase position, and the ultimate kinetic energy approximates to the work done by the corresponding constant field during the time of flight of the particle.

Having considered the fundamental principles involved in the acceleration of charged particles by a longitudinal electric wave of constant phase velocity, one wishes to see the scope of such devices. For the purpose of order of magnitude estimates let us assume that the amplitude of the electric field is so high that $(\lambda_0/2\pi)eE_0 \simeq 50$ keV. (Working at a 'guided' wavelength $\lambda_0 \simeq 100$ mm, this would require an r.f. field amplitude of $E_0 \simeq 3$ kV mm^{-1}, which is quite feasible.) Thus, *for electrons*, the parameter value $(\lambda_0/2\pi)(eE_0/m_0 c^2) \simeq 0{\cdot}1$. This is the value used in constructing the contour lines in Figures 46 and 47; in the former $v_0 = \frac{1}{2}c$, in the latter $v_0 = c$. These diagrams show that electrons with moderately high initial energy can be trapped in both cases. If the phase velocity is $\frac{1}{2}c$, an initial energy of a few tens of thousands of electron volts is sufficient, but the maximum energy obtainable is limited as already explained. Considering a travelling wave of the same amplitude parameter but travelling with the velocity of light, the injection energy has to be higher in order to obtain trapping over the same range of phases. However the required initial energy, of the order of a million electron volts, is readily available as the output energy of a d.c. high-voltage accelerator. Using such an accelerator as an injector, the electrons trapped in the travelling wave will gain an additional energy approaching 3 MeV per metre length of flight. To utilize the trapping phase range as well as possible,

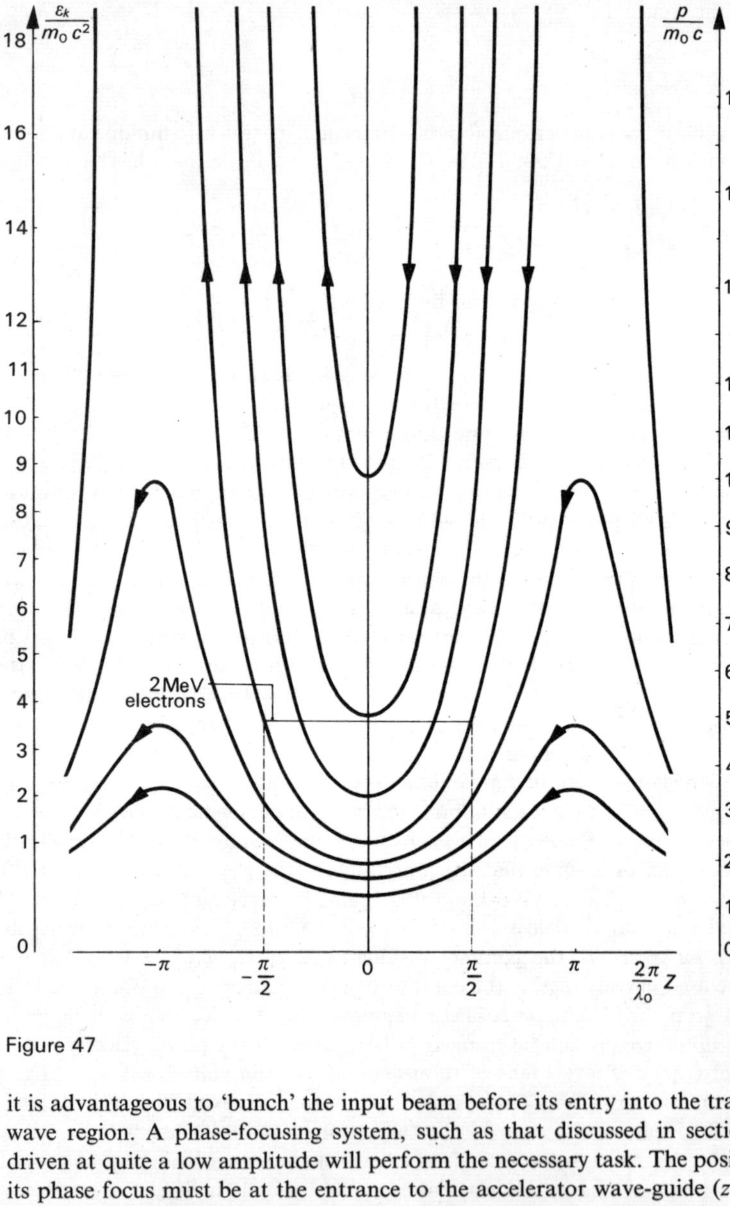

Figure 47

it is advantageous to 'bunch' the input beam before its entry into the travelling wave region. A phase-focusing system, such as that discussed in section 3.3, driven at quite a low amplitude will perform the necessary task. The position of its phase focus must be at the entrance to the accelerator wave-guide ($z = z_0$), and it must be phased in such a manner that at entry ($t = t_0$) the centres of the bunches satisfy the condition $2\pi Z/\lambda_0 = -(\omega/c)(z_0 - ct_0) = 0$.

Although the same principles are applicable to charged particles of any mass, in practice one cannot accelerate particles of large rest mass, for example, protons, with the aid of travelling waves of constant phase velocity. At feasible amplitudes, the parameter $\lambda_0 eE_0/m_0 c^2$ is very small, say 10^{-3} or less, and for phase velocities $v_0 < c$, the closed contour lines in the phase diagram are squashed in the vertical direction. Thus the maximum energy gain in the course of a phase oscillation becomes small relative to the required injection energy. Similarly, if the travelling wave velocity is $v_0 = c$, either the required field amplitude, E_0, or the necessary injection energy, is far beyond the limits of technical possibility.

Acceleration of heavy particles can be obtained only if the phase velocity of the travelling wave itself increases along the length of the wave-guide. Such a wave can also be described in terms of a constant frequency and a wavelength which varies along the direction of propagation. If we recall our accelerator scheme (c), described earlier in this section, with drift tubes of increasing length, we recognize that the field in such a system can be described as the superposition of travelling waves which have the same frequency as the field in the gaps, but whose wavelengths, λ_0, $\frac{1}{2}\lambda_0$, $\frac{1}{3}\lambda_0$, etc., vary along the accelerator axis. A single travelling wave with varying phase velocity (that is, wavelength) can be obtained in a wave-guide in which there are apertures with non-uniform spacing between them.

Considering a travelling wave of this kind, the field intensity must be described in the form

$$E = E_0 \sin \omega \left[t - \int \frac{ds}{v_0(s)} \right]$$

instead of as in equation **4.2**. The quantitative analysis of the motion of a charged particle in such a field is much more involved than in the case of constant phase velocity, discussed previously. With some restrictive assumptions, however, the same general approach gives much useful information, and – as before – it hinges on the existence of an equilibrium phase of the field to which a particle can be locked. The motion of other particles is described with reference to this equilibrium phase.

Let us suppose that at an instant t the particle is at $z_0(t)$, and therefore it is acted upon by a field E, defined above. This particle is locked at a constant phase of the field provided the same field acts upon it at all subsequent times, that is,

$$t - \int \frac{dz_0}{v_0(z_0)} = \text{constant} \equiv t_0.$$

$v_0(z_0)$ is determined by the structure of the wave-guide, and therefore the equation above describes the position of this particle, z_0, as a function of time, provided the constant t_0 is given. This constant too can be found by applying the condition that the particle 'sees' a constant field. Suppose that the wave propagating in the wave-guide has a smaller acceleration than the particle would acquire if subjected to the peak field intensity E_0. In this case the constant t_0 is defined by the condition that the acceleration of particles subject to the constant field $E_0 \sin \omega t_0$ shall be equal to the acceleration of the field.

Once the equilibrium phase ωt_0 is established, it may be used as the origin of a coordinate system for the description of the motion. Thus one is led to a problem analogous to that discussed in the case of constant phase velocity and it can be treated by similar methods, although one cannot find a rigorously valid Hamiltonian in such a simple form as expression **4.5**. Nevertheless if the considerations are restricted to such a narrow range ΔZ about the equilibrium phase, that within this range the phase velocity of the wave can be considered constant, an approximate description takes a relatively simple form. The essential conclusion is that there are particles whose representative points in the phase plane describe closed curves about the equilibrium phase. As before, this means that if particles are injected within a finite range of phase and energy, they are trapped in the field and oscillate about a stable phase equilibrium. While doing so, their energy varies periodically about the energy corresponding to that of a particle which is at rest in the equilibrium phase, but on the average they gain the same amount of kinetic energy.

While this result is similar to that obtained in the case of waves of constant phase velocity $v_0 < c$, the acceleration has the important effect that the *phase oscillations are damped*, that is, during the process of acceleration, the particles are concentrated in increasingly tight bunches about the stable equilibrium phase. Without attempting a quantitative determination of the damping of phase oscillations, a simple treatment will show why this damping takes place. It has been pointed out previously that the vertical dimension of a closed orbit in phase space, characteristic of a particle trapped in a wave, depends on its phase velocity, say $v_1 < c$. Keeping all other parameters constant, the height of such an orbit for a wave of another phase velocity v_2 will be larger if $v_2 > v_1$. In fact the phase orbit will grow taller in a continuous fashion, provided the acceleration takes place so *slowly* that the phase velocity of the wave can be considered constant while the representative point in phase space describes a number of cycles. At the same time the horizontal dimension of the phase orbit changes too.

To see how this process takes place, let us recall the 'adiabatic theorem' (section 1.2.2) which states that, if the Hamiltonian is a constant of motion except for a slowly changing parameter, a closed phase orbit stays closed and its shape changes in such a manner that the phase integral is conserved. In the present context, it can be asserted that if the phase velocity of the travelling wave changes slowly, it leaves the area within a closed phase orbit invariant. Hence an increase in the dimension of the phase orbit, parallel to the p-axis must be accompanied by a decrease in its dimension parallel to the Z-axis. This indicates how the amplitude of the oscillation about the stable equilibrium position is reduced in the course of the acceleration.

So far we have studied the motion of particles along a straight line which was chosen as the z-axis of a coordinate system. The assumption that on this axis the electric field has only a z-component implies that the field is symmetrical about this axis. Thus a particle, which is moving on this axis initially, will continue to do so; that is, the z-axis constitutes an equilibrium trajectory. It remains to be seen, however, whether this equilibrium is a stable or an unstable one. We must consider what happens to a particle which is slightly displaced from the z-axis. The answer to this question can be derived from first principles.

Let us reconsider the motion of a particle on the z-axis, but this time let us describe it as it appears to an observer who is moving with the travelling wave, located at the point of equilibrium phase. If the phase velocity of the wave is $v_0 < c$, this can be readily done by performing a Lorentz transformation. For our present purpose, however, it is quite sufficient to carry out some qualitative considerations. One can say with certainty that particles which are trapped in the field according to our previous analysis will also appear trapped to the observer stationed at the point of equilibrium phase. He will describe the electric field as a static one, and will observe that the trapped particles oscillate about an equilibrium position. We may add that in his description the equilibrium position is defined by the minimum of the (scalar) potential distribution along the z-axis, which is a symmetry axis of the potential distribution. If he considers the motion of particles displaced from the z-axis, he will discover that they must be subject to radial forces directed away from the z-axis. The reason is that in empty space the scalar potential must satisfy Laplace's equation, and hence one can show that if the potential distribution along one direction *has* a minimum at a particular point, then at the same point but in a direction perpendicular to the previous one, the distribution cannot have a minimum. Hence when a particle is subject to a quasi-elastic restoring force in the $\pm Z$ direction, it is, at the same time, repelled from the z-axis once it suffers any small initial displacement. Regarding the effect as it appears in the laboratory system, one can conclude that the region where phase stability occurs has a radial defocusing effect and *vice versa*. In other words, where electrons are compressed into bunches along the z-axis, the bunches flow apart in the radial direction. In order to combat the defocusing effect, some special arrangement is necessary, based essentially on static geometrical electron-optic principles. In electron accelerators, an axial magnetic field, or a series of magnetic quadrupoles (with alternate focusing–defocusing planes), can be used. In ion accelerators, one can introduce 'grids' in front of the drift tubes. Without considering the details of their effect, we mention only that on such grids charges and currents are induced, and the contribution to the field by these sources can sufficiently modify the field distribution to over-compensate the radial defocusing effect without destroying the phase focusing.

4.3 Cyclic accelerators

The most conspicuous feature of all cyclic accelerators is that the particles are guided on a curved orbit by means of a magnetic field. With the exception of the betatron, the acceleration of the particles is achieved by a radio-frequency electric field, which is synchronized to the rotational frequency of the particles. The temporal and spatial distribution of such a field can be described as the superposition of travelling waves, one of which rotates at the same speed as the particle. Thus cyclic accelerators, with the one exception mentioned, can be looked upon as 'coiled-up' linear accelerators, in the sense that the fundamental processes are essentially the same as those in a linear accelerator, although the equilibrium orbit of the particles, which was a straight line in linear accelerators, is in this case a plane curve. We shall classify these accelerators according to the main features of the equilibrium orbits.

The equilibrium orbit in accelerators of the *cyclotron* family is either a plane spiral, on which the accelerating particle moves outwards, or a family of closed osculating curves (usually circles), and the particle follows paths of increasing radius as its energy increases. In accelerators of the *synchroton* family, the equilibrium orbit is always a well-defined closed curve: a circle or a race-track shape, for example. In the *betatron*, the equilibrium orbit is again a circle but the mechanism of the acceleration is entirely different from that outlined above, and it will be convenient to describe it before discussing synchrotons. As in linear accelerators, the problem of maintaining a high-intensity beam is solved by static electron-optical methods. In the case of cyclic accelerators, this is achieved by the suitable choice of the distribution of the magnetic guiding field, and the main difference between various types of accelerators within one family appears in the different approach to this electron-optic problem.

4.3.1 *Accelerators of the cyclotron family*

The classical cyclotron, historically the first cyclic accelerator, is designed to accelerate particles in the non-relativistic range of velocities, that is, up to energies which are still low in comparison with their rest energies. Therefore its practical significance lies in the acceleration of heavy particles such as protons and other ions.

A uniform magnetic field of flux density B is maintained between the pole faces of an electromagnet, and in this field there are two electrodes formed by the halves of a drum-shaped conductor, split along its diameter (see Figure 48). (Because of their shapes the electrodes are usually called the dees of the cyclotron.) If an alternating potential difference is applied between the dees, there will be a field $E = E_0 \sin \omega t$ essentially confined to the gap between the dees, perpendicular to the magnetic field, the regions inside the dees being field free. Particles emitted by a source at the centre of the system are subject to the electric and the magnetic fields while they pass the gap, and at other times they are subject to the magnetic field only. Therefore they move on a trajectory which consists of semi-circular arcs; the angular velocity of the orbiting particle is

$$\omega_c = \frac{eB}{m} \qquad \textbf{4.15}$$

and the radius of the orbit is

$$R = \frac{p}{eB}, \qquad \textbf{4.16}$$

where p is the momentum of the particle.

In the non-relativistic range of velocities, the particle has a constant mass, and therefore the angular velocity ω_c is also constant. Thus, if the frequency of the alternating field is equal to the cyclotron frequency $\omega = \omega_c$ particles which leave

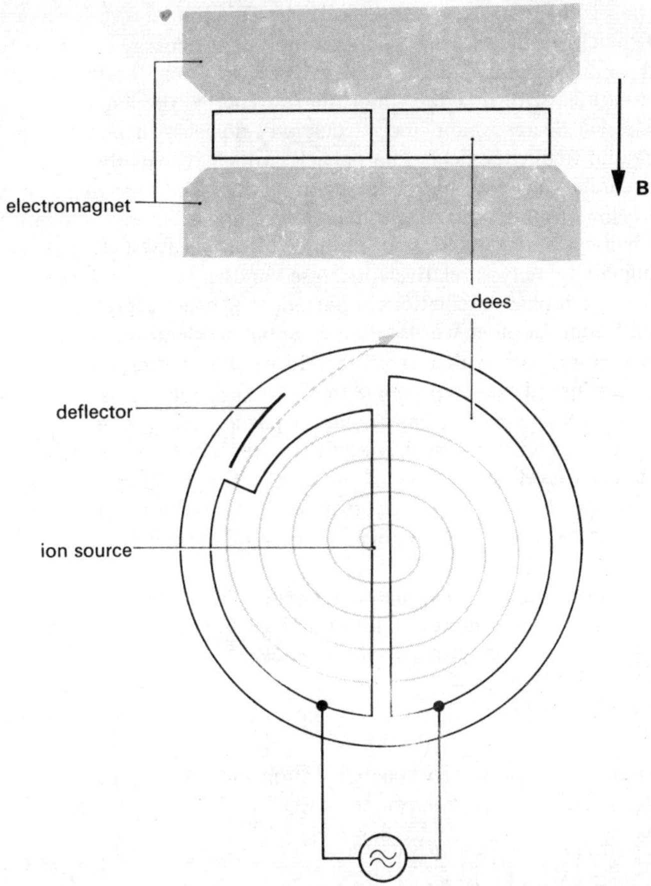

Figure 48

the source during the accelerating half-cycle of the field will always cross the gap between the dees when the field is such as to accelerate them. Note that the phase at which the particles cross the gap is not critical at all. One may say that the phase equilibrium is a neutral one. The amount of energy the particle acquires during a passage through the gap does, of course, depend on the phase at which the crossing takes place, but all the particles which eventually move on an orbit of radius R have the same momentum $p = eBR$, and hence a kinetic energy $(eBR)^2/2m$.

Equation **4.15** is correct relativistically if m signifies the relativistic mass $m = m_0(1-v^2/c^2)^{-\frac{1}{2}}$. Multiplying the numerator and denominator of the right-hand side by c^2, we may write

$$\omega_c = \frac{eBc^2}{W}, \qquad \textbf{4.17}$$

where $W = mc^2$ is the total energy of the particle. This equation shows how the operation of the classical cyclotron reaches a limit as the energy of the particle increases. At a kinetic energy W_k, the total energy is $W = m_0 c^2 + W_k$ and, if W_k becomes comparable with or larger than the rest energy, the angular velocity of the particle will decrease, and the particle will not stay in phase with the accelerating radio-frequency field. The particles which repeatedly crossed the field with a definite phase during the accelerating half-cycle $0 < \omega t_0 < \pi$ will start lagging behind as their kinetic energy increases, and eventually slip into the decelerating half-cycle $-\pi < \omega t_0 < 0$. Because of the energy loss they suffer here, their angular velocity is relatively increased and as the process continues, the particles carry out phase oscillations about $\omega t_0 = 0$, which represents a phase of stable equilibrium. In other words, after an initial acceleration, owing to the relativistic change of mass with increasing velocity, the particles are bunched about the equilibrium phase. In principle they can keep this phase indefinitely, orbiting at a constant energy. Further increase of particle energy can be obtained by decreasing the frequency of the driving field at the same rate as the angular velocity of the particles drops because of the mass variation. A cyclotron which overcomes the difficulty arising from the relativistic change of mass by synchronizing the driving field with the angular velocity of the particles is called a '*synchro-cyclotron*'.

At a given value of the angular frequency ω of the electric 'driving' field the total energy of a particle in the equilibrium phase is obtained from equation **4.17** by putting $\omega = \omega_c$, or in terms of the rest energy

$$\frac{W}{m_0 c^2} = \frac{eB}{m_0 \omega}. \qquad \textbf{4.18a}$$

On the other hand, the relation between momentum and orbit radius, as given by equation **4.16**, is relativistically correct, and therefore

$$\frac{W}{m_0 c^2} = \left[1+\left(\frac{p}{m_0 c}\right)^2\right]^{\frac{1}{2}} = \left[1+\left(\frac{eBR}{m_0 c}\right)^2\right]^{\frac{1}{2}}. \qquad \textbf{4.18b}$$

Equating the right-hand sides of equations **4.18a** and **b**, one obtains the relation between driving frequency ω and the radius of tne equilibrium orbit R:

$$\left(\frac{eB}{m_0 \omega}\right)^2 = 1+\left(\frac{eBR}{m_0 c}\right)^2. \qquad \textbf{4.19}$$

Equation **4.18b** also shows that the kinetic energy of the particle, $W_k = W - m_0 c^2$, is uniquely determined by the radius of the equilibrium orbit R, and therefore the ultimate limit of the kinetic energy attainable at a given flux density B is set by the size of the electromagnet which produces the magnetic guiding field.

From the qualitative description of the way phase oscillations develop about a phase of stable equilibrium, it can be seen that the situation is quite similar to that in a linear accelerator. Indeed the quantitative analysis of the phase oscillations, conditions of stability etc., which will not be carried out here, can follow the same arguments and give analogous results to the case of linear accelerators.

Particles whose momentum (equilibrium orbit radius) is within a finite range about that of the particle in equilibrium phase will stay in its neighbourhood. Thus the operation of the synchro-cyclotron can be summarized as follows. Initially the particles are accelerated as in a classical cyclotron. The driving frequency, however, is gradually reduced, and the equilibrium orbit radius R grows according to the relation **4.19**. Particles whose instantaneous orbit radii r are within a small range about the nominal equilibrium radius R will be trapped in a state of stable phase oscillations and will move out with the equilibrium orbit as it increases, picking up kinetic energy all the time.

Next we should consider the problem of orbit stability, that is, what happens if a particle is slightly displaced from the equilibrium orbit. In the initial stages of the acceleration, in a classical cyclotron as well as in a synchro-cyclotron, the force acting on the particle during its passage between the dees is of great importance, because at relatively low energies, a field component perpendicular to the plane of the equilibrium orbit can produce significant deflection. From inspection of the field pattern during the accelerating phase (see Figure 49), one can expect that the resulting effect is a focusing one, tending to keep the particles

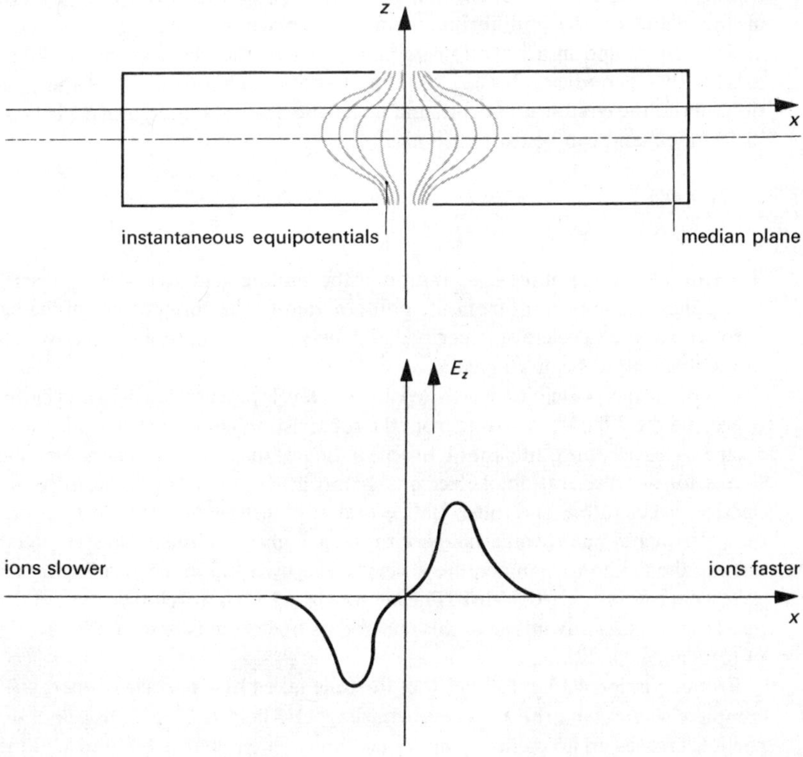

Figure 49

in the vicinity of the plane of the equilibrium orbit. This result is confirmed by quantitative arguments. At more advanced stages of the acceleration, however, the effect of the accelerating field from the focusing point of view is negligible, and the confinement of the beam depends entirely on the magnetic guiding field. Considering a particle moving on an equilibrium orbit of radius R = constant, the behaviour of particles moving with the same momentum on adjacent trajectories can be discussed in terms of geometric electron optics, and we can refer to the results obtained in section 2.6. It was seen there that in order to obtain a restoring force directed towards the plane of the equilibrium orbit, one requires a rotationally symmetrical magnetic field which is also symmetrical with respect to the plane of the equilibrium orbit, and whose intensity in this plane falls with increasing radius. Taking the z-axis of a cylindrical polar coordinate system perpendicularly to the plane of the equilibrium orbit at the centre of the orbit, the required field distribution in the plane $z = 0$ is

$$\mathbf{B} = \mathbf{e}_z B_0 \left(\frac{r}{R}\right)^n \qquad (-1 < n < 0).$$

In this case the particles will oscillate about the equilibrium orbit. The oscillations in the plane of the equilibrium orbit take place with a spatial periodicity $2\pi R(1+n)^{-\frac{1}{2}}$, and in a direction perpendicular to the plane of the equilibrium orbit with a periodicity $2\pi R|n|^{-\frac{1}{2}}$. The quantity n, characteristic of the oscillations about the equilibrium orbit, and hence the beam stability, is often called the 'field index' and can be expressed as

$$n = \frac{(\partial B/\partial r)_R}{B_R/R} \quad \text{or} \quad n = \frac{\partial(\ln B)}{\partial(\ln r)}. \qquad \textbf{4.20}$$

The careful shaping of the distribution of the guiding field, to match a prescribed field index n, is important from the point of view of the confinement of the beam throughout the acceleration period, and hence from the point of view of the attainable output beam intensity.

Although the principles which led to the development of synchro-cyclotrons overcome the difficulties arising from the relativistic change of mass with velocity, such devices become quite impracticable if the rest mass of a particle is very small, that is, for the acceleration of electrons. In any attempt to design a cyclotron-type electron accelerator, one must realize that it is technically feasible to produce radio-frequency microwave fields with such high amplitudes that an electron passing the field once can acquire a kinetic energy equal to, or even higher than, its rest energy, $m_0 c^2 \simeq 0{\cdot}5$ MeV. The *microtron*, which is a cyclotron-type electron accelerator, takes advantage of this fact and its operation is based on a novel type of resonance condition.

From equation **4.17**, it follows that the time taken by a particle of energy W to complete a circular orbit in a uniform magnetic field is $T = 2\pi W/eBc^2$. If the particle crosses an accelerating field once on each circular orbit, and its kinetic energy is increased by eU_0 each time, then after n passages its total energy is

$W = m_0c^2 + neU_0$ and its period of orbital rotation is

$$T_n = (2\pi/eBc^2)(m_0c^2 + neU_0).$$

Hence the difference in the period of orbital rotation on two subsequent orbits is

$$\Delta T = T_n - T_{n-1} = \frac{2\pi U_0}{Bc^2}.$$

Observe that ΔT depends on the energy gained by the particle, and on the flux density of the guiding field only. The particle will cross the accelerating field always in the same phase, if the angular frequency ω of the field satisfies the relation $\Delta T = 2\pi k_1/\omega$, where k_1 is an integer number, that is,

$$\frac{U_0}{Bc^2} = \frac{k_1}{\omega}. \qquad \textbf{4.21a}$$

In addition to this condition, one must ensure that the particle also spends an integral number of periods of the driving field in the first orbit. Denoting the initial kinetic energy by eU_i this condition is

$$\frac{m_0c^2 + eU_i + eU_0}{eBc^2} = \frac{k_2}{\omega}, \qquad \textbf{4.21b}$$

where k_2 is an integer, independent of k_1. From equations **4.21a** and **b**, one obtains the required driving frequency and energy increment per turn, in terms of the injection energy, the flux density of the guiding field and the arbitrary parameters k_1, k_2:

$$\omega = \frac{eB}{m_0}\left[1 + \left(\frac{eU_i}{m_0c^2}\right)^2\right]^{-1}(k_2 - k_1),$$

$$eU_0 = (m_0c^2 + eU_i)k_1(k_2 - k_1)^{-1}, \qquad \textbf{4.22}$$

where both k_1 and k_2 are integers subject to the restriction $k_2 > k_1$.

The simplest mode of operation arises if the electrons start from rest, $eU_i = 0$, and one chooses $k_1 = 1$, $k_2 = 2$, yielding $\omega = eB/m_0$, which is the cyclotron frequency corresponding to the rest mass, and $eU_0 = m_0c^2$, which is the rest energy. One can see that in this mode of operation, and for convenient flux densities, the frequency of the driving field falls in the microwave region. Thus the accelerating field can be maintained by a suitably shaped resonant cavity. The required kinetic energy increment per turn is fairly high and therefore the cavity must be excited at a high power level, but the requirements are well within technical feasibility. A simple microtron system is schematically shown in Figure 50. The electron source is one of the lips of the cavity resonator, from which electrons are drawn out by the high instantaneous field as a 'field emission' current.

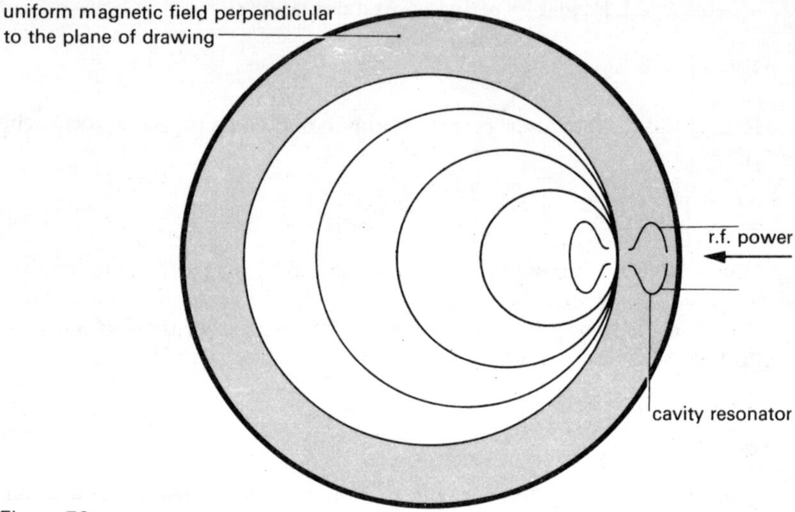

Figure 50

If it is assumed that the transit time in the accelerating cavity of width d is negligible, the energy increment per turn depends on the instant of transit

$$e\,U(t) = edE_0 \sin \omega t$$

and the amplitude of the field must be sufficiently high to supply the increment $eU_0 = edE_0 \sin \omega t_0$ at a phase $t_0 \neq \frac{1}{2}\pi$. We leave it to the reader to show that phase stability requires that $\frac{1}{2}\pi < \omega t_0 < \pi$.

The problem of beam stability is not simple. The radio-frequency field itself contributes to the focusing particularly during the first few orbits. The focusing effect produced by the magnetic guiding field cannot be described in simple terms. Since the orbits are osculating circles, they cannot all be centred on the symmetry axis of a magnetic field, and thus the results obtained in section 2.6, are not directly applicable. The total number of orbits, however, is usually small, and therefore the conditions to maintain orbit stability are, on the whole, less stringent than in a large synchro-cyclotron. The effect of electric focusing on the inner orbits, combined with the magnetic focusing of the outer orbits arising from the slow radial decrease of the guiding field, gives satisfactory beam stability.

4.3.2 *The betatron*

The betatron is an electron accelerator based on a principle which is entirely different from those discussed previously. Briefly, one can say that the betatron is a transformer whose secondary coil is replaced by a doughnut-shaped vacuum chamber, and the conduction current, which is driven in the secondary circuit of a conventional transformer by the induced electromotive force, is replaced by a beam of free electrons accelerated by the induced electric field.

The induction law relates the line integral of the electric field **E**, calculated along a closed path, to the rate of change of the total magnetic flux linked by this curve:

$$\oint \mathbf{E}.d\mathbf{s} = -\frac{\partial}{\partial t}\int_S \mathbf{B}.d\mathbf{S} = -\frac{\partial f}{\partial t}, \qquad \textbf{4.23}$$

where S is an arbitrary surface bounded by the curve along which the line integral is taken. Let us consider now a magnetic field which is rotationally symmetrical about the z-axis, and in which the $z = 0$ plane is a plane of mirror symmetry. If the magnetic field is constant, an electron has a circular equilibrium orbit of radius R in the $z = 0$ plane, centred on the symmetry axis, provided that its momentum is tangential to the circle, $\mathbf{p} = p\mathbf{e}_\theta$ and it satisfies the relation

$$p = -eR\,B(R). \qquad \textbf{4.24}$$

If the magnetic field is time dependent, an electron at any radial position R in the $z = 0$ plane is subject to an electric field $\mathbf{E} = E\mathbf{e}_0$, the magnitude of which follows from equation **4.23**, if the integration is carried out with respect to a circle of radius R in the $z = 0$ plane:

$$2\pi RE = -2\pi\int_0^R r\frac{dB(r)}{dt}\,dr. \qquad \textbf{4.25}$$

The electron will stay on the equilibrium orbit of constant radius R if $dp/dt = eE$, that is, from equations **4.24** and **4.25**,

$$\frac{d}{dt}B(R) = \frac{1}{R^2}\int_0^R r\frac{dB(r)}{dt}\,dr.$$

This condition is satisfied if

$$B(R) = \frac{1}{R^2}\int_0^R r\,B(r)\,dr. \qquad \textbf{4.26}$$

With a given flux-density distribution $B(r)$ in the $z = 0$ plane, the mean flux density within the equilibrium orbit is

$$\overline{B}_R = \frac{2}{R^2}\int_0^R r\,B(r)\,dr$$

and therefore equation **4.26** can be written as

$$B(R) = \tfrac{1}{2}\overline{B}_R. \qquad \textbf{4.27}$$

Thus the flux density at any point of the equilibrium orbit must be equal to half of the mean flux density within the orbit. This condition can be met if the flux density decreases with radial distance (see Figure 51).

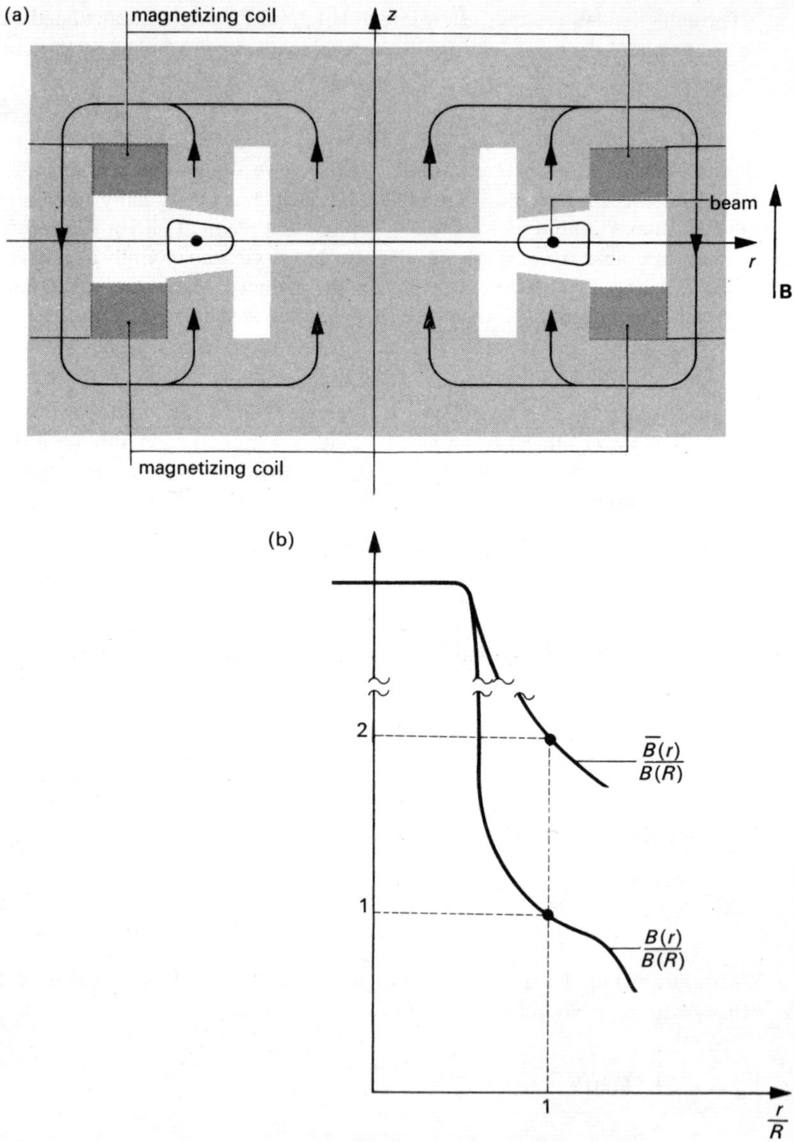

Figure 51

Once equation **4.27** is satisfied, the momentum of an electron orbiting on a circle of radius R will increase in time as $B(R)$ increases (see equation **4.24**). Note, however, that, unlike the accelerators discussed previously, the acceleration process in the betatron provides no mechanism for phase focusing.

The details of the guiding field distribution required in the betatron are not specified by equation **4.27**, and this freedom enables one to choose the field distribution in the region of the equilibrium radius R in such a manner that it should provide beam stability, that is, electrons slightly displaced from the equilibrium orbit should carry out harmonic oscillations about the equilibrium orbit.

The problem of orbit stability is greatly simplified by the fact that although the magnetic field is time dependent, it changes slowly in comparison with the period of the orbital motion. The momentum of the electron moving on the equilibrium orbit is given by equation **4.24**, and hence its total energy is

$$W = [(m_0 c^2)^2 + (ecRB)^2]^{\frac{1}{2}}.$$

The linear velocity of the orbiting electron is v, where

$$\frac{v}{c} = \frac{pc}{W} = \frac{eRB}{m_0 c}\left[1+\left(\frac{eRB}{m_0 c}\right)^2\right]^{-\frac{1}{2}}, \qquad \textbf{4.28}$$

and its angular velocity is

$$\omega = v/R \rightarrow c/R \qquad \text{when } v \rightarrow c,$$

that is, if $eRB \gg m_0 c$.

A typical betatron can be driven at mains frequency (50 Hz), and the accelerating period is that during which the flux-density distribution is not distorted by the saturation of the core. The energy gain per orbital rotation is quite small, say tens of electron volts, but while the flux density $B(R)$ reaches its peak value, a few kilogauss, the particle performs several hundred thousand orbital rotations. During most of its motion, the electron travels with very nearly the velocity of light.

This outline of the betatron operation indicates that it is justifiable to assume that during an orbital revolution the guiding field and the particle momentum are approximately constant. For this reason, the beam stability can be discussed with the assumption that the electron is in a stationary state of motion. In other words, the results of static electron-optic considerations are applicable. Quoting therefore the results of section 2.6, one can say that the electrons will perform stable oscillations in the vicinity of the equilibrium orbit if, in this region, the field distribution is again

$$B(r) = B_0\left(\frac{r}{R}\right)^n \qquad (-1 < n < 0).$$

Having assumed a stationary state of motion, the results can give no information about the effect of the slow change of energy on the oscillations. If the equations of motion are set up in a manner which allows for such a slow variation of the energy, it is found that the amplitude of the oscillation is damped in inverse proportion to the square root of the momentum.

It is of some interest to point out the most conspicuous difference between the microtron and the betatron, arising from the different mechanisms of acceleration rather than from the different technical facilities necessary for their operation. In the microtron, the energy gain per orbital rotation is comparable with the rest energy of the electron, and as a result the neighbouring orbits are well separated in the region diametrically opposite to the osculating point of the orbits. Consequently a target can be placed to intersect the outermost orbit without interfering with the acceleration process. Furthermore, the guiding field can be slightly distorted, in such a manner that the electrons leave the outermost orbit tangentially in a well-shaped beam, without the other orbits suffering significant perturbation. In the betatron, the beam is held tightly bound within a small region about a well-defined equilibrium orbit, and it is a very delicate problem to decide how to bring the electrons on to a target at the end of the accelerating cycle, without obstructing the passage of the electrons during the acceleration cycle. The extraction of the beam presents an even more complex problem.

4.3.3 *Synchrotrons*

The operation of the cyclotron and the synchro-cyclotron was based on a resonance mechanism by which the momentum of the particles was increased and from the relation $p = eBR$ it follows that in the *constant* guiding field the equilibrium orbit radius must also increase. This relation, however, shows that particles of increasing momentum can stay on an equilibrium orbit of constant radius R if the flux density along the prescribed equilibrium orbit, $B(R)$, is decreased at the same rate as the momentum is increasing. The resonance accelerator which is designed on this principle is the synchrotron. In order to maintain the resonant interaction between the radio-frequency accelerating field and the orbiting particles, the frequency of the driving field in the synchro-cyclotron is matched to the instantaneous total energy W of the particle in accordance with the relation $\omega = ec^2B/W$. If the total energy is expressed in terms of the momentum, one obtains a relation between the instantaneous driving frequency and the instantaneous flux density of the guiding field, along the equilibrium orbit of constant radius:

$$\omega = \frac{e\,B(R)}{m_0}\left[1+\left(\frac{eR\,B(R)}{m_0\,c}\right)^2\right]^{-\frac{1}{2}}. \qquad \textbf{4.29}$$

This equation defines the essential conditions which determine the design features of a synchrotron. As we shall see shortly, equation **4.29** gives different requirements depending on whether the particle is initially moving with a velocity much less than the velocity of light, or whether the particle is injected at nearly this velocity. The latter is quite feasible if the task is the acceleration of electrons. Therefore there are design problems which are different in the case of electron synchrotrons and proton synchrotrons respectively. Before turning to any specific aspect which distinguishes the two types of synchrotrons, let us see their common features.

The most conspicuous feature which distinguishes the synchrotons from the accelerators in the cyclotron family appears in the structure of the electromagnet. Cyclotrons require an approximately uniform magnetic field extending over an area somewhat larger than that enclosed by the ultimate equilibrium orbit. Since

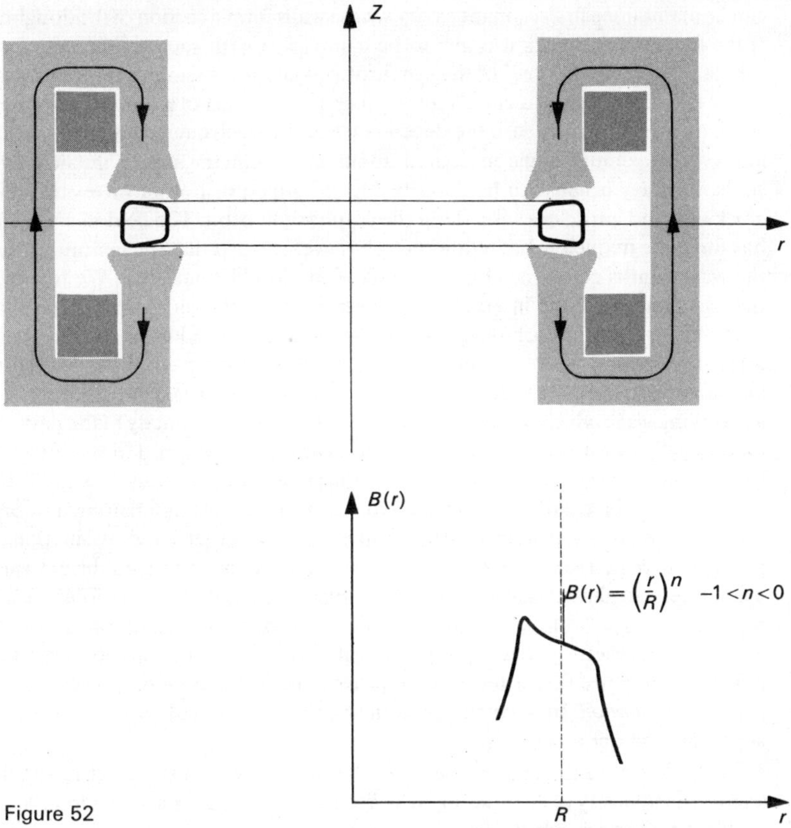

Figure 52

in the synchrotron the equilibrium orbit has a constant radius, it is sufficient to maintain the guiding field over a relatively narrow annular area, the mean radius of which equals the constant equilibrium orbit radius (see Figure 52). Thus at very high energies, where the size of a solid magnet becomes prohibitive on economic grounds, the construction of an annular magnet is still possible. In order to achieve beam stability, the flux density distribution about the equilibrium orbit radius has to be shaped to satisfy the relation $B(r) = B_0(r/R)^n$ with $-1 < n < 0$. The fact that such a field distribution allows stable oscillations for particles slightly displaced from the equilibrium orbit, as in betatrons, is explained by the fact that the time-dependent parameters again vary slowly relative to the period of orbital rotation.

It must be emphasized that synchrotrons are resonance accelerators and the acceleration process is accompanied by a phase-focusing effect. The technical realization of an accelerating radio-frequency field can take a variety of forms, depending on the frequency range required in a particular accelerator. The vacuum chamber of the synchrotron is doughnut-shaped, and the radio-frequency can be maintained in a resonant cavity whose walls form a section of the doughnut. If the frequency range is too low to be compatible with such a solution, a bent tubular electrode (the 'cee' of the synchrotron) and the vacuum chamber can act as the two electrodes of a capacitance which forms a part of a lumped-parameter resonant circuit. In any case the dependence of the electromagnetic field on time, and on the azimuth angle measured about the symmetry axis of the field from some arbitrary origin, can be described as the superposition of waves travelling clockwise and anti-clockwise along the equilibrium orbit. The field in each wave has the same frequency but a different phase velocity, so that integer multiples of the wavelengths equal the circumference of the equilibrium orbit. We recognize that the analysis of the interaction between such a field and particles travelling along the equilibrium orbit is analogous to that considered in the context of the linear accelerator, with the equilibrium orbit playing the part of the accelerator axis. Thus particles of a well-defined momentum can be locked at a fixed phase of a travelling wave which has the same phase velocity as the velocity of the particles. Other particles with slightly deviating momenta can be trapped in the travelling wave in such a manner that they perform phase oscillations about the equilibrium phase. It must be recalled that in the synchrotron, particle momentum and orbit radius are uniquely related to one another. On the other hand, oscillations in phase involve periodic variations of the momentum about an equilibrium value. Therefore the phase oscillations are coupled with radial oscillations, usually referred to as *synchrotron oscillations.* This effect has to be distinguished from the essentially geometrical electron-optic effect, governed by the distribution of the guiding field. Since this latter effect appears in the betatron too, it is often called *betatron oscillation.* In a synchrotron the *two types* of radial oscillations appear superimposed upon one another.

Let us return to the relation between the frequency ω of the accelerating field and the flux density of the guiding field $B(R)$ given by equation **4.29**. Observe that at kinetic energies which are very small compared with the rest energy, or $eRB \ll m_0 c$, equation **4.29** reduces to the linear relationship $\omega \simeq eB/m_0$. At the other extreme, when the particles are moving with very nearly the velocity of light, $eRB \gg m_0 c$ and $\omega = c/R =$ constant. In the transition between the completely non-relativistic and the extreme relativistic situation, there is no simplifying condition which would allow an approximation to equation **4.29**, and this makes the matching of the driving frequency to the guiding flux density a very serious technical problem.

If the task is the acceleration of electrons, this problem can be avoided because electrons can be accelerated to relativistic velocities by some simpler method, and can be injected into the synchrotron, when it can operate at a constant radio frequency. A particularly interesting solution is based on the fact that the betatron also operates with an equilibrium orbit of constant radius. Thus it is possible to build a combined machine, which at first operates as a betatron, accelerating the electrons to relativistic energies, and then the synchrotron operation takes over when the radio-frequency field is switched on. Note that in the *betatron*, the equilibrium orbit radius R_b is uniquely determined by the field distribution, and the angular velocity of the electrons varies as $\omega_b = v/R_b = (v/c)(c/R_b)$.

Let the *synchrotron* operation be designed for an equilibrium orbit radius $R = R_b + \Delta R$, that is, the constant frequency of the accelerating field is

$$\omega = c(R_b + \Delta R)^{-1};$$

$\Delta R \ll R_b$. A smooth transition is achieved between the betatron operation and the synchroton operation, if the change-over takes place (by switching on the radio-frequency field) when $\omega = \omega_b$, that is, when $v/c = (1+\Delta R/R_b)^{-1}$. This condition determines the flux density on the betatron orbit at the instant of switch-over, since

$$\frac{v}{c} = \frac{pc}{W} = \frac{eR_b B(R_b)}{m_0 c}\left[1+\left(\frac{eR_b B(R_b)}{m_0 c}\right)^2\right]^{-\frac{1}{2}}.$$

Under these conditions, once the radio-frequency field is switched on, the mechanism of the acceleration is analogous to that in a linear accelerator, in which the phase velocity of the accelerating field is equal to the velocity of light. During the acceleration by the betatron effect, there is no phase focusing and therefore only those electrons, which enter the radio-frequency field within a finite range about the equilibrium phase, can be accelerated further. The width of this range, remember, depends on the electron energy at the time of the transition from betatron to synchrotron operation. It is also good to remember that, under these circumstances, no phase oscillations take place; the phases are approaching monotonically an asymptotic value, and the radius of the orbit is approaching R asymptotically.

Conditions for orbit stability in the electron synchrotron are the same as in the betatron. If the annular magnetic field is described by $B = B_0(r/R_0)^n$ where $-1 < n < 0$, both the betatron and the synchrotron operation allows the electrons to oscillate about the equilibrium orbits of radius R_b and R respectively, provided that both these orbits are sufficiently close to the circle of radius R_0. That is why, in considering the transition from betatron to synchroton operation, the condition $R - R_b \ll R_b \simeq R_0$ is an essential one, and an error in timing the transition leads to excessively large amplitudes in the oscillation of the electrons about the synchrotron orbit.

The solution of certain technical problems becomes easier if, instead of a closed annular magnet, identical sections are built with gaps between them, so that the closed equilibrium orbit consists of circular arcs in the magnetic field and straight flight paths in field-free regions, which join smoothly. In such an arrangement, the straight sections make it easier to insert ports for injecting and extracting the beam, and junctions for the supply of the radio-frequency accelerating power, pumping connexions, etc. Considering a straight section of length L and a magnet section of arc length $2\pi R/N$, with flux density distribution $B = B_0(r/R)^n$, forming a 'unit cell' of the guiding field, one can still maintain stable betatron oscillations about the equilibrium orbit. The conditions for stability follow from the arguments presented in section 2.8.

The matrix representing the field is of the type given by equation **2.62**, and the matrix representing the field-free region is of the type given by equation **2.62a**. Thus the diagonal elements of the matrix representing the 'unit cell' are

$$m_{11} = \cos\frac{2\pi}{N}k_j R - k_j L \sin\frac{2\pi}{N}k_j R,$$

$$m_{22} = \cos\frac{2\pi}{N}k_j R,$$

where in the case of displacements in the equilibrium plane one must substitute

$$k_1 = R^{-1}(1+n)^{\frac{1}{2}}$$

and for displacements in the direction perpendicular to the plane of the equilibrium orbit,

$$k_2 = R^{-1}(-n)^{\frac{1}{2}}.$$

The condition of stability is

$$\tfrac{1}{2}|\text{tr M}| = \cos\frac{2\pi}{N}k_j R - \frac{LN}{4\pi R}\frac{2\pi}{N}k_j R \sin\frac{2\pi}{N}k_j R < 1.$$

Given the number of unit cells N, the inequality above, written for k_1 and k_2 separately, permits the determination of the range of the field index n within which oscillations in both directions are stable.

In the design and development of *proton synchrotrons*, more serious technical difficulties have to be faced because, unlike electrons, heavy particles cannot be injected at relativistic speeds. Thus the frequency of the accelerating field, and the flux density of the guiding field, must be varied in such a manner that condition **4.29** is satisfied throughout the accelerating cycle. If we denote

$$\frac{eR}{m_0 c}B(R) = \tan x,$$

then from equation **4.29** one finds the requirement

$$\omega = \frac{c}{R}\sin x.$$

Thus the necessary relationship can be maintained, for example, by a sophisticated electronic control system, which detects at some point of the guiding field an instantaneous flux density B proportioned to $B(R)$, calculates

$$x = \tan^{-1}[eR\,B(R)/m_0 c],$$

calculates $\sin x$, and produces a signal which sets the angular frequency of a power oscillator to $(c/R)\sin x$. To appreciate the difficulties, it is well to realize that the radio-frequency amplitude must be sufficient to produce a kinetic energy gain of about a thousand electron volts per turn. Thus, if the output energy is to be a thousand million electron volts, the particles have to describe during the accelerating cycle a million orbits, while the driving frequency has to increase by an order of magnitude.

So long as equation **4.29** is satisfied, the mechanism of the acceleration is analogous to that outlined in the context of the interaction between charged particles and an electromagnetic wave travelling with increasing phase velocity in a linear accelerator. The main difference arises again from the fact that, in the present case, the equilibrium orbit is a circle of constant radius instead of a straight line. As in similar situations encountered before, particles will be accelerated if their energy on entering the radio-frequency accelerating field is higher than a threshold value, and their entrance phase is within a finite range about an equilibrium phase. As pointed out earlier, the particles which are trapped in the travelling wave and perform phase oscillations carry out radial oscillations too.

Any deviation from equation **4.29**, in the relation between the driving frequency and the flux density of the guiding field, has the effect of displacing the equilibrium phase and changing the equilibrium orbit radius. This causes excessive synchrotron and betatron oscillations, and reduces the number of particles trapped in the travelling wave. To appreciate the stringency of the conditions on the oscillation amplitudes, one should consider that typical proton synchrotron orbit radii are of the order of ten metres, thus during the accelerating cycle particles cover a total flight path of several hundred miles, and will hit the wall of the vacuum chamber if they deviate from the nominal equilibrium orbit by a foot or so.

3.4 *Strong focusing*

The desire to accelerate particles to higher and higher energies means building synchrotrons of ever-increasing diameter. To maintain the intensity of the output beam, the cross-section of the vacuum chamber has to be increased correspondingly, as shown below.

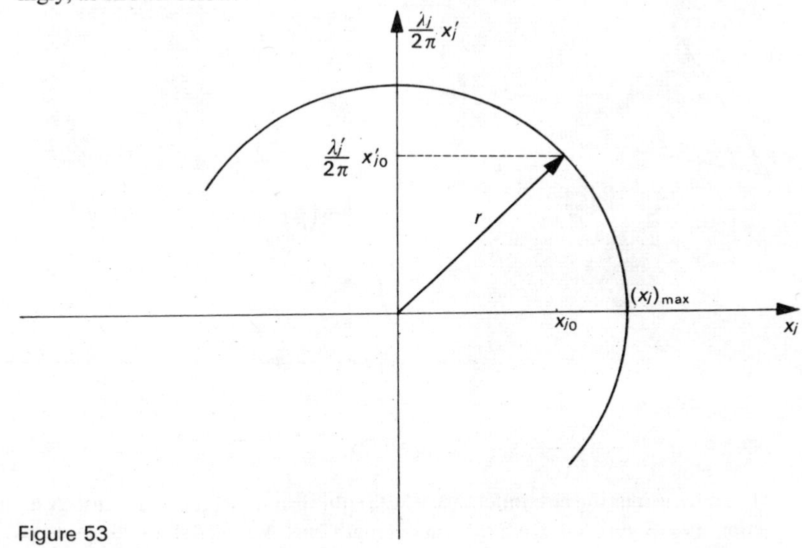

Figure 53

If the trajectories adjacent to the equilibrium orbit are sinusoidal curves, the displacement from the equilibrium orbit can be readily displayed in the (x_j, x'_j) plane. The explicit equations obtainable from the much used matrix relation, $\mathbf{X} = \mathbf{MX}_0$, indicate that one can proceed as in sections 2.2.1 and 2.9. One plots, along two mutually perpendicular axes, the displacement x_j from the equilibrium orbit and $(\lambda_j/2\pi)x'_j$. where λ_j is the spatial periodicity of the trajectories; the subscript $j = 1$ refers to motion in the equilibrium plane and $j = 2$ refers to motion in the perpendicular direction (see Figure 53). If x_{j_o} and x'_{j_o} are the initial coordinate and trajectory slope respectively, the displacement is given by the projection of a vector of length

$$r = \left[x_{jo}^2 + \left(\frac{\lambda_j}{2\pi} x'_{jo}\right)^2\right]^{\frac{1}{2}}$$

on the x_j-axis; as the particle progresses, the vector rotates about the origin of the coordinate system. Thus, with the given initial conditions, the maximum displacement is r, which decreases with decreasing λ_j. In section 2.6, it was shown that, if the flux-density distribution of the guiding field is $B(r) = B_0(r/R)^n$, the oscillations are stable in both directions only if $-1 < n < 0$, and in this case the spatial periodicity is given by $\lambda_1 = 2\pi R(1+n)^{-\frac{1}{2}}$ and $\lambda_2 = 2\pi R|n|^{-\frac{1}{2}}$, in the two directions respectively. Thus to accommodate a beam of a given angular aperture, the cross-section of the vacuum chamber must be increased in proportion to the area defined by the equilibrium orbit. The purpose of 'strong focusing' is to reduce the spatial period of oscillations for a given equilibrium orbit radius, and it is achieved by using a guiding field of *alternating* flux-density *gradient*.

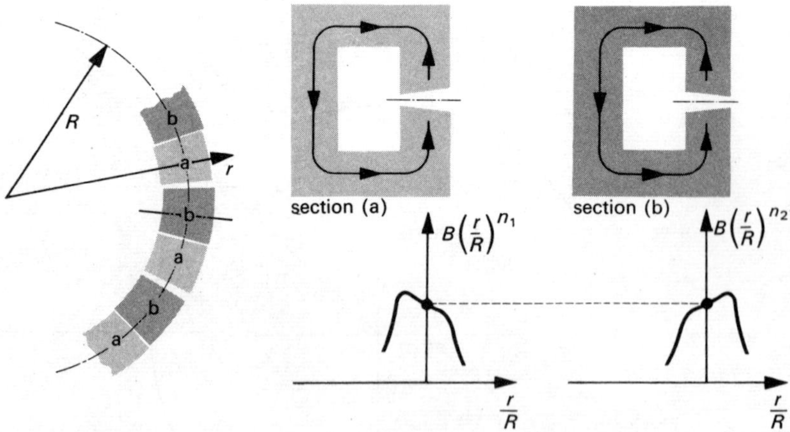

Figure 54

Let us consider the circumference of the equilibrium orbit divided into N equal sections (see Figure 54). Each section of length $2\pi R/N$ is further divided into parts of length $s_1 < 2\pi R/N$ and $s_2 = 2\pi R/N - s_1$, in which the flux-density distribution is

$$B_1 = B_0\left(\frac{r}{R}\right)^{n_1} \qquad \text{if } 0 \leqslant s \leqslant s_1,$$

$$B_2 = B_0\left(\frac{r}{R}\right)^{-n_2} \qquad \text{if } s_1 \leqslant s < \frac{2\pi R}{N},$$

respectively, with $n_1 \gg 1$, $n_2 \gg 1$.

Observe that between adjoining sections the direction as well as the magnitude of the field gradient is alternating. Nevertheless along a circle of radius R, the flux density is the same, and this fact leads to the existence of an equilibrium orbit of radius R. Since the magnetic guiding field consists of identical 'unit cells', the considerations of section 2.8 are applicable, and it will be shown that the field indices n_1 and n_2 can be chosen in such a manner that particle trajectories are confined to the neighbourhood of the equilibrium orbit. Furthermore, it will be seen that the spatial period of the oscillations becomes smaller than in the case of a guiding field of uniform gradient.

To obtain a beam confined to a small region about the equilibrium orbit, the matrix M, characteristic of the unit cell, has to satisfy the condition $|\text{tr M}| < 2$. If m_{jk}^+ and m_{jk}^- are the elements of the matrices M^+ and M^- respectively, defined by equation **2.62**, the stability condition becomes

$$|\text{tr}(M^+M^-)| = |m_{11}^+ m_{11}^- + m_{12}^+ m_{21}^- + m_{21}^+ m_{12}^- + m_{22}^+ m_{22}^-| < 2. \qquad \textbf{4.30}$$

For simplicity's sake, let us consider a unit cell which consists of sections of equal length, $s = \pi R/N$; one with field index n_1, and the other with $-n_2$. Then from equations **2.59a** and **b**, one obtains

for section 1:

$$Rk_1^{(1)} = (1+n_1)^{\frac{1}{2}}$$
$$Rk_2^{(1)} = (-n_1)^{\frac{1}{2}}$$

for section 2:

$$Rk_1^{(2)} = (1-n_2)^{\frac{1}{2}}$$
$$Rk_2^{(2)} = n_2^{\frac{1}{2}}$$

and with the notation

$$\phi_1 = \frac{\pi}{N}(n_1+1)^{\frac{1}{2}} \qquad \psi_1 = \frac{\pi}{N}(n_2-1)^{\frac{1}{2}}$$

$$\phi_2 = \frac{\pi}{N} n_2^{\frac{1}{2}} \qquad \psi_2 = \frac{\pi}{N} n_1^{\frac{1}{2}}$$

equation **4.30** yields

$$\left|\cos\phi_j \cosh\psi_j + \frac{1}{2}\left(\frac{\psi_j}{\phi_j} - \frac{\phi_j}{\psi_j}\right)\sin\phi_j \sinh\psi_j\right| < 1, \qquad \textbf{4.31}$$

where $j = 1$ refers to oscillations *in* the plane of the equilibrium orbit, and $j = 2$ to oscillations in the *perpendicular* direction.

Note that if $n_1 \gg 1$ and $n_2 \gg 1$, one finds

$$\phi_1 \simeq \psi_2 \qquad \text{and} \qquad \phi_2 \simeq \psi_1.$$

Let us denote for sake of brevity

$$\cos \pi \frac{n_1^{\frac{1}{2}}}{N} \cosh \pi \frac{n_2^{\frac{1}{2}}}{N} - \frac{1}{2} \frac{n_1 - n_2}{(n_1 n_2)^{\frac{1}{2}}} \sin \pi \frac{n_1^{\frac{1}{2}}}{N} \sinh \pi \frac{n_2^{\frac{1}{2}}}{N} \equiv F\left(\frac{n_1}{N^2}, \frac{n_2}{N^2}\right)$$

and plot the curves

$$F\left(\frac{n_1}{N^2}, \frac{n_2}{N^2}\right) = 1 \qquad \text{and} \qquad F\left(\frac{n_1}{N^2}, \frac{n_2}{N^2}\right) = -1$$

in a Cartesian coordinate system, where the parameters n_1/N^2 and n_2/N^2 are measured along the two axes. Points between the two curves represent values of these parameters, which are compatible with *stable* oscillations *in the plane* of the equilibrium orbit.

If the variables $n_1^{\frac{1}{2}}/N$ and $n_2^{\frac{1}{2}}/N$ are interchanged, one obtains a function, which can be denoted by G, and the curves

$$G\left(\frac{n_1}{N^2}, \frac{n_2}{N^2}\right) = \pm 1$$

define the areas within which the pairs of parameters $n_1^{\frac{1}{2}}/N$ and $n_2^{\frac{1}{2}}/N$ are compatible with *stable* oscillations in the direction *perpendicular* to the equilibrium orbit. Note that the way the function G is formed out of F indicates that, once F is plotted, the G-curves can be obtained by taking the mirror image of the F-curves about the line $n_1 = n_2$ (see Figure 55a). Overlapping areas of the two

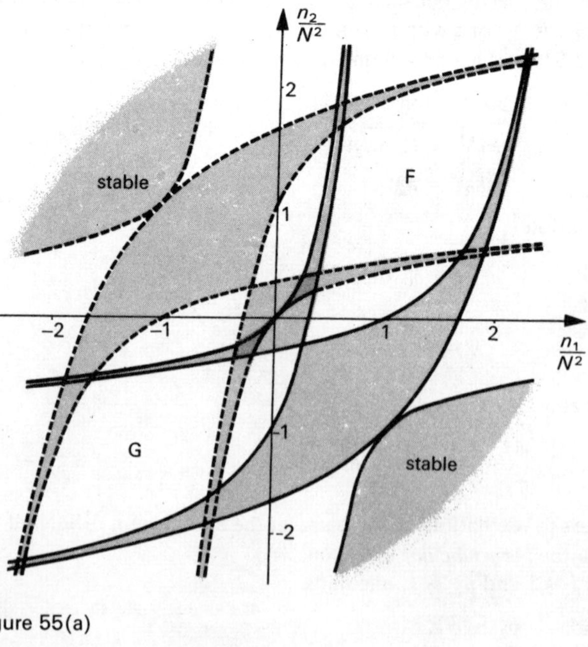

Figure 55(a)

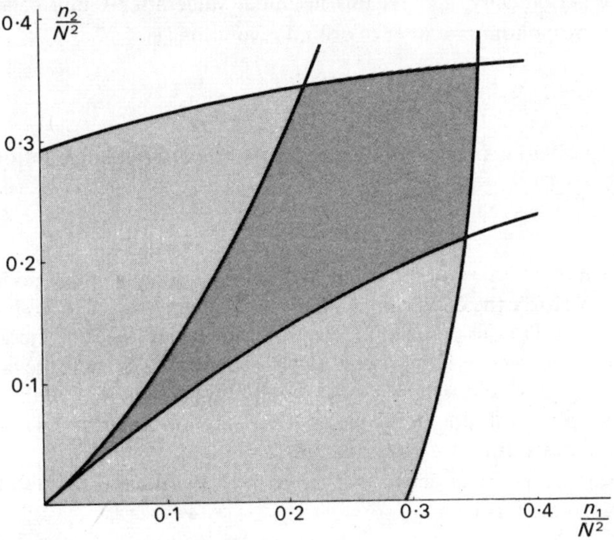

Figure 55(b)

stability regions define the range of the parameter values $n_1^{\frac{1}{2}}/N$ and $n_2^{\frac{1}{2}}/N$ which lead to a complete confinement of the beam to a finite neighbourhood of the equilibrium orbit. The lowest stability region (shown on an enlarged scale in Figure 55b), covering the largest range of the parameters, has a 'necktie' shape, and is often referred to by this name

Having established the condition of stable oscillations about the equilibrium orbit, it still remains to be seen that the alternating-gradient guiding field achieves its original purpose, that is, it increases the frequency of orbital oscillations.

It was pointed out in section 2.8 that if the stability conditions are satisfied, the trajectories adjacent to the equilibrium orbit are periodic curves, but not sinusoidal. Without a knowledge of the precise shape of the trajectories within the unit cell, some important conclusions can be reached by the following argument. If equation **4.30** is satisfied, one is entitled to write

$$\tfrac{1}{2}\,\mathrm{tr}\,(\mathrm{M}^{+}\mathrm{M}^{-}) = \cos\mu \qquad (\mu \neq 0, \pi, 2\pi \ldots)$$

and express the matrix $\mathrm{M} = \mathrm{M}^{+}\mathrm{M}^{-}$ in the form of equation **2.82**:

$$\mathrm{M} = \mathrm{I}\cos\mu + \mathrm{J}_1 \sin\mu,$$

where I is the unit matrix, and J_1 has the properties $\det \mathrm{J}_1 = 1$, $\mathrm{tr}\,\mathrm{J}_1 = 0$ and $\mathrm{J}_1^2 = -\mathrm{I}$. Therefore, whatever happens within a unit cell, if the position of the particle and the trajectory slope, which are contained in the vector

$$\mathbf{X} = \begin{bmatrix} x \\ x' \end{bmatrix},$$

are determined at intervals equal to the length of the unit cell, $S = 2\pi R/N$, it is found to vary periodically, and regains its initial value after k unit cells, if $k = 2\pi/\mu$. The corresponding number of orbital revolutions is

$$\nu = \frac{k}{N} = \frac{2\pi}{N\mu},$$

or the number of oscillations about the equilibrium orbit per orbital revolution is

$$\frac{1}{\nu} = \frac{N\mu}{2\pi}. \qquad \textbf{4.32}$$

Now if the ratios n_j/N^2, characteristic of the alternating gradients of the guiding field, are chosen to satisfy the condition **4.31**, μ is fixed, but equation **4.32** shows that the number of oscillations per orbital revolution can be increased by increasing N, provided n_j is increased to keep n_j/N^2 = constant. In our previous notation $n_1/N^2 = (\psi_2/\pi)^2$ and $n_2/N^2 = (\phi_2/\pi)^2$, and, in general, the number of orbital oscillations per revolution in the plane of the equilibrium orbit, $1/\nu_1$, and in the perpendicular direction, $1/\nu_2$, are different.

As a simple example, however, let $n_1 = n_2 \equiv n_0 \gg 1$. In this case, the stability condition **4.31** takes the form

$$\tfrac{1}{2}(\text{tr } \mathrm{M}^+\mathrm{M}^-) \equiv \cos\phi \cosh\psi = \cos\mu.$$

Choosing, for example, $n_0/N^2 = \frac{1}{4}$, then $\phi = \frac{1}{2}\pi$, and hence $\cos\mu = 0$, that is, $\mu = \frac{1}{2}\pi$. Thus equation **4.32** yields

$$\frac{1}{\nu} = \frac{N}{4} = \tfrac{1}{2}n_0^{\frac{1}{2}} \qquad (n_0 \gg 1).$$

This result should be compared with the corresponding one in the case of a constant guiding field gradient; here the number of oscillations per orbital revolution is

$$\frac{2\pi R}{\lambda} = \begin{cases} n^{\frac{1}{2}} & \text{perpendicular to the plane of the equilibrium orbit,} \\ (1-n)^{\frac{1}{2}} & \text{in the plane of the equilibrium orbit,} \end{cases}$$

where $-1 < n < 0$.

It has been pointed out repeatedly that, in the case of stable orbits, the trajectories adjacent to the equilibrium orbit are periodic but not sinusoidal. Therefore our previous considerations do not give direct information about the maximum displacement of a particle from the equilibrium orbit. To obtain such information, the differential equation of the particle trajectories must be considered in detail, as indicated in section 2.8. This task is beyond the scope of this book and let it suffice to state the result of such considerations.

At the beginning of this subsection, it was seen that the maximum displacement r from the equilibrium orbit depends in a simple way on the spatial periodicity λ_j of the trajectories. In the case of alternating-gradient field, one *under*estimates the maximum displacement if one simply replaces $2\pi R/\lambda_j$ by $1/\nu$.

The discrepancy depends on the location of the parameter values n_1/N^2 and n_2/N^2 in the stability diagram. In the case considered above, $n_1 = n_2 = n_0$, the discrepancy decreases with increasing number of sections, and becomes insignificant if $n_0/N^2 \ll 1$.

It must be emphasized that the preceding considerations referred to what were termed 'betatron oscillations', and no attempt was made to consider quantitatively the effect of phase oscillations. On this topic the reader must consult specialist literature.

In conclusion one should realize that, under practical circumstances, the distribution of any guiding field has some deviations from the nominal one, because the individual magnets, which constitute the complete system, can be built only approximately to the specifications of the design. Since the particles perform a large number of orbital revolutions, the deviations from the nominal flux-density distribution result in periodic perturbations, and harmful resonance effects can occur. Perturbations due to errors in the field gradient have the result that within the necktie-shaped stability region of the $(n_1/N^2, n_2/N^2)$ diagram there are belts, so-called stop-bands, where the betatron oscillations become unstable. Another source of perturbation is the fact that magnetic imperfections destroy the strict mirror symmetry of the field with respect to a plane. Hence the equilibrium orbit is not a plane curve, and in this case there is a coupling between the oscillations in the assumed plane of the equilibrium orbit and in the perpendicular direction. This coupling can also lead to a resonance which makes the beam unstable in certain belts within the nominally stable region of the $(n_1/N^2, n_2/N^2)$ diagram.

4.4 Limits of the energy attainable by particle accelerators

It is well known from the classical theory of electromagnetic fields that an accelerating charge loses energy by the emission of electromagnetic radiation. This process must set an ultimate limit attainable by an accelerator, because if the particle gains by acceleration a certain amount of energy in unit time (say $d\varepsilon/dt$) and there is an energy loss by radiation at a rate $-dw/dt$, the ultimate energy of the particle is reached once the rate of energy gain equals the rate of energy loss. The purpose of this section is to investigate the consequences of this statement in quantitative terms.

If the momentum of the particle is $\mathbf{p} = \gamma m_0 \mathbf{v}$, and its total energy is

$$W = (p^2c^2 + m_0^2c^4)^{\frac{1}{2}} = \gamma m_0 c^2,$$

with the usual notation $\gamma = (1 - v^2/c^2)^{-\frac{1}{2}}$, the energy lost by radiation in unit time is given by

$$-\frac{dw}{dt} = \frac{2}{3}\frac{r_0}{m_0 c}\gamma^2\left[\left(\frac{d\mathbf{p}}{dt}\right)^2 - \frac{1}{c^2}\left(\frac{dW}{dt}\right)^2\right]. \qquad 4.33$$

The quantity r_0 is the 'classical electron radius', which can be expressed in terms of other universal constants in S.I. units:

$$r_0 = (4\pi\varepsilon_0)^{-1}\frac{e^2}{m_0 c^2}.$$

If one substitutes the momentum **p** and energy W in terms of the velocity, in equation **4.33**, and carries out the differentiation, one is led to the equation which is usually presented in textbooks:

$$-\frac{dw}{dt} = \frac{2}{3}\frac{r_0 m_0}{c}\gamma^6\left[\dot{v}^2 - \frac{1}{c^2}(\mathbf{v}\times\dot{\mathbf{v}})^2\right] \qquad \mathbf{4.34}$$

and hence one obtains, as a non-relativistic approximation ($v/c \ll 1$), the more familiar expression

$$-\frac{dw}{dt} = \frac{2}{3}\frac{r_0 m_0}{c}\dot{v}^2. \qquad \mathbf{4.34a}$$

In discussing the radiation loss from particles in high-energy accelerators, one obviously must use the relativistically correct equations.

Considering the case of a particle in a linear accelerator, it must be remembered that the total momentum is parallel to the accelerator axis ($\mathbf{p} = p\mathbf{e}_z$) and therefore one finds, using the energy–momentum relation,

$$\frac{dp}{dt} = \frac{dW/dt}{dW/dp} = \frac{1}{v}\frac{dW}{dt}.$$

By substituting this expression, equation **4.33** yields

$$-\frac{dw}{dt} = \frac{2}{3}\frac{r_0}{m_0 c}\frac{1}{v^2}\left(\frac{dW}{dt}\right)^2. \qquad \mathbf{4.35}$$

This equation gives the rate of the radiation loss, and can easily be converted into the radiation loss per unit length of flight path. Since $v = dz/dt$, equation **4.35** can be rewritten as

$$-\frac{dw}{dz} = \frac{2}{3}\frac{r_0}{m_0 c^2}\frac{c}{v}\left(\frac{dW}{dz}\right)^2.$$

Taking the extreme relativistic limit $v \to c$, one finds

$$\frac{-dw/dz}{dW/dz} = \frac{2}{3}\frac{r_0}{m_0 c^2}\frac{dW}{dz}. \qquad \mathbf{4.36}$$

On the left-hand side, we see the ratio of the rate of the radiation loss to the rate of the energy gain in the accelerator. According to equation **4.36**, this ratio is proportional to the rate of energy gain itself, and it becomes unity only if the energy gain of the particle, over a flight path equal to the classical electron radius, becomes larger than the rest energy of the particle, that is, $r_0(dW/dz) = (3/2)m_0 c^2$. Since $r_0 \simeq 2{\cdot}82 \times 10^{-15}$ m, this rate of acceleration is far beyond technical feasibility even for electrons ($m_0 c^2 \simeq 0{\cdot}5$ MeV). In other words, the limitation to the attainable energy, which is inherent in the mechanism of linear acceleration, has no practical significance.

The situation is quite different in the case of cyclic accelerators. Let us consider the case when the particle is moving on an equilibrium orbit of constant radius in a betatron or as a synchrotron. In this case, the acceleration of the particle is perpendicular to the orbital velocity and $\dot{v} = v^2/R$. Thus equation **4.34** yields

$$-\frac{dw}{dt} = \frac{2}{3}\frac{r_0 m_0}{c}\gamma^4\frac{v^4}{R^2}.$$

Hence the energy loss by radiation during one orbital rotation is

$$w_1 = -\frac{2\pi R}{v}\frac{dw}{dt} = \frac{4\pi}{3}\frac{r_0}{R}\left(\frac{v}{c}\right)^3 m_0 c^2\left(\frac{W}{m_0 c^2}\right)^4.$$

Here use has been made of the relation $\gamma = W/m_0 c^2$. In the extreme relativistic case ($v \simeq c$),

$$\frac{w_1}{m_0 c^2} = \frac{4\pi}{3}\frac{r_0}{R}\left(\frac{W}{m_0 c^2}\right)^4. \qquad \textbf{4.37}$$

An alternative expression is obtained if the orbit radius is expressed in terms of the flux density of the guiding field: $1/R = ecB/pc$; since in the extreme relativistic approximation $pc \simeq W$, one obtains

$$w_1 = \frac{4\pi}{3} r_0 ecB\left(\frac{W}{m_0 c^2}\right)^3. \qquad \textbf{4.37a}$$

The equations above show that the radiation loss per orbital revolution increases rapidly with the energy of the particle. Although the classical electron radius is very small, the radiation loss of electrons becomes very substantial at high energies. Considering, for example, a typical electron synchrotron with a guiding field $B \simeq 1$ Wb m^{-2}, designed for an output energy $W \simeq 500$ MeV (the corresponding equilibrium orbit radius is $R \simeq 1{\cdot}66$ m), one finds $w_1 \simeq 3{\cdot}5$ keV. In order to reach the output energy specified the energy gain ε_1 per orbital revolution must be $\varepsilon_1 \geqslant w_1$.

Equation **4.37** is, of course, valid for heavy particles too. In practical terms, however, the limitation imposed by the radiation loss is not significant, since the particles' rest energy is so much higher. For example, using a guiding field $B \simeq 1$ Wb m^{-2}, $W \simeq 50$ GeV protons ($W/m_p c^2 \simeq 50$) have a radiation loss $w_1 \simeq 0{\cdot}45$ eV per revolution.

The characteristic properties of the radiation emitted by the accelerating electron, its directional pattern and its spectral distribution, can be determined by classical (that is, non-quantum-theoretical) arguments. Even a summary of these considerations goes beyond the scope of this book; let it suffice to say that the directional pattern of the radiation is very strongly forward peaked, and at energies above 30 MeV, considerable energy is radiated in the visible spectrum.

The spectrum is continuous and as the electron energy increases the peak of the distribution becomes increasingly pronounced, shifting through the ultra-violet into the X-ray region. The experimental investigation of the radiation of electrons accelerated in synchrotrons has confirmed the theoretical predictions, and has led to an interesting optical technique of studying the instantaneous position and cross-section of the electron beam, during the acceleration process. For further information on the theoretical and the experimental investigation of the synchrotron radiation, the reader should consult specialist literature.

Further reading

J. C. SLATER, *Microwave Electronics*, van Nostrand, 1950.

Advanced theory of the electromagnetic field in wave-guides and cavity resonators, and of the interaction of such fields with free electrons. Chapters 11 and 14 are devoted to the theory of resonance accelerators. Outstandingly stimulating reading.

S. FLUGGE (ed.), *Encyclopedia ('Handbuch') of Physics*, Vol. 44: *'Nuclear Instrumentation*, Springer, Berlin 1959.

Collection of review articles for the specialist, written at an advanced level, on the theory and the design of different types of accelerators, some described in technical details.

R. KOLLATH (ed.), *Particle Accelerators*, Pitman, 1967.

Collection of review articles on the theory and design of accelerators of different types. The theory is summarized rather than developed in detail; description and design parameters of a number of up-to-date machines are given.

Chapter 5
Electronic Amplifiers and Oscillators

That the different modes in which electrical excitement takes place will some day or other be reduced under one common law can hardly be doubted, though for the present we are bound to admit distinctions. It will be a great point gained when these distinctions are, not removed, but understood.

M. Faraday

5.1 Introduction

The function of most electronic amplifiers, and of all oscillators, is to convert direct-current energy into radio-frequency energy. Out of the very large variety of devices, we shall consider electronic valves only, that is, devices in which this energy conversion takes place by virtue of the interaction between *free* electrons, moving in vacuo and electromagnetic fields. Solid-state devices, whose operation depends upon the properties of semiconductors, form a class of similar importance but will be completely ignored.

It was shown in section 3.1 that if electrons pass through a radio-frequency electromagnetic field, an energy conversion process takes place and electromagnetic radiation is generated at the expense of the *mean* kinetic energy of the electrons. It was also mentioned that this is the fundamental process on which the operation of most valve amplifiers and oscillators rests. The study of this process is the subject of the subsequent discussion.

We recognize that the process in question is the reverse of that operating in an accelerator, and it is tempting to repeat the considerations followed in the analysis of the properties of accelerators. However, it can readily be seen that since the function of accelerators on the one hand, and that of amplifiers and oscillators on the other, are so very different, it is advisable to approach the study of these devices in a somewhat different way. Until now, our attention has been concentrated on the motion of electrons, or other charged particles, in well-defined radio-frequency fields, but how these fields are produced or maintained has not so far been investigated. More precisely, the effect of moving electrons on the properties of the resonant cavities, that is, the reaction of the r.f. system, was completely disregarded. Although these problems are of great importance in the design of the r.f. system of the accelerators, they could be neglected in a first review of the mechanism of their operation. When we turn to problems concerning amplifiers and oscillators, the situation is reversed. The purpose of these devices is to deliver r.f. power to some load; the electrodes between which (or the resonant cavity within which) the electrons move, and the load to which power has to be delivered, are all parts of the same network. Thus the primary question is how the electron motion affects the properties of this network. The answer to this question will reveal the working mechanism of these devices. Whatever the details of this mechanism turn out to be, it is plausible that in order to obtain high gain in amplification, or to generate oscillations at substantial power levels, high electron-beam intensities are required. To take proper account of the mutual interaction between electrons, we shall in some cases sacrifice some of the information concerning the motion of individual electrons, and shall analyse the operation of amplifiers and oscillators in terms of the properties of space-charge flow.

For the purpose of practical applications, valves can be specified by quantities which characterize their behaviour as circuit elements. Once these characteristic quantities are known, the valves can be represented by 'equivalent networks', and the design of an amplifier, the investigation of the conditions under which self-excited oscillations can be obtained, the description of the properties of the output signal etc., become tasks which can be handled by methods of network analysis, without any reference to the physical process inside the valve. In the following considerations, the relation between the interaction of free electrons with electromagnetic fields within a valve, and the properties of valves as circuit elements, will be outlined.

A first step in this direction is to recognize that if a charge is moving between two electrodes connected by a conductor, there will be current flowing in this conductor *while* the charge is in motion between the electrodes (see Figure 56).

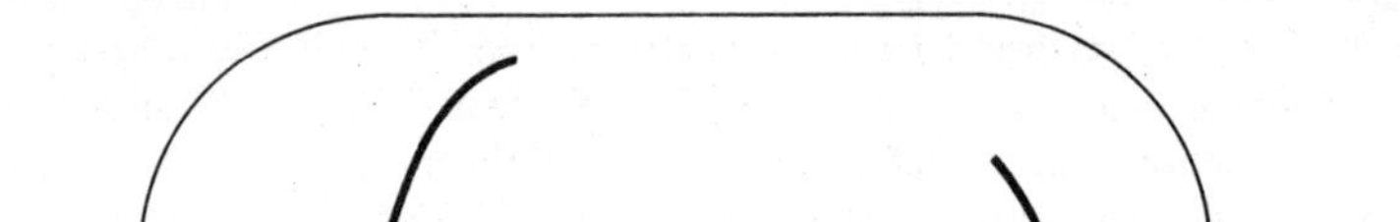

Figure 56

To see the validity of this statement, let us recall that if there is a stationary charge between the electrodes, it will induce, on the electrode surfaces, charges which will take up a static equilibrium distribution, determined by the condition that no electric field can exist inside the conductor. If the free charge is displaced, that induced on the electrodes must be redistributed, and this process is necessarily accompanied by the flow of conduction current in the lead between the electrodes. This picture shows that the flow of conduction current in the connecting lead is a consequence of the motion of the charge between the electrodes, and not of charge leaving or entering the electrodes.

To determine the conduction current in the external lead, one can use simple energy considerations. Suppose the two electrodes were connected to a battery, maintaining a constant potential difference U_0 between the electrodes. In this case, there will be a constant electric field $\mathbf{E}_0$ between the electrode, such that if A_1 and A_2 are points on the two electrodes then

$$\int_{A_1}^{A_2} \mathbf{E}_0 \cdot d\mathbf{s} = U_0.$$

If a charge e is moving in the field $\mathbf{E}_0$ with a velocity $\mathbf{v}$, the work done on the charge by the field in the time dt is $e\mathbf{E}_0 . \mathbf{v}\, dt$. This work is supplied by the battery which drives a conduction current i in the external circuit connecting the electrodes, and the work done by the source during the same time dt is $iU_0\, dt$. Energy balance is obtained if

$$i = \frac{e(\mathbf{E}_0 . \mathbf{v})}{U_0}. \qquad \textbf{5.1}$$

Note that the vector $\mathbf{E}_0/U_0$, which has the dimension of reciprocal length, is independent of the potential difference, but is a function of the coordinates, and is uniquely determined by the electrode geometry.

In the case of plane parallel electrodes a distance d apart, $E_0 = U_0/d$, and if the velocity component parallel to the field is denoted by v_E, equation **5.1** is reduced to

$$i = \frac{ev_E}{d}. \qquad \textbf{5.1a}$$

The equations above are in agreement with our previous conclusion that the conduction current i flows only while the electron is in motion ($v \neq 0$) with the additional condition that the velocity is not perpendicular to the field, $\mathbf{E}_0 . \mathbf{v} \neq 0$.

If an electron moves in a field defined by an arbitrary number of electrodes $1, 2, \ldots, k$, at potentials $U_1, U_2, \ldots, U_k$, there will be a current $i_1, i_2, \ldots, i_k$ induced in the external circuit connected to each electrode (see Figure 57). For example, an electron moving in a multi-grid valve will, in general, induce current in each grid circuit. In such a case, each current i_l is determined by an equation

$$i_l = \frac{e(\mathbf{E}_l . \mathbf{v})}{U_l},$$

where v is the velocity of the particle in the field, maintained by the electrodes at potentials $U_1, U_2, \ldots, U_k$, and $\mathbf{E}_l$ is the field if electrode number l is at the potential U_l while all the other electrodes are at zero potential.

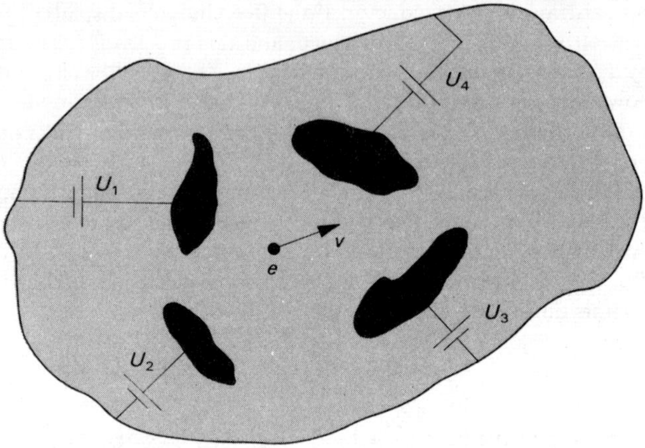

Figure 57

If there are a number of isolated charges, $q_1, q_2, \ldots,$ in the field moving with velocities, $v_1, v_2, \ldots,$ the conduction current in the external circuit is the sum of the contributions from each single charge. It will be seen that considering a continuous space-charge distribution of density ρ is often preferable to dealing with a system of individual electrons. In this case, the charge e in equation **5.1** can be replaced by the charge contained in an infinitesimal volume element dV, and the current in the external circuit becomes

$$i = U_0^{-1} \int \rho \mathbf{v}.\mathbf{E}\, dV,$$

or introducing the current density $\mathbf{j} = \rho \mathbf{v}$,

$$i = U_0^{-1} \int \mathbf{j}.\mathbf{E}\, dV. \qquad \mathbf{5.2}$$

Let us now determine the current in the external circuit if the electron motion takes place in a time-dependent field. To enable us to follow arguments similar to those used earlier, the time rate of change of the fields must be restricted: the free-space wavelength of the field in question must be large compared with the linear dimensions of the region in which the electron motion takes place.* If this condition is met, the argument followed above can be repeated to give the rate of change of the energy of the moving charges as $\int \mathbf{j}.\mathbf{E}(t)\, dV$, where the field is a function of time. This expression, however, does not describe the complete rate of energy change in the volume considered. Since the field is time dependent, the energy stored in it, $\frac{1}{2}\varepsilon_0 \int E^2\, dV$, changes too at a rate $\varepsilon_0 \int \mathbf{E}.\dot{\mathbf{E}}\, dV$, and this change must be taken into consideration. If the external source maintains a time-

*If this condition is not satisfied, one cannot consider the electrodes as equipotential surfaces and the electric field cannot be derived from a scalar potential, as assumed in the argument given above.

dependent potential difference $U(t)$ between the electrodes, and the current in the external lead is denoted by $i(t)$, the energy balance requires that

$$i(t)\,U(t) = \int \mathbf{j}.\mathbf{E}\,dV + \varepsilon_0 \int \mathbf{E}.\dot{\mathbf{E}}\,dV. \tag{5.3}$$

Compare equation **5.3** with equation **3.7**, which reads

$$-\int_S (\mathbf{E}\times\mathbf{H}).d\mathbf{S} = \int_V \mathbf{j}.\mathbf{E}\,dV + \frac{\partial}{\partial t}\int_V \frac{1}{2}(\mathbf{E}.\mathbf{D}+\mathbf{B}.\mathbf{H})\,dV. \qquad [3.7]$$

On the right-hand side, the first term is the same in both equations and, according to equation **3.7**, it represents the *sum* of the rate of change of the energy *stored* as kinetic energy of the electrons present in the interaction region, and the net energy *carried away* by the electrons leaving this region in unit time. The second terms in equations **5.3** and **3.7** look different but they have the same meaning: they represent the rate of change of the energy stored as electromagnetic energy in the interaction region. The difference in appearance arises because the assumption made in the derivation of equation **5.3** allowed us to consider a pure electric field.

The left-hand sides of the two equations are significantly different. In equation **3.7**, which was derived under the most general conditions, we have the rate at which energy is radiated (or absorbed) to balance the rate at which energy changes within the interaction region. If equation **3.7** is evaluated for a quasi-stationary system, one must recognize that, although the Poynting vector is in general different in each volume element, it falls off at least as the inverse fifth power of distance, and therefore its surface integral is negligible. In other words, under such circumstances no radiation leaves (or enters) the interaction region, and the rate of change of the energy in this region is balanced by the power delivered (or absorbed) by the source driving the current in the external circuit.

The mean power delivered by the external source, if the time-dependent quantities vary sinusoidally with a frequency ω, is, from equation **5.3**,

$$-\frac{\omega}{2\pi}\int_0^{2\pi/\omega} i(t)\,U(t)\,dt = \frac{1}{2\pi}\int_0^{2\pi}\int_V \mathbf{j}.\mathbf{E}\,dV\,d(\omega t), \tag{5.4}$$

as the average of the second term over a whole period vanishes. In section 3.1 we have seen that the right-hand side is the mean value of the net energy carried away by the electrons during a period of $2\pi/\omega$.

Before considering examples which demonstrate the processes considered above, let us see how one can translate the relations obtained so far into the language of network analysis. Let us write the external current as the sum of two terms

$$i = i_1 + i_2.$$

To satisfy equation **5.3**, the two components are defined by the relations

$$-Ui_1 = \int \mathbf{j}.\mathbf{E}\,dV, \tag{5.5a}$$

$$-Ui_2 = \frac{\partial}{\partial t}\int \frac{\varepsilon_0}{2}E^2\,dV. \qquad \textbf{5.5b}$$

Equation **5.5b** shows that i_2 is independent of the flow of electrons in the field. The right-hand side of this equation gives the rate of change of the energy stored in the field. If the electrodes have a constant capacitance C and the potential difference between them is U, equation **5.5b** can be written as $Ui_2 = d/dt(\frac{1}{2}CU^2)$, and hence

$$i_2 = C\frac{dU}{dt}.$$

This is the current flowing through a condenser of capacitance C, when connected to a potential difference $U(t)$; it arises purely by virtue of the capacitance of the electrodes which form the boundaries of the interaction region.

The current component in the external circuit, which is generated by the interaction between electrons and field, is i_1 and can be written as

$$i_1 = -U^{-1}\int \mathbf{j}.\mathbf{E}\,dV. \qquad \textbf{5.6}$$

$U(t)$ and $\mathbf{j}(t)$ have, in general, both constant and time-dependent components, and therefore the same can be said about i_1:

i.e. $U = U_0 + U_1(t)$,
$i_1 = I_0 + I_1(t)$.

To describe the effect of the interaction between electrons and the field on the a.c. properties of the network connected to the electrodes, the interaction region

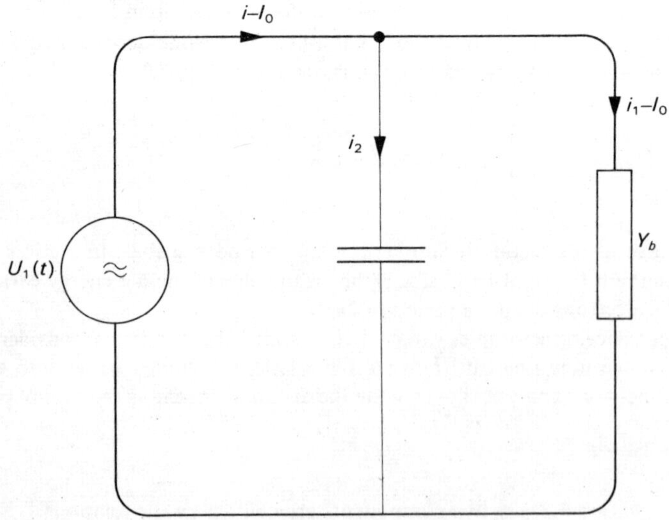

Figure 58

can be represented by an equivalent circuit element which has an admittance

$$Y_b = \frac{i_1 - I_0}{U - U_0}. \qquad \textbf{5.7}$$

The behaviour of the valve in the absence of an electron current, is of course represented by the electrode capacitances. The 'beam admittance' Y_b is the quantity by which the dynamic properties of electrons can be introduced into the analysis of valve circuits (see Figure 58).

5.2 Some features of the operation of classical thermionic valves

The characteristic properties of thermionic valves such as diodes, triodes and multi-electrode valves, depend on the geometry of the electrodes. If, however, one is satisfied with understanding the basic features of the operation of such valves, without aspiring to a quantitative description of the various characteristic properties, one is entitled to consider the simplest possible geometry compatible with the problem. That is why, in most cases, we shall consider valves with electrodes arranged in parallel planes.

5.2.1 *Diodes at high frequencies neglecting space-charge effects*

Let us consider a diode consisting of a plane cathode and a parallel plane anode. An external source maintains a potential difference

$$U = U_0 + U_1 \sin \omega t$$

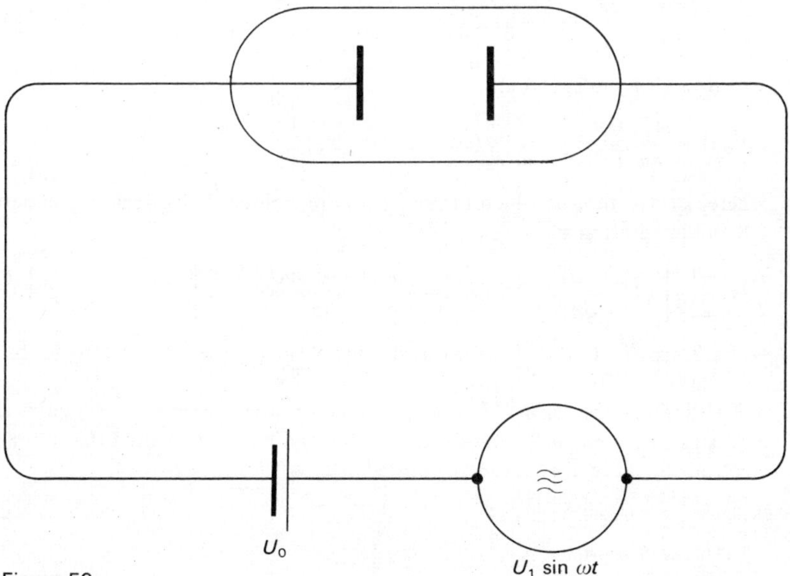

Figure 59

between the electrodes (see Figure 59). $U_1 < U_0$, so that electrons emitted by the cathode are always subject to a net accelerating force, and keep moving towards the anode. We assume that the inter-electrode distance is much smaller than the free-space wavelength of the electric field, so that quasi-stationary conditions apply. This, however, does not preclude the possibility that the time of flight of the electrons between the electrodes is comparable with the period $2\pi/\omega$. We assume, further, that the cathode emits electrons with zero initial velocity at a uniform rate, and we wish to determine the current i induced by the moving electrons in the external circuit.

If the constant emission current is denoted by I_0, the charge emitted between t_0 and t_0+dt_0 is $I_0\,dt_0$, and the current induced by this charge in the external circuit is, from equation **5.1a**,

$$di_1 = \frac{v(t)I_0\,dt_0}{d},$$

where d is the inter-electrode distance and v is the instantaneous velocity of the electrons under consideration. The total current in the external circuit is induced by all the electrons moving between the electrodes at this instant. If the electron transit time between the electrodes is T, the total current at time t is

$$i_1 = \int_{t-T}^{t} di_1 = \frac{I_0}{d}\int_{t-T}^{t} v\,dt_0. \tag{5.8}$$

The instantaneous velocity v at the time t is determined by the equation of motion:

$$\frac{dv}{dt} = \frac{eU_0}{md}\left(1+\frac{U_1}{U_0}\sin\omega t\right),$$

which yields by integration

$$v(t_0, t) = \frac{eU_0}{md}\left[t-t_0-\frac{U_1}{\omega U_0}(\cos\omega t-\cos\omega t_0)\right], \tag{5.9}$$

where t_0 is the time at which the electrons are emitted. Substitution in equation **5.8** and integration gives

$$\frac{i_1}{I_0} = \frac{1}{2T_0^2}\left[\tfrac{1}{2}T^2-\frac{U_1}{\omega U_0}\left(T\cos\omega t-\frac{\sin\omega t-\sin\omega(t-T)}{\omega}\right)\right],$$

where $T_0 = d(2eU_0/m)^{-\frac{1}{2}}$ is the transit time of the electrons in the absence of the r.f. field ($U_1 = 0$).

To eliminate the term $\frac{1}{2}T^2$ from the equation above one must recognize that electrons which reach the anode at an arbitrary instant t covered the full inter-electrode distance d if they were emitted at $t_0 = t-T$:

$$\int_{t-T}^{t} v(t_0, t')\,dt' = d.$$

Integrating equation **5.9** and substituting $t_0 = t-T$ one finds

$$1 = \frac{1}{2T_0^2}\left[\tfrac{1}{2}T^2 - \frac{U_1}{\omega U_0}\left(\frac{\sin \omega t - \sin \omega(t-T)}{\omega} - T\cos \omega(t-T)\right)\right]$$

and therefore

$$\frac{i_1}{I_0} - 1 = \frac{U_1}{U_0}\left(\frac{T}{T_0}\right)^2\left(2\frac{\sin \omega t - \sin \omega(t-T)}{(\omega T)^2} - \frac{\cos \omega t + \cos \omega(t-T)}{\omega T}\right). \qquad \textbf{5.10}$$

In general T depends on the time of emission, t_0. If, however, $U_1 \ll U_0$ then the transit time is essentially determined by the steady component of the field, and the alternating component gives a small t_0-dependent correction:

$$T = T_0 + \tau(t_0),$$

where $\tau(t_0) \ll T_0$ and $\omega\tau \ll 1$ for any t_0. In this case a first-order approximation is permissible, and terms containing τ^2 and $\tau U_1/U_0$ can be neglected. The result of this approximation has the same form as equation **5.10**, but T is replaced by T_0; after some rearrangement of the equation one finds

$$i_1 = I_0 + 2\frac{I_0}{U_0}U_1[F_1(\omega T_0)\sin \omega t + F_2(\omega T_0)\cos \omega t], \qquad \textbf{5.11}$$

where
$$F_1(\omega T_0) = (\omega T_0)^{-2}(1 - \cos \omega T_0 - \tfrac{1}{2}\omega T_0 \sin \omega T_0),$$
$$F_2(\omega T_0) = (\omega T_0)^{-2}[\sin \omega T_0 - \tfrac{1}{2}\omega T_0(1 + \cos \omega T_0)]. \qquad \textbf{5.12}$$

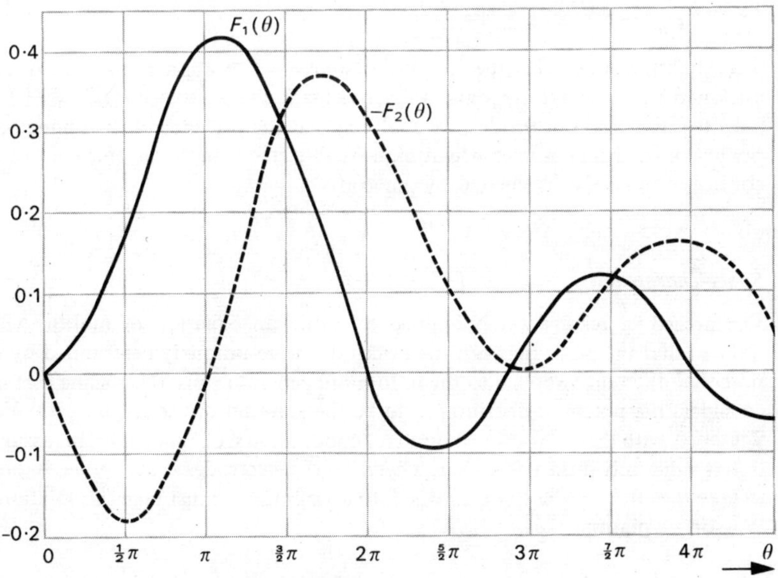

Figure 60

The functions F_1 and F_2 are plotted in Figure 60, and a glance at these curves allows one to draw the following conclusions.

(1) If the time of flight is small compared with the period of the alternating potential, $\theta = \omega T_0 \ll \frac{1}{2}\pi$, then $|F_1| \simeq |F_2| \ll 1$, and the second term in equation **5.11** becomes negligible, so that $i_1 \simeq I_0$. This means that the current in the external circuit is equal to the emission current, and is not affected by the alternating component of the field. In other words, under these conditions, the current in the external circuit can be changed only by varying the emission current. It must be remembered that the interaction between electrons was disregarded, that is, space-charge effects were not taken into account in the equation of motion. It will be seen later that if such effects are included, the emission current itself becomes a function of the applied voltage. Alternatively, the emission current can be varied by the introduction of an auxiliary electrode, whether space-charge effects are significant or not. Triodes are valves where the emission current is controlled by an auxiliary electrode, called the grid.

(2) If the time of flight of the electrons is comparable with the period of the alternating field, there will be two alternating components in the current induced in the external circuit. One component is in phase, the other is in quadrature, the behaviour of the valve as a circuit element changes drastically. If the time dependence of the respective quantities is written in the complex notation, as is customary, the beam impedance, defined by equation **5.7**, can be obtained from equation **5.11** in the form

$$Y_b = 2\frac{I_0}{U_0}(F_1 + iF_2). \qquad \textbf{5.13}$$

This equation shows that the a.c. power source will 'see' an ohmic load of conductance $1/R = 2(I_0/U_0)F_1$, and a reactive load of susceptance $2(I_0/U_0)F^2$.

In the discussion of triodes, we shall see that the presence of the ohmic component of the beam admittance at high frequencies reduces the efficiency of the control of the emission current by the grid.

5.2.2 *Space-charge flow*

The preceding results were obtained by using an equation of motion which implied that the fields between the electrodes were uniquely determined by the potential difference applied to them. In more general terms, this means that one considers the potential distribution to be the solution of the Laplace equation, $\nabla^2\Phi = 0$, with the appropriate boundary conditions (see Figure 61). This assumption is valid only if the space charge between the electrodes is negligible. In order to take the effect of the space charge into account, one must take the solution of Poisson's equation

$$\nabla^2\Phi = -\frac{\rho}{\varepsilon_0} \qquad \textbf{5.14}$$

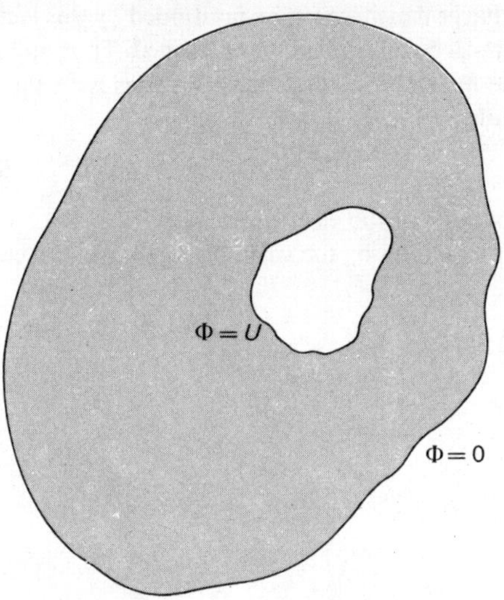

space-charge density between electrodes = ρ

Figure 61

with the appropriate boundary conditions; ρ is the space-charge density.

To calculate the current density

$$\mathbf{j} = \rho \mathbf{v},$$

it is sufficient to know that the magnitude of the velocity is determined by the energy relation

$$\tfrac{1}{2}mv^2 = -e\Phi$$

provided that the electrons leave the cathode with zero velocity. Certainly equation **5.14**, and the two other relations above, are sufficient to find some general relationship between current density and potential at a given point in the field. To obtain such a relationship, it must be recognized that the *shape* of an electron trajectory depends only on the potential *distribution*, if the electrons have zero velocity where $\Phi = 0$. In other words, if the potential is raised at every point by a factor of α, the shape of the trajectory remains the same*. Therefore multiplying the potential at every point by a factor α results in multiplying the velocity vector $\mathbf{v}$ by a scalar factor $\alpha^{\frac{1}{2}}$. Equation **5.14** shows that in order to multiply the

*This is best shown by the principle of least action. As seen in section 2.1, for motion in a static field, this can be written in the form

$$\delta \int_{(1)}^{(2)} 2m(W-e\Phi)^{\frac{1}{2}}\, dl = 0,$$

where W is the total energy of the particle, and dl is the line element of the trajectory. If the initial velocity of the particle is zero, W is scaled up or down, together with the potential. The trajectory along which the above equation is satisfied remains unchanged, if $W - e\Phi$ is multiplied by a constant.

potential by a factor α, the space-charge density must be multiplied by this factor. Consequently the current density **j** will be multiplied by a factor $\alpha^{\frac{3}{2}}$. Thus one can conclude that, at any point in the field, the current density varies with the $\frac{3}{2}$th power of the potential at that point:

$$\mathbf{j} \propto \Phi^{\frac{3}{2}}\mathbf{e}_v, \qquad 5.15$$

where $\mathbf{e}_v$ denotes the unit vector parallel to the electron velocity.

The current induced in the external circuit by the space-charge flow is given by equation **5.5a**:

$$i = -U^{-1}\int \mathbf{j}.\mathbf{E}\,dV = U^{-1}\int \mathbf{j}.\nabla\Phi\,dV.$$

With the aid of the vector identity

$$\nabla.(a\mathbf{b}) \equiv \mathbf{b}.\nabla a + a\nabla.\mathbf{b},$$

the equation above gives

$$i = U^{-1}\int \nabla.(\Phi\mathbf{j})\,dV - U^{-1}\int \Phi(\nabla.\mathbf{j})\,dV.$$

The second term vanishes because, from the Maxwell equations, in a static or quasi-stationary field, the divergence of the convection current density vanishes. Applying Gauss's theorem to the first integral, one finds that

$$i = U^{-1}\int_S \Phi\mathbf{j}.d\mathbf{S}$$

where the integration extends over the surface enclosing the interaction region. Since the potential of the cathode is 0, and that of the anode is U, substituting into equation **5.15** gives

$$i \propto U^{\frac{3}{2}}\int_S \mathbf{e}_v.d\mathbf{S}. \qquad 5.16$$

Thus one finds that the current induced by the space-charge flow in the external circuit of the diode, under static or quasi-stationary conditions, is proportional to the $\frac{3}{2}$th power of the anode voltage. The integral in equation **5.16** represents a 'form factor', which depends on the electrode geometry only.

The explicit determination of **j** can be rigorously carried out if the electrode geometry is simple, for example, in the case of plane parallel electrodes whose linear dimensions are very large compared with the inter-electrode distance, so that fringing fields at the electrode boundaries can be disregarded. In this case, the electrons, which leave the cathode with zero initial velocity, move parallel to the field, and the whole problem reduces to a one-dimensional one. The divergence condition reduces to $dj/dx = 0$; that is, the convection current density in the interaction space is constant, and the potential is the solution of the differential equation **5.14**, which now reduces to

$$\frac{d^2\Phi}{dx^2} = -\frac{\rho}{\varepsilon_0}. \qquad 5.17$$

If one introduces $j = \rho v$, and $v = (2e\phi/m)^{\frac{1}{2}}$, equation **3.11** can be written

$$\frac{d^2\Phi}{dx^2} = \frac{j}{\varepsilon_0}\left(\frac{m}{2e}\right)^{\frac{1}{2}}\Phi^{-\frac{1}{2}}$$

or $$\frac{1}{2}\frac{d}{dx}\left(\frac{d\Phi}{dx}\right)^2 = \frac{j}{\varepsilon_0}\left(\frac{m}{2e}\right)^{\frac{1}{2}}\frac{d}{dx}\Phi^{\frac{1}{2}}$$

and by integration one obtains

$$\left(\frac{d\Phi}{dx}\right)^2 = \frac{j}{\varepsilon_0}\left(\frac{m}{2e}\right)^{\frac{1}{2}}\Phi^{\frac{1}{2}} + E_0^2, \qquad \textbf{5.18}$$

where E_0 is the field intensity at $x = 0$, that is, on the cathode, where $\Phi(0) = 0$.

Equation **5.18** shows that the presence of a current density, j, modifies the field distribution, so that it is no longer uniform ($d\Phi/dx \neq$ constant). Its value on the cathode, E_0, decreases with increasing current density. If the initial velocity of the electrons is zero, they can leave the cathode only if $E_0 \geqslant 0$. The current density will in fact increase until the condition $E_0 = 0$ is reached. Using this condition, equation **5.18** becomes

$$\frac{d\Phi}{\Phi^{\frac{1}{4}}} = \left(\frac{j}{\varepsilon_0}\right)^{\frac{1}{2}}\left(\frac{m}{2e}\right)^{\frac{1}{4}} dx.$$

Integration, followed by a small rearrangement, gives

$$j = \frac{4}{3}\varepsilon_0\left(\frac{2e}{m}\right)^{\frac{1}{2}} x^{-2}\Phi^{\frac{3}{2}}(x) = \frac{4}{3}\varepsilon_0\left(\frac{2e}{m}\right)^{\frac{1}{2}}\frac{U^{\frac{3}{2}}}{d^2}. \qquad \textbf{5.19}$$

Equation **5.19** is the Child–Langmuir law, which relates the current density of space-charge flow in a plane parallel diode to the constant anode voltage and the inter-electrode distance. One can readily see from equation **5.19**, how the space charge 'distorts' the potential distribution:

$$\Phi(x) = U\left(\frac{x}{d}\right)^{\frac{4}{3}}.$$

The current in the external circuit is, as expected,

$$i = -U^{-1}\int \mathbf{j}.\mathbf{E}\, dV = A\frac{j}{U}\int\frac{d\Phi}{dx}dx = Aj, \qquad \textbf{5.20}$$

where A is the surface of the electrodes. Equation **5.19** is in agreement with the more general result, $i = Aj \propto U^{\frac{3}{2}}$. It must be emphasized once again that all the relations obtained are applicable only under static or quasi-static conditions. Furthermore, it was assumed that electrons leave the cathode with zero initial velocity. In practice, electrons emitted by thermionic cathodes have finite initial velocities, which follow a Maxwellian distribution. As a result, electrons can leave the cathode in opposition to a retarding field. Thus in a stationary state of the space-charge limited flow, the field at the cathode surface is $E_0 < 0$, and a potential minimum is formed in front of the cathode usually referred to as a 'virtual cathode'.

5.2.3 *Amplification by a triode*

The process of amplification will be discussed qualitatively only, as a simple example of some of the ideas developed so far. This will suffice to point out the limitations of triodes when operated at high frequencies.

In a triode, between the cathode and the anode, there is a 'grid' which is usually a wire helix surrounding the cathode. The field distribution between the cathode and anode is affected by the grid potential, and the grid affords a convenient way of controlling the space-charge flow.* Since the grid is much closer to the cathode than to the anode, the electron current density is more sensitive to the variation of the grid voltage than to that of the anode voltage. The relation between the current induced in the anode circuit and the electrode voltages has, over a considerable range, a form similar to the Child–Langmuir law:

$$i_a \propto (U_a+\mu U_g)^n,$$

where U_a and U_g are the anode and grid voltages respectively, n is a constant which is, to a very good approximation, equal to $\frac{3}{2}$, and μ is a numerical factor, called the 'amplification'.

The grid is usually negative with respect to the cathode, so electrons cannot reach the grid and thus, under quasi-stationary conditions, no current flows in the grid circuit. Consequently, the space-charge flow can be controlled, for example sinusoidally modulated, without drawing appreciable energy from the source connected to the grid.

Considering the simple triode circuit, shown in Figure 62, if the grid is always negative with respect to the cathode:

$$U_g = -U_{g0}+U_{g1}\sin \omega t, \qquad U_{g0} \ll U_{g1}.$$

If the electron current is never cut off, it will contain an alternating component, and can be written in the form

$$i_a = I_0+I_1 \sin(\omega t+\phi), \qquad I_1 \ll I_0. \tag{5.21}$$

Restricting our considerations to quasi-stationary conditions, $\phi = 0$, the current induced in the external circuit is the same as the electron current. On account of the voltage drop across the resistance R, the potential difference between anode and cathode is always smaller than the battery voltage:

$$U_a = U_b-Ri_a = (U_b-RI_0)-RI_1 \sin \omega t.$$

Note that the modulation of the anode voltage is in anti-phase with respect to that of the anode current.

*If the grid is sufficiently negative, the field in front of the cathode is a retarding one, even if the anode is positive. In such a situation, the flow of electrons is completely cut off. In a certain range of the grid voltage, there are zones of the cathode surface in front of which there is a retarding field, and others in front of which the field is accelerating. The ratio of the two areas depends on the grid voltage. Under these conditions, the grid voltage can control the electron flow even if the space-charge density is negligible.

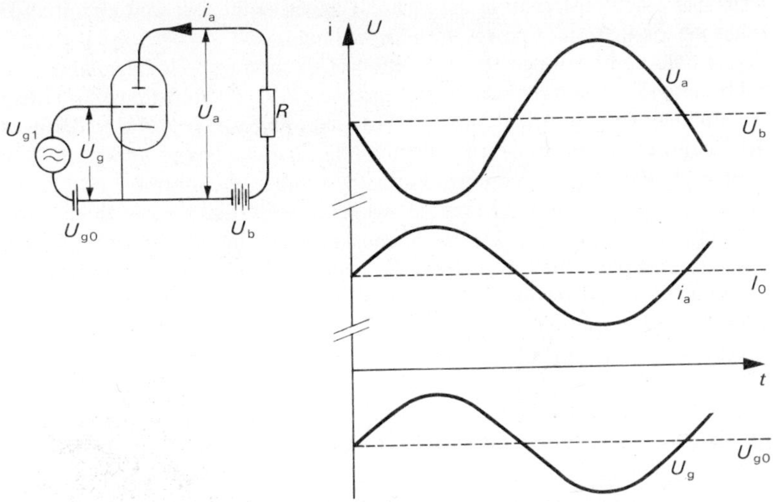

Figure 62

The net energy carried away by the electrons per cycle is, from equation **5.4**,

$$\overline{U_a i_a} = \overline{U_b i_a} - R\overline{i_a^2} = U_b I_0 - R(I_0^2 + \tfrac{1}{2}I^2)$$

and hence the mean a.c. power delivered to the load is

$$\overline{P_\sim} = \tfrac{1}{2}I_1^2 R = U_b I_0 - \overline{U_a i_a} - I_0 R^2.$$

The source of this power is evidently the d.c. power $U_b I_0$ delivered by the battery but the conversion of this power into a.c. power is not perfect. Some of the power, $\overline{i_a U_a}$, is carried away by the electrons and is dissipated as heat in the anode which intercepts the electron beam. $I_0^2 R$ is a d.c. loss which occurs in the load resistance itself. The d.c. losses can be reduced if the grid is biased so much that the anode current is allowed to flow only during a fraction of the cycle. In this case, the anode current consists of periodically occurring pulses yielding a lower mean current, and the a.c. components of the current and voltage will no longer be purely sinusoidal.

If the external circuit consists of a parallel resonant circuit and the useful load is coupled to this, say inductively, one can tune the whole system to resonate at the frequency ω. In this case, the external network behaves as a short circuit, at all frequencies outside the neighbourhood of the resonant frequency. At the resonance frequency it presents an ohmic load which is a parallel combination of two resistances, one representing the inherent losses of the resonant circuit and the other the dissipation of power in the actual load.

It should be mentioned that a fraction of the output power available can be fed back to the grid to replace the externally applied control voltage, and thereby the amplifier can be turned into a self-excited oscillator. In order to discuss the details of the properties of a triode amplifier, the conditions under which self-excited oscillations are maintained etc., the process of conversion of d.c. power into a.c. power is taken for granted; with this tacit assumption, the triode can be considered as an *a.c. generator* of constant electromotive force and constant internal resistance. The e.m.f. of this 'equivalent' generator is proportional to the a.c. voltage impressed on the control grid. To study the development of the equivalent network representing the triode, and its applications, the reader should consult a textbook on valve circuits.

It must be stressed that all that has been said about the triode is valid only as long as *quasi-stationary* conditions are maintained, *and* the time of flight of the electrons between the electrodes is much shorter than the period of the a.c. component of the fields. If the electric field varies at a high frequency, it is no longer true that there is no current induced in the grid circuit unless the grid intercepts the beam. As in the case of the diode, if the time of flight of the electrons between the electrodes is comparable with the period of the alternating field, in the external circuit there *is* a net induced current flowing, one component of which is in phase with the grid voltage. This component presents an apparent ohmic load connected to the grid voltage generator. The anode current will contain a component which is in anti-phase with the grid voltage, as before, but it will also contain a component in time quadrature. In terms of the quantities used in the analysis of valve circuits, the conventional valve parameters (internal resistance, amplification factor, mutual conductance), which are real quantities at low frequencies, now become complex quantities. The most significant effect, which impairs the operation of the triodes at very high frequency, is the ohmic load 'seen' by the grid voltage generator, in consequence of which the efficiency of the control of the emission current by the grid is greatly reduced. For this reason, valves designed to operated at very high frequencies require some alternative method for the control of the electron current density.

One can readily see that there is a variety of possibilities available. Consider n electrons passing through a surface element dS, somewhere in the valve, between times t and $t+dt$. These electrons constitute a convection current of density $\mathbf{j}$, defined by $\mathbf{j}.d\mathbf{S}\,dt = en$. These electrons left a surface element of the cathode, dS_0, during an interval between earlier instants t_0 and t_0+dt_0. The current density at the cathode, $\mathbf{j}_0$, must satisfy the charge conservation relation

$$\mathbf{j}_0.d\mathbf{S}_0\,dt_0 = \mathbf{j}.d\mathbf{S}\,dt$$

and hence the current density in the direction of the unit vector $\mathbf{n}$ normal to the surface element dS is

$$\mathbf{j}.\mathbf{n} = \mathbf{j}_0.\frac{d\mathbf{S}_0}{dS}\frac{dt}{dt_0}. \tag{5.22}$$

This equation shows that the current density, at the surface element dS, can be varied by changing any one of the three factors on the right-hand side*. Under the usual operating conditions of triodes, the grid voltage controls directly the emission current density j_0. When the anode current is nearly completely cut off, and the highly negative potential of the grid allows the emission of electrons from restricted zones of the cathode, small changes of the grid voltage in effect control the ratio dS_0/dS. This method may be called 'concentration modulation' because it effects the area from which the electrons are collected on to a given surface element. The ratio dt_0/dt can be varied by means of velocity modulation, and this is the method which has become of utmost importance in achieving current-density modulation, in valves designed for very high frequency operation.

5.3 Klystrons and reflex klystrons

Klystrons are thermionic valves in which the electron current-density modulation is achieved by velocity modulation accompanied by phase focusing, as described in section 3.3. Klystrons can be used at extremely high frequencies (microwaves) as amplifiers, harmonic frequency generators or oscillators. Reflex klystrons are oscillators. The principle of amplification by klystrons will first be outlined. Consider the system shown schematically in Figure 63. Electrons are emitted by a cathode; a suitable electrode system (electron gun) accelerates them and produces a beam of electrons of energy eU_0. The beam passes through gaps defined by plane parallel electrodes, with each pair connected to a parallel resonant circuit tuned to the same resonant frequency ω. Each circuit is connected to the same d.c. potential, say ground, and therefore in the region between the two gaps, there is no field. If the first resonant circuit is excited by an external signal of frequency ω, there will be an alternating potential difference between the electrodes which define the first gap,

$$U_1 = U_{10} \cos \omega t. \qquad \textbf{5.23}$$

Let us suppose that the time of flight τ of the electrons in the gaps is very small compared with the period of the alternating field ($\omega \ll 1$) and $U_{10} \ll U_0$. In this

*Over the surface of a cathode, the effective emission-current density can vary for two reasons. Either the emission-current density of a surface element dS_0 changes, or the place of origin of the current which 'feeds' the surface element dS changes. The latter is the case if one considers electrons which reach the surface element dS after passing through a time-dependent deflecting field. In general, the emission current varies at a rate

$$\frac{dj_0}{dt} = \frac{\partial j_0}{\partial t} + \mathbf{u}.\nabla j_0.$$

The first term gives the rate of change of the emission-current density of a surface element; $\mathbf{u}$ is the velocity at which the surface element acting as source sweeps across the cathode, and ∇j_0 is the rate at which the emission current density varies as a function of the position on the cathode surface. 'Chopping' the beam by sweeping it in front of an aperture is the simplest form of this type of intensity modulation.

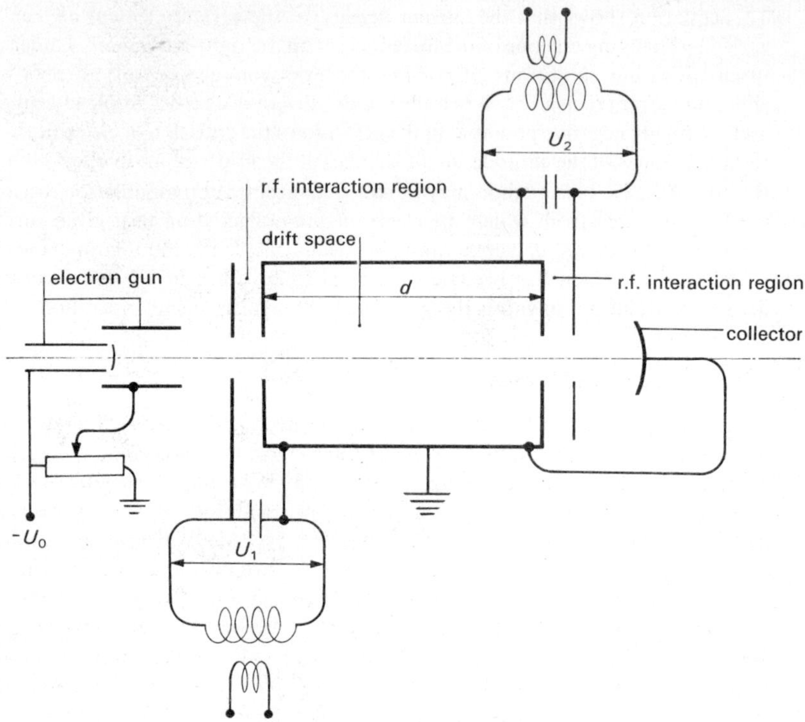

Figure 63

case, the considerations described in section 3.3 are applicable, and the electron current at the second gap, I, is given by equation **3.36**:

$$\frac{I}{I_0} = 1+2J_1(\xi)\sin\omega\left(t-\frac{d}{u}\right)+\text{terms of higher harmonics,} \qquad \mathbf{5.24}$$

where I_0 is the emission current,

$$u = \left(\frac{2eU_0}{m}\right)^{\frac{1}{2}}$$

is the velocity of the electrons at the entrance to the first gap,

$$\xi = \frac{1}{2}\frac{d}{u}\frac{U_{10}}{U_0}\omega$$

is the 'bunching parameter', and J_1 is the first-order Bessel function of the first kind.

Because of the small transit time, the mean power drawn by the electrons from the external circuit is negligible in comparison with the power dissipated in the loss resistance of the resonant circuit. Thus the generator which supplies the signal U_1 is loaded mainly by the resonant circuit, and the velocity modulation requires no substantial power. Note that since the electrons are accelerated before the first gap, the transit phase angle in the gap, $\omega\tau$, can always be kept small by applying sufficiently high accelerating voltage. At the same time, by increasing the length of the drift space, d, the bunching parameter can be chosen to obtain the required high 'modulation depth' $J_1(\xi)$. Note that in triodes the electrons in the cathode–grid section are necessarily slow, and that is why the transit time in this region cannot be reduced sufficiently to eliminate their adverse effects at high frequencies.

In the external circuit connected to the electrodes which define the second gap, the modulated electron current I induces a current $i = I$. The resonant circuit acts as a short circuit for the constant component I_0 and, since its resonance frequency is the same as the fundamental frequency of the alternating component, there will be a potential difference across the resonant circuit:

$$U_2 = IR_s = 2I_0 R_s J_1(\xi) \sin \omega\left(t-\frac{d}{u}\right) \equiv U_{20} \sin \omega\left(t-\frac{d}{u}\right). \qquad 5.25$$

The alternating potential difference, $-U_2$, maintains a field which opposes the motion of the electrons and the mean a.c. power delivered by the electron is

$$\overline{P_\sim} = \overline{IU_2} = I_0 U_{20} J_1(\xi) = 2I_0^2 R_s J_1^2(\xi). \qquad 5.26$$

In both the equations above R_s is the shunt resistance of the resonant circuit. In general, R_s is the parallel combination of a resistance r representing the inherent losses of the resonant circuit, and a resistance R, which is induced in the resonant circuit as a result of the power dissipation in the useful load. Thus $1/R_s = 1/r + 1/R$, and the maximum useful power, delivered if $r = R$, is half of that given by equation **5.26**.

Without analysing the details of the klystron amplifier, a few general remarks will be useful. The above description of the operation of the klystron amplifier shows that the source of a.c. power output is the kinetic energy of the electrons, acquired during the acceleration in a constant field. In extreme situations the operation of the klystron can be one of two types. For a given accelerating voltage, frequency and drift space the bunching parameter becomes $\xi \ll 1$, if the voltage amplitude U_{10} is sufficiently small. In this case $J_1(\xi) \simeq \xi$ and the amplitude of the alternating component of the beam current at the second gap, as well as that of the output voltage across the load, becomes proportional to the amplitude of the input signal. This may be readily seen from equations **5.24** and **5.25**. In this case, from the power output described by equation **5.26**, the efficiency of the d.c. to a.c. power conversion becomes

$$\frac{\overline{P_\sim}}{I_0 U_0} = \frac{U_{20}}{U_0} J_1(\xi) \qquad 5.27$$

and is very small. Under these conditions one speaks about 'low-level' operation.

$J_1(\xi)$, however, reaches a maximum of 0·58 when $\xi \simeq 1{\cdot}84$ and therefore the a.c. current reaches $1{\cdot}16 I_0$ for an input signal amplitude which makes a bunching parameter equal to 1·84. Klystron operation in the region of $\xi \simeq 1{\cdot}84$ is known as 'high-level' operation. Equation **5.27** seems to indicate that the maximum efficiency is 58 per cent, if one assumes that the highest permissible voltage amplitude across the output gap is $U_{20} \simeq U_0$. This is the condition under which the beam can still pass the gap. This estimate, however, is over optimistic. If the electrons are significantly slowed down in the gap, the increased time of flight in the gap will reduce the efficiency of the energy transfer, in much the same way as was seen in the case of conventional diodes. An estimate taking this 'beam loading' into account gives a maximum efficiency of about 34 per cent, a figure which has actually been obtained with klystrons in practice.

The previous considerations are applicable in describing the klystron's performance as a harmonic frequency generator. If the resonant circuit connected to the second gap is tuned to the frequency $n\omega$, it will respond to this frequency component of the beam current. Equation **3.36** shows that the amplitude of the nth harmonic is $2I_0\, J_n(n\xi)$, and therefore, in quantitative considerations, the properties of the nth-order Bessel function have to be used, while the qualitative features remain unchanged.

The preceding considerations also seem to be strictly applicable only at low frequencies, since they are based on a picture involving lumped-parameter circuits. We can, however, readily convince ourselves that the basic processes operate at any frequency, provided that the external 'network' is suitably modified. A lumped-parameter resonant circuit would radiate at very high frequencies. To avoid the radiation loss, they must be replaced by resonant cavities, as shown in Figure 64. By choosing the shape of the cavity appropriately, one can ensure that in the region where the beam passes, there is practically an electric field only, and the magnetic field inseparably coupled to it appears in the region which is free of charge. In the example shown in the figure, the latter is the region between the two coaxial conductors. Thus the cavity resonators make it possible, at very high frequencies, to meet all the conditions under which the interaction between electrons and fields converts d.c. power into a.c. power.

It is a different matter to show how the properties of cavity resonators, waveguides etc. are described quantitatively, and how output power, amplifier gain etc. are expressed in practical terms; this is the subject of the theory of microwave networks. The usual procedure is to define 'equivalent' network parameters in terms of quantities which are meaningful at extremely high frequencies. For example, the line integral of the instantaneous field between the opposite faces of the gap of the cavity can play the part of the instantaneous voltage. An equivalent capacitance can be defined by the relation between the voltage and the energy stored in the electric field, the Q-value is defined in terms of the energy dissipated in a cycle and the energy stored in the cavity, and the conventional definition for resonant circuits in terms of circuit parameters allows the introduction of an equivalent loss resistance etc. These problems are, of course, of utmost import-

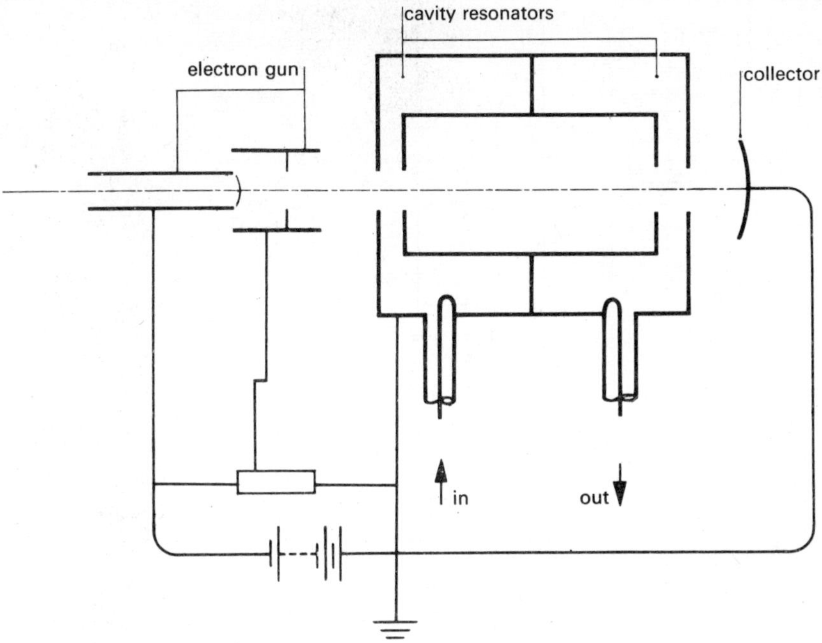

Figure 64

ance for the users of klystrons, but they are outside the scope of this book.

The klystron amplifier can readily be converted into an oscillator, by using a fraction of the output power for the excitation of the first cavity. To obtain self-excited oscillations, the feedback loop must always satisfy the condition that the total phase shift, around the complete feedback loop, is an integral multiple of 2π. In klystrons there is a remarkably simple method for establishing a feedback mechanism using the electron beam itself. A retarding field can be applied to reflect the electrons after their passage through a cavity, and those electrons which pass through the same gap a second time deliver energy to the same cavity which produced the velocity modulation during the first passage. Klystrons which generate r.f. power by this method are called *reflex klystrons*.

5.3.2 To describe the operation of the reflex klystron, an expression is required for the electron current observed at the gap as a function of time. This information can be obtained by considering the electron motion in a system schematically shown in Figure 65. Electrons are emitted and accelerated by an electron gun to a kinetic energy eU_0. If there is an alternating field $E(t)$ in the gap, the electron velocities will be modulated as described in section 3.3. After crossing the modulating gap, the electrons enter a constant *uniform retarding field* of intensity $\mathbf{E}_r$, which forces them to return to the interaction gap and cross it once more.

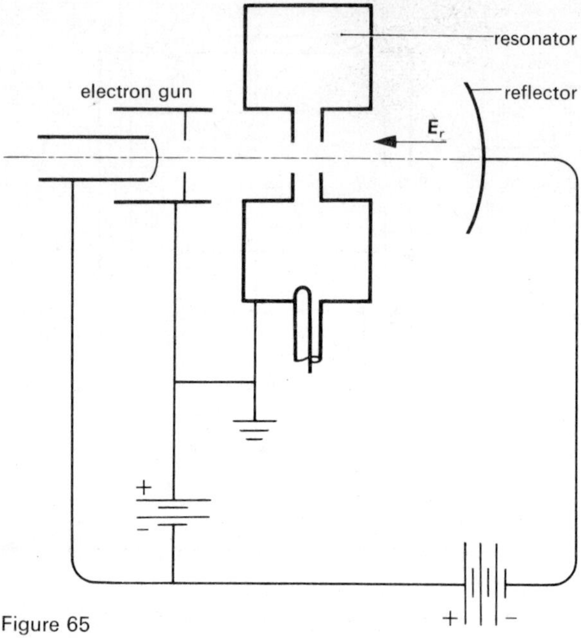

Figure 65

If the equivalent potential difference across the gap is

$$U = U_1 \cos \omega t = \int_0^d E(t)\, ds, \qquad \textbf{5.28}$$

electrons which enter the gap at $t = t_0$ with velocity $u = (2eU_0/m)^{\frac{1}{2}}$ will leave the gap with a velocity given by equation **3.25**. To a first-order approximation:

$$v_1 \simeq u\left(1+\frac{1}{2}\frac{U_1}{U_0}\cos \omega t\right),$$

provided that $U_1 \ll U_0$. In the retarding field, at time t, these electron have a velocity

$$v = v_1 - \frac{eE_r}{m}(t-t_0)$$

and arrive back at the gap when $v = -v_1$, that is, at the time

$$t = t_0 + \frac{2m}{eE_r} v_1 = t_0 + \frac{2m}{eE_r} u\left(1+\frac{1}{2}\frac{U_1}{U_0}\cos \omega t_0\right). \qquad \textbf{5.29}$$

If the phase angle corresponding to the mean time of flight for the round trip of the electrons is denoted by

$$\theta = \frac{2m}{eE_r} u\omega \qquad \textbf{5.30}$$

and the parameter

$$\xi = \frac{1}{2}\theta\frac{U_1}{U_0} \qquad 5.31$$

is introduced, then equation **5.29** can be written in the form

$$\omega t - \theta = \omega t_0 + \xi \cos \omega t_0. \qquad 5.32$$

This equation is equivalent in its role to equation **3.34**, which appeared in the discussion of the current modulation caused by velocity modulation in a field-free drift space.

To obtain the electron current in the beam, one can follow the same arguments as in section 3.3. If the emission-current intensity is I_0, charge conservation requires that, for any plane, $I = I_0(dt_0/dt)$; for the current at the gap due to the reflected electrons, one finds an expression similar to equation **3.36**:

$$\frac{I}{I_0} = 1 - 2J_1(\xi)\sin(\omega t - \theta) + \text{terms of higher harmonics.} \qquad 5.33$$

It should be noted that the direction of this current is opposite to the direction of the initial flow, because of the reversal of the electron motion in the retarding field. Since the beam passes through the gap of the same resonant cavity, we shall be concerned only with the effects caused by the fundamental frequency component of the beam current.

The most convenient approach to the analysis of the reflex oscillator is to describe the interaction between the electron beam and the field in the resonator gap, in terms of the concepts of network analysis. For this reason, let us consider the time-dependent 'voltage' (**5.28**) and the alternating component of the current (**5.33**) as the real parts of complex functions of time:

$$\boldsymbol{U} = U_1 \exp(\mathrm{i}\omega t),$$
$$\boldsymbol{I} - I_0 = 2I_0 J_1(\xi) \exp[\mathrm{i}(\omega t - \theta + \tfrac{3}{2}\pi)]. \qquad 5.34$$

Thus one obtains for the beam admittance, defined by equation **5.7**,

$$-\boldsymbol{Y} = \frac{I_0}{U_0}\theta\frac{J_1(\xi)}{\xi}\exp[-\mathrm{i}(\theta - \tfrac{3}{2}\pi)]. \qquad 5.35$$

Hence the mean value of the power delivered to the cavity is

$$P = \tfrac{1}{2}\mathscr{R}[\boldsymbol{U}^*(\boldsymbol{I} - I_0)] = I_0 U_1 J_1(\xi)\cos(\theta - \tfrac{3}{2}\pi)$$
$$= 2I_0 U_0 \xi J_1(\xi)\theta^{-1}\cos(\theta - \tfrac{3}{2}\pi), \qquad 5.36$$

where U_1 was eliminated by using equation **5.31**. Thus it can be seen that the

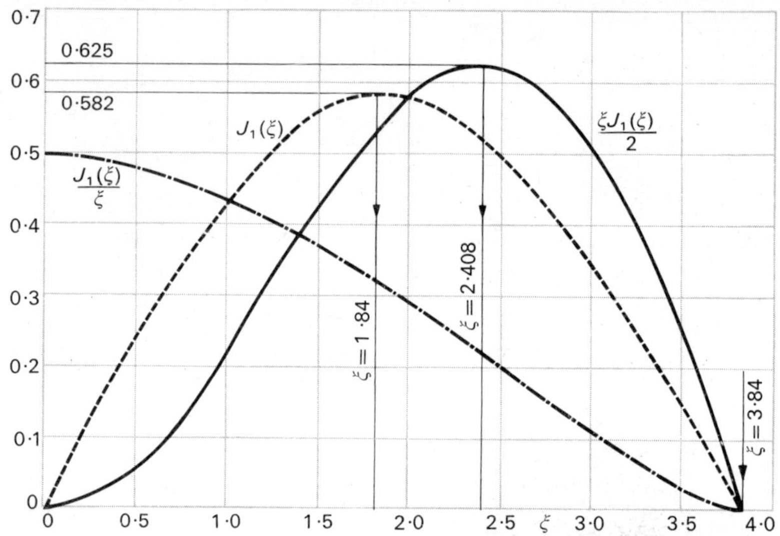

Figure 66

alternating current, the admittance and the output power are proportional to the functions $J_1(\xi)$, $\xi^{-1}J_1(\xi)$ and $\xi\, J_1(\xi)$ respectively. We shall frequently refer to these functions; they are plotted in Figure 66 in the range between $\xi = 0$ and $\xi = 3{\cdot}84$, where all three functions vanish for the first time if $\xi > 0$.

If one represents the resonant cavity by a resonant circuit which has the same resonant frequency and the same Q-value as the resonant cavity, and the beam is represented by the admittance given by equation **5.35**, the reflex oscillator can be represented by an equivalent circuit shown in Figure 67. Here R takes into account the resultant of the inherent losses of the cavity, and those due to some external load coupled to the cavity. Now, from the network equations (Kirchoff's equations), one can readily find that oscillations with a constant amplitude are established if the total admittance of the network is zero.* The implications of this condition can be brought to light by a graphical method.

Let us assume that the a.c. voltage is infinitesimally small ($U_1 \rightarrow 0$). In this case $\xi \rightarrow 0$ and hence $\xi^{-1}J_1(\xi) \rightarrow \frac{1}{2}$, so that equation **5.35** reduces to

$$-\boldsymbol{Y} \equiv -(g+\mathrm{i}b) = \frac{1}{2}\frac{I_0}{U_0}\,\theta\,\exp\left[-\mathrm{i}(\theta-\tfrac{3}{2}\pi)\right]. \qquad \mathbf{5.37}$$

*The physical content of this condition is as follows. At a given conductance G, the susceptance B is characteristic of the phase shift between voltage and current. Since the phase shift around the complete circuit must be zero, the resultant susceptance must vanish. The resultant conductance must vanish too, because if there is an excess conductance energy is dissipated, and the oscillation amplitude decays exponentially. If there is a 'defect' in the total conductance, instead of a dissipation of energy, the energy present at a given instant will grow indefinitely, giving oscillations with exponentially growing amplitude.

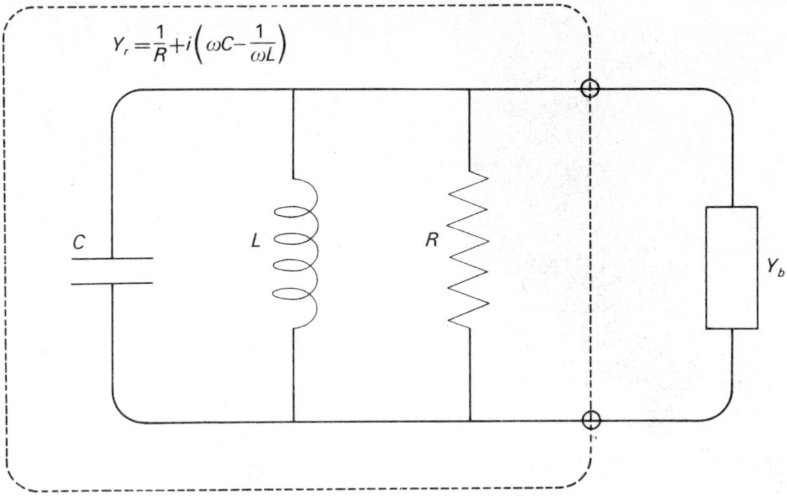

Figure 67

This equation defines an Archimedean spiral in the complex plane (g, ib), winding clockwise (see Figure 68) as the parameter θ, that is, the phase angle of transit in the retarding field increases. Next one can plot the admittance of the resonant circuit

$$Y_r = \frac{1}{R} + i\left(\omega C - \frac{1}{\omega L}\right)$$

in the same plane, taking the frequency as a variable parameter. Thus a straight line is obtained, running parallel to the imaginary axis. The intersection with the real axis occurs at $g = 1/R$, corresponding to the resonant frequency

$$\omega = \omega_0 = (LC)^{-\frac{1}{2}}.$$

A glance at this diagram shows that if the beam admittance is represented by a point on the spiral left of the straight line, no stable oscillation can exist, because the beam conductance cannot cancel the conductance of the external circuit. If the time of flight in the reflector space is increased by gradually decreasing the intensity of the retarding field, the beam admittance eventually reaches the point marked A on the diagram. Here the beam and circuit admittances cancel, and oscillations can start, although the conditions as set out give infinitesimally small amplitudes. If θ is increased further and the point A′ is reached, there will be an excess of negative conductance available, and the oscillation amplitude starts to increase exponentially. As a result – other parameters staying constant – the

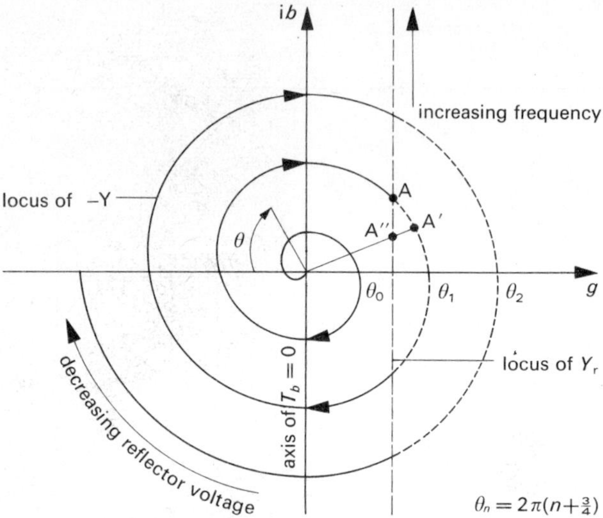

Figure 68

bunching parameter increases, yielding $\xi^{-1}J_1(\xi) < \frac{1}{2}$. Now equation **5.35** shows that as a consequence the point A′, representing the beam admittance, moves along the radius vector (since θ is fixed) towards the origin. When the representative point in the diagram reaches A″ the beam and circuit admittances are balanced and the oscillator settles down at a constant amplitude. The frequency of the oscillation will be that appropriate to the point A″ on the circuit admittance line.

The most conspicuous properties of the reflex oscillator can be seen from equations **5.34–6**, and the general features of the diagram developed.

(1) Equation **5.34** shows that, at a given value of the bunching parameter ξ the alternating current has a maximum if

$$\theta - \tfrac{3}{2}\pi = 2n\pi \qquad \text{or} \qquad \theta = 2\pi(n+\tfrac{3}{4})$$

where n is an *integer*. In this case, the a.c. current is in anti-phase with the a.c. voltage. It is left to the reader to show by kinematic arguments, that this relation is equivalent to saying that the centre of a bunch of electrons, produced by phase focusing, is subject to maximum deceleration when it crosses the gap on its return trip.

(2) The diagram shows that stable oscillations occur in a number of *ranges* of the parameter θ, that is, in different ranges of the retarding field intensity. Each such range is in the neighbourhood of $\theta \simeq 2\pi(n+\frac{3}{4})$, where each value of n defines a different 'mode' of the oscillation.

(3) If the retarding field intensity is varied *within* the stable range of oscillation, point A″ shifts along the vertical straight line, indicating that the frequency of the stable oscillation will depend on the setting of the retarding field. In other words, the oscillator can be 'tuned' electronically, that is, purely by controlling the electron motion. Observe that the oscillation frequency is equal to the resonance frequency of the resonator if, and only if, $\theta = 2\pi(n+\frac{3}{4})$. In higher modes of oscillation (larger n), the frequency range of stable oscillations is larger.

(4) If the retarding field is swept through the range which gives stable oscillation, the output power passes through a maximum when $\theta = 2\pi(n+\frac{3}{4})$. This can be seen from the following argument. Equation **5.36** shows that the power generated is proportional to the quantity $\xi\, J_1(\xi)$ and depends only on this quantity, if θ has a fixed value. Since, as one moves from a point of the zero-power spiral along a radial vector (θ = constant) towards the origin, the quantity $\xi^{-1}J_1(\xi)$ decreases from $\frac{1}{2}$ to 0, one can conclude from Figure 66 that at the same time $\xi\, J_1(\xi)$ increases from zero, goes through a maximum of 0·625 and then falls again to zero. Thus along any radial line there will be one point which represents an admittance $g+ib$, to which there corresponds a relative power maximum. On the other hand, the value of this relative maximum is proportional to $\cos(\theta-\frac{3}{2}\pi)$, and therefore it is higher for $\theta = 2\pi(n+\frac{3}{4})$ than for any other θ in the neighbourhood of this specific value. Thus it becomes plausible that for a given mode of oscillation, n = constant, there exists one point on the g-axis of the admittance plane where the power has an absolute maximum, and the loci of the points where the power

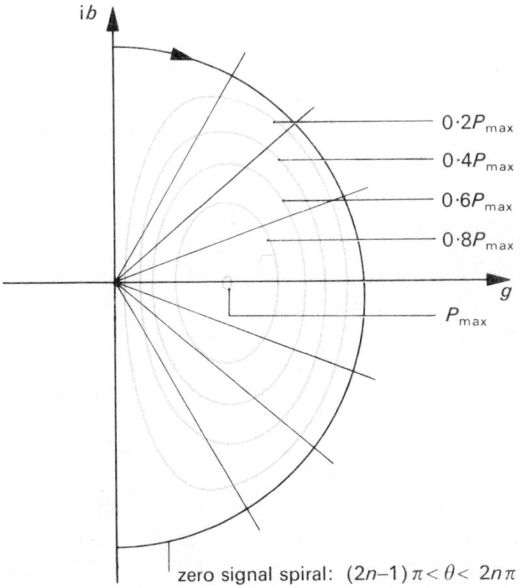

Figure 69

is constant are closed loops encircling this point (see Figure 69). Detailed calculations confirm the validity of this conclusion. Because of the factor θ^{-1} in equation **5.36**, under otherwise fixed conditions a reflex oscillator generates a lower maximum power in higher modes of oscillations, that is, for large values of n.

5.3.3 In the frequency range where conventional valves can operate, the gain of amplifiers is increased by connecting single amplifier stages in cascade and using the output signal of one stage as input signal for the next one. The possibility of connecting klystron amplifiers in cascade leads naturally to the idea of coupling the consecutive amplifier stages purely electronically. This means taking a sequence of identical cavity resonators and letting the electron beam pass through all of them. The input signal can be applied to the first cavity, and the output can be coupled out of the last one. The intermediate cavities are excited by the beam, and one expects that the energy stored in these cavities will enhance the bunching of the beam, so that the over-all gain of the system as a whole is increased. This idea is, indeed, feasible. The next possibility is to increase the number of cavities over a fixed length of a beam, and this process eventually leads to a continuous system carrying an electromagnetic wave, the amplitude of which is expected to increase in the direction of propagation. In the previous discussions there are several clues which suggest that such a system should be feasible. One can suggest that the operating conditions of a linear accelerator should be reversed: let bunches of electrons travel in synchronism with an electromagnetic wave in such a manner that the bunches are subject to a constant retarding field. In this case, the energy lost by the electrons will be fed into the electromagnetic field. That an initial velocity modulation produces the right kind of electron beam for such a purpose is indicated by equation **3.36**, which shows that the alternating components of the beam *intensity*, as functions of time and distance measured along the beam, are given by expressions typical of *travelling waves*.

To obtain a clear picture of 'beam type' amplifiers, a general knowledge of the properties of space-charge waves is required.

5.4 Space-charge waves and cyclotron waves

Let us consider an electron beam of constant space-charge density ρ, moving with a constant velocity $\mathbf{u}_0$, in a region where there is no externally applied field. At a plane perpendicular to the beam axis, which is taken as the z-axis of our coordinate system, let some perturbation alter the velocity of the electrons. This can be achieved by means of a velocity modulator gap of the kind discussed in the preceding section. As a result of the perturbation of the electron velocities, a sequence of events will occur. It was seen before, that the space-charge density distribution will change, and consequently the potential distribution will change. The excessive space-charge concentration will produce a force which tends to restore the equilibrium state. An observer, riding with the beam with the velocity u_0, would find that, subject to the restoring force, the electrons oscillate about their equilibrium position. For the observer fixed in the laboratory, it appears that the velocity of the electrons, the space-charge density, and the electromagnetic field generated by it vary in space and time as travelling waves. The time variation in the waves is determined by the initial perturbation.

In studying the properties of a space-charge wave, the main problems are to determine its 'propagation constant' or spatial periodicity, and the phase relations between the 'velocity wave' and the 'current-density wave'. The electron beam which carries the space-charge wave is usually of high current density and of an approximately uniform cross-section, maintained by magnetic confinement as outlined at the end of section 2.8.

Due to the periodic perturbation of the electron velocity at $z = 0$, the velocity of the electrons and the space-charge density will acquire time-dependent and z-dependent components:

$$\mathbf{v} = \mathbf{u}_0 + \mathbf{u}_1(\mathbf{r}, z, t), \qquad \mathbf{5.38}$$
$$\rho = \rho_0 + \rho_1(\mathbf{r}, z, t),$$

where $\mathbf{r}$ is the position vector in the plane perpendicular to the z-axis. Assuming that the amplitudes of the time-dependent components are small compared with the steady components, the current density will be, to a first-order approximation,

$$\mathbf{j} = \mathbf{j}_0 + \mathbf{j}_1(\mathbf{r}, z, t) = \rho\mathbf{v} = \rho_0\mathbf{u}_0 + \rho_0\mathbf{u}_1 + \rho_1\mathbf{u}_0. \qquad \mathbf{5.39}$$

The space charge and current densities, together with the electromagnetic fields produced by them, must satisfy Maxwell's equations; the electron velocity $\mathbf{v}$ and the fields are related by the equations of motion, and as charge is conserved, the charge density and current density are related by the equation of continuity. The rigorous way of handling the problem of space-charge waves is to derive from Maxwell's equations, in the usual manner, a wave equation, and to seek a solution of the form $f(\mathbf{r}) \exp[i(\omega t - \beta z)]$ for each time-dependent and coordinate-dependent quantity, under the appropriate boundary conditions. The task is greatly simplified if the beam is assumed to be initially uniform and of infinite cross-section. In this case their spatial distribution is one dimensional: all the quantities are functions of z and time only.

The result obtained for the one-dimensional problem gives a good approximation to the description of the properties of beams confined to a finite cross-section, by means of a strong magnetic field. As pointed out in section 2.8, in such a beam the electrons have practically no radial motion, and the instantaneous electron velocity is the same everywhere in a cross-section of the beam. Thus $\mathbf{v}$ and ρ are functions of z and t only. This situation requires that the electric field has no radial component, so the Poisson equation reduces to $\partial E_z/\partial z = -\rho/\varepsilon_0$, where E_z is a function of z and t alone. In the subsequent calculations, as already mentioned, we shall be concerned with functions of the form $\exp[i(\omega t - \beta z)]$, where we take ω as given, and β must be determined. It should be understood that the observable physical quantities are described by the real part of the complex quantities. Using complex exponentials, the operators $\partial/\partial t$ and $\partial/\partial z$ become multiplying factors of $i\omega$ and $i\beta$ respectively.

The z-dependent and time-dependent quantities u_1, ρ_1, j_1 and E_z, are related to one another as follows. Since

$$\frac{d\mathbf{v}}{dt} = \frac{\partial \mathbf{v}}{\partial t} + (\mathbf{v}.\nabla)\mathbf{v},$$

the equation of motion $m(d\mathbf{v}/dt) = e(\mathbf{E}+\mathbf{v}\times\mathbf{B})$ gives

$$u_1 = -\mathrm{i}\frac{(e/m)E_z}{\omega-\beta u_0}. \tag{5.40}$$

The equation of continuity, $\nabla.\mathbf{j}+\partial\rho/\partial t = 0$, gives

$$\rho_1 = \frac{\beta}{\omega} j_1. \tag{5.41}$$

Finally, substituting equations **5.40** and **5.41** into $j_1 = u_0\rho_1 + u_1\rho_0$ (see equation **5.39**), one finds

$$j_1 = -\mathrm{i}\omega\frac{(e/m)\rho_0 E_z}{(\omega-\beta u_0)^2}. \tag{5.42}$$

and from equation **5.41**

$$\rho_1 = -\mathrm{i}\beta\frac{(e/m)\rho_0 E_z}{(\omega-\beta u_0)^2}. \tag{5.43}$$

To determine the propagation constant β, it must be recalled that the *total* current density is the sum of the convection and displacement current densities, that is,

$$\mathbf{j}_{\text{tot}} = \mathbf{j}+\varepsilon_0\frac{\partial \mathbf{E}}{\partial t}$$

and that by virtue of the Maxwell equations, the divergence of the total current density must vanish. In the present case

$$j_{\text{tot}}\ j_0+j_1+\varepsilon_0\frac{\partial E_z}{\partial t} = j_0+\mathrm{i}\omega\varepsilon_0\left[1-\frac{(\rho_0/\varepsilon_0)(e/m)}{(\omega-\beta u_0)^2}\right]E_z.$$

The first term is constant, and its divergence vanishes. But $\partial E_z/\partial t \neq 0$, and therefore the condition $\partial j_{\text{tot}}/\partial z = 0$ is satisfied only if

$$(\omega-\beta u_0)^2 = \frac{\rho_0}{\varepsilon_0}\frac{e}{m}. \tag{5.44}$$

The quantity

$$\left(\frac{\rho_0}{\varepsilon_0}\frac{e}{m}\right)^{\frac{1}{2}} \equiv \omega_{\mathrm{p}}, \tag{5.45}$$

determined by the mean charge density ρ_0, is called the 'plasma frequency', and in terms of this, the propagation constant becomes, from equation **5.44**,

$$\beta = u_0^{-1}(\omega \pm \omega_p). \qquad \textbf{5.46}$$

The spatial period of the wave, that is, its wavelength is $\lambda = 2\pi/\beta$.

The phase velocity w of a travelling wave of frequency ω and propagation constant β is $w = \omega/\beta$. Thus it can be seen that, in space-charge waves, the quantities in question propagate in the shape of *two* travelling waves, which have different phase velocities. One is

$$w_s = u_0 \frac{\omega}{\omega + \omega_p} < u_0, \qquad \textbf{5.47a}$$

$$w_f = u_0 \frac{\omega}{\omega - \omega_p} > u_0. \qquad \textbf{5.47b}$$

The first type of wave is called a 'slow' wave, and the second one a 'fast' wave for obvious reasons.* The energy in a wave is transported with a velocity known as the 'group velocity' and defined as $w_g = d\omega/d\beta$. In the present case, one finds that both the slow and fast waves have the same group velocity, $w_g = u_0$.

Now we are ready to establish explicit expressions for the various waves, which are *superpositions* of a slow and a fast component. The *velocity* wave takes the form

$$u_1(z, t) = a_s \exp\left[i\left(\omega t - (\omega + \omega_p)\frac{z}{u_0}\right)\right] + a_f \exp\left[i\left(\omega t - (\omega - \omega_p)\frac{z}{u_0}\right)\right], \qquad \textbf{5.48}$$

where a_s and a_f are constants determined by initial conditions. Suppose that a sinusoidal velocity modulation is produced by an infinitesimally narrow gap located at $z = 0$, so that the electrons leaving the gap have a velocity

$$v = u_0 + u_1(0, t) = u_0 + u_{10} \cos \omega t.$$

In this case one finds $a_s = a_f = \frac{1}{2}u_{10}$, and hence

$$u_1(z, t) = \tfrac{1}{2}u_{10}\left[\exp i\left(\omega t - (\omega + \omega_p)\frac{z}{u_0}\right) + \exp i\left(\omega t - (\omega - \omega_p)\frac{z}{u_0}\right)\right]. \qquad \textbf{5.49}$$

*It is interesting to note that in a coordinate system which moves along the z-axis with the mean velocity of the electrons, u_0, the two waves appear to propagate in opposite directions, with different velocities:

$$w_+ = w_f - u_0 = \frac{u_0 \omega_p}{\omega - \omega_p} \qquad w_- = w_s - u_0 = \frac{-u_0 \omega_p}{\omega + \omega_p}.$$

The wavelengths, when measured in the laboratory system, and in the frame moving with the velocity u_0 (or any other constant velocity, for that matter), are the same for the corresponding waves. It takes two simultaneous measurements, in each system, to establish the distance between points where a periodically varying quantity has the same value at the same time. Under non-relativistic conditions, the distance between such points is the same in any inertial coordinate system. Therefore $\beta_f = \beta_+$ *and* $\beta_s = \beta_-$. Hence the frequency of the waves, measured by the observer moving with velocity u_0, will be $\omega_\pm = \beta_+\omega_+ = \beta_-\omega_- = \pm\omega_p$. Thus it can be said that the plasma frequency is the frequency of the collective oscillation of the electrons about their equilibrium positions. The reader can readily show that the difference in the frequencies apparent to an observer in the laboratory frame and one in the moving frame is the result of the Doppler shift.

To determine the *current-density* wave, note that according to equations **5.42** and **5.40**

$$j_1 = u_1 \rho_0 \frac{\omega}{\omega - \beta u_0}.$$

This relation, of course, is valid for the slow and the fast wave *separately*. Substituting u_1 and the appropriate propagation constant, one finds by superimposing the two components,

$$j_1(z, t) = \tfrac{1}{2} u_{10} \rho_0 \frac{\omega}{\omega_p} \left[\exp \mathrm{i}\left(\omega t - (\omega + \omega_p)\frac{z}{u_0}\right) - \exp \mathrm{i}\left(\omega t - (\omega - \omega_p)\frac{z}{u_0}\right)\right]. \quad \textbf{5.50}$$

Note that $j_1(0, t) = 0$, that is, the alternating *current* density is *zero at the gap where* the initial *velocity modulation is produced.*

If one studies the behaviour of the velocity wave and of the current-density wave further downstream, one recognizes that changes take place in both because the *slow and fast components get out of phase* on account of the different phase velocities. Taking out the common exponential factor, $\exp[\mathrm{i}\omega(t - z/u_0)]$, equations **5.49** and **5.50** can be readily rewritten

$$u_1(z, t) = u_{10} \cos \frac{\omega_p}{u_0} z \exp\left[\mathrm{i}\omega\left(t - \frac{z}{u_0}\right)\right], \quad \textbf{5.49a}$$

$$j_1(z, t) = -\mathrm{i} u_{10} \rho_0 \frac{\omega}{\omega_p} \sin \frac{\omega_p}{u_0} z \exp\left[\mathrm{i}\omega\left(t - \frac{z}{u_0}\right)\right]. \quad \textbf{5.50a}$$

Equations **5.49a** and **5.50a** show 'spatial beats' arising from the superposition of two travelling waves of identical frequency but different wavelengths (see Figure 70). The *amplitudes* of the alternating components of the electron velocity

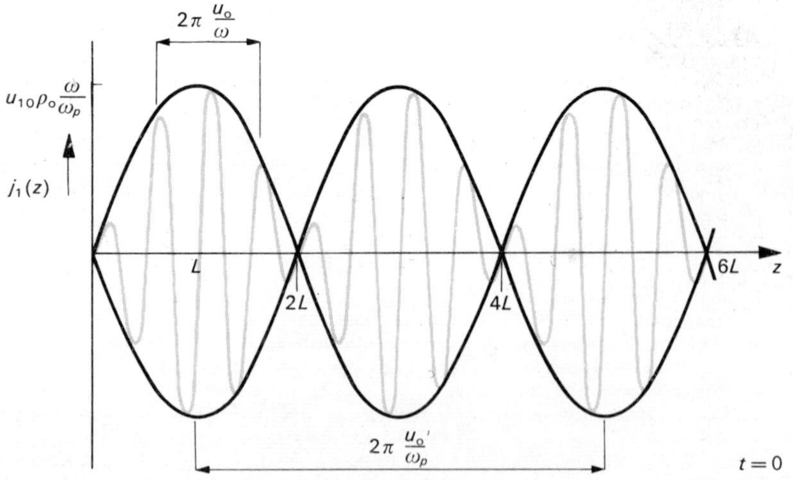

Figure 70

and of the current density depend on the position z at which these quantities are observed. It should be noted that the amplitudes vary sinusoidally with z, and that there is a $\frac{1}{2}\pi$ phase shift between velocity and current-density waves. Furthermore, the *travelling waves* of velocity and current density are also shifted in phase by $\frac{1}{2}\pi$, as shown by the factor $-\mathrm{i}$, which which occurs in equation **5.50a** but not in equation **5.49a**.

Recalling what was said about amplifier klystrons in the preceding section, one finds that the optimum position of the output cavity can be readily predicted from equations **5.49a** and **5.50a**. The optimum position is that at which the alternating current (density) has a maximum amplitude, that is, where

$$\frac{\omega_p}{u_0} z = (k+\tfrac{1}{2})\,\pi.$$

At the same position, the electrons move with their mean velocity, since at this position $u_1(t) = 0$. Note that this position depends on the mean charge density through ω_p.

Next we wish to determine the time average of the *kinetic energy density* of the electrons in a beam carrying a slow or fast wave. Since the number of electrons per unit volume is ρ/e, and each has a kinetic energy $\frac{1}{2}mv^2$, the time average of the kinetic energy density is

$$\frac{1}{2}\frac{\overline{mv^2\rho}}{e} = \frac{m}{2e}\overline{jv} = \frac{m}{2e}(j_0 v_0 + \overline{j_1 v_1}). \qquad \textbf{5.51}$$

Equations **5.49** and **5.50** show that both the slow and the fast waves have the same amplitude, and that the *slow* velocity and current-density waves are in *antiphase*, whereas the *fast* velocity and current-density waves are *in phase*. Thus $(\overline{j_1 v_1})_s < 0$, $(\overline{j_1 v_1})_f > 0$, and therefore the *kinetic energy density in the fast wave is higher*, and *in the slow wave it is smaller than in the unmodulated beam*.

It is customary to define the density of the 'kinetic power flow' as the *time average* of the kinetic energy of the electrons crossing unit area in unit time at a fixed cross-section. The mean velocity u_0 is taken as the uniform drift velocity of all the electrons. This amounts to saying that the velocity of the energy flow in both waves is taken as the common group velocity $w_g = u_0$. Thus the kinetic power flow, which is the mean velocity times the kinetic energy density, is smaller for the slow wave, and larger for the fast wave, than the power flow in the unmodulated beam. This state of affairs is often referred to by the slightly unfortunate statement that the slow wave is a negative energy carrier, and the fast wave is a positive energy carrier. It is interesting to note that if the interaction between a field and an electron beam takes place without a net energy exchange, slow and fast waves of equal amplitudes are excited. This is the case in the interaction between a beam of uniform charge density and the field in an infinitesimally small gap. On the other hand, if the mean kinetic energy of the electrons is reduced in the interaction, the slow wave is enhanced and the fast wave is damped. If the mean kinetic energy is increased, the two waves show the opposite behaviour.

In the next section, we shall refer to these processes in the interpretation of 'travelling-wave' amplifiers.

The space-charge waves described previously are longitudinal waves rather like pressure waves in a gas stream. The cause is a periodic perturbation applied parallel to the beam axis, and the waves are maintained by space-charge restoring forces. Space-charge waves reminiscent of transverse waves require transverse restoring forces. If the beam is immersed in a magnetic field, this can supply the required restoring force. Waves excited under these conditions are usually called cyclotron waves for reasons which will become apparent shortly. There are many ways of exciting cyclotron waves, and we shall consider one example below.

Let electrons flow parallel to a uniform magnetic field, $\mathbf{B} = B\mathbf{e}_z$, and pass a uniform electric field which is perpendicular to the magnetic field and which varies sinusoidally in time: $\mathbf{E} = \mathbf{e}_x E \cos \omega t$. If the time τ spent by the electrons in the electric field $\mathbf{E}$ is short compared with the period of the alternating field ($\omega\tau \ll 2\pi$),

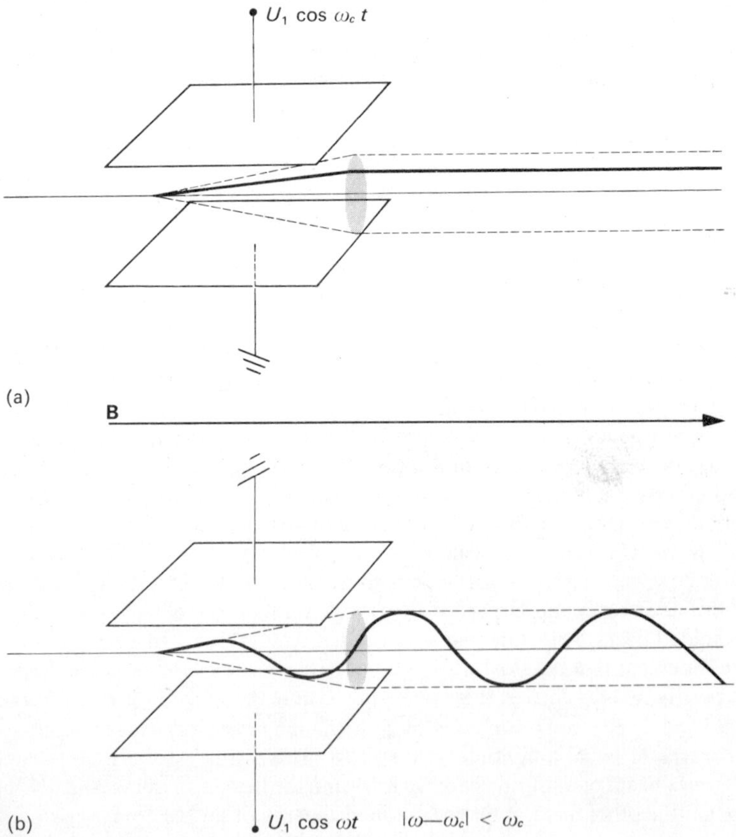

Figure 71

each electron acquires a transverse velocity corresponding to the instantaneous field, and starts spiralling around the direction of magnetic field at the cyclotron frequency $\omega_c = eB/m$. If the electrons have a common drift velocity u_0 parallel to the z-axis, the beam will spread and becomes refocused at intervals of $2\pi u_0/\omega_c$.

If the time of flight is so long that $\omega\tau \gg 2\pi$, *and* the frequency of the deflecting field is in the neighbourhood of the cyclotron frequency, $\omega \simeq \omega_c$, the situation will be quite different. If one treats the electric field **E** as the superposition of two fields which rotate in opposite directions, one of these will rotate very nearly in synchronism with the electrons, and the other component will rotate in the opposite sense. Thus the effect of the first component will be cumulative and will make the radius of the cyclotron orbit increase, whereas the other will cause some periodic perturbation, which becomes negligible as time goes on. If the deflecting field frequency is equal to the cyclotron frequency, one obtains a rotating beam which sweeps a conical surface, with the cone's apex at the point where the electrons enter the deflecting field, and the cone's axis lying along the magnetic field. When the beam leaves the deflecting field, the radius of the cyclotron orbit stays constant, and a straight beam is obtained which rotates at the cyclotron frequency, sweeping a cylindrical surface (see Figure 71a). The motion of the individual electrons does not change significantly if the frequency of the deflecting field does not coincide exactly with the cyclotron frequency, provided that $|\omega_c - \omega|\tau \ll 2\pi$, but the shape of the beam as a whole is distorted. The shape, in fact, becomes a helix wound on a cone inside the deflecting field and on a matching cylinder after leaving the deflecting field (see Figure 71b). The reason is that if $\omega \neq \omega_c$, the individual electrons keep orbiting at the frequency ω_c, but the point where the beam crosses a fixed plane moves with an angular velocity ω. This can be predicted qualitatively by the following argument. If an electron which entered the deflecting field at an instant t_0 reaches the exit plane of the deflecting field at a particular point, the next electron which reaches this point will have had to 'see' the same field throughout its passage, and therefore it must have entered at an instant $t_0 + n(2\pi\omega)$, where n is an integer.

One can readily describe the motion of the electron in quantitative terms. Denoting the transverse velocity components by u_x and u_y, the equation of motion gives

$$\frac{du_x}{dt} = \frac{e}{m} E \cos \omega t + \omega_c u_y,$$

$$\frac{du_y}{dt} = -\omega_c u_x.$$

Considering u_x and u_y as the real and imaginary parts of a complex number, $u = u_x + iu_y$, the above equations can be combined:

$$\frac{du}{dt} - i\omega_c u = \frac{e}{2m} E[\exp(i\omega t) + \exp(-i\omega t)].$$

After multiplying both sides by $\exp(i\omega_c t)$, the equation can be readiiy integrated over the time of flight, that is, between t_0 and $t_0+\tau$; after some rearrangement one obtains

$$u(t_0+\tau)-u(t_0)\exp(-\mathrm{i}\omega_c\tau) = \frac{e}{2m}E\left[\exp(\mathrm{i}\omega t_0)\frac{\exp(\mathrm{i}\omega\tau)-\exp(-\mathrm{i}\omega_c\tau)}{\mathrm{i}(\omega_c+\omega)}+\right.$$

$$\left.+\exp(-\mathrm{i}\omega t_0)\exp(-\mathrm{i}\omega_c\tau)\frac{\exp[\mathrm{i}(\omega_c-\omega)\tau]-1}{(\omega_c-\omega)}\right]. \qquad \mathbf{5.52}$$

At the electron's entry $u(t_0) = 0$. Taking into account the conditions $\omega \simeq \omega_c$, $\omega_c\tau \gg 2\pi$, but $|\omega_c-\omega|\tau \ll 2\pi$, the fraction in the first term inside the brackets becomes approximately $\omega_c^{-1}\sin\omega_c\tau$, and the fraction in the second term becomes τ in first-order approximation. Thus equation **5.52** reduces to

$$u(t_0+\tau) = \frac{e}{2m}E\tau\left[\exp(\mathrm{i}\omega t_0)\frac{\sin\omega_c\tau}{\omega_c\tau}+\exp(-\mathrm{i}\omega t_0)\exp(-\mathrm{i}\omega_c\tau)\right].$$

In view of the condition $\omega_c\tau \gg 2\pi$, the first term in the brackets can be neglected, and hence

$$u(t_0+\tau) = \frac{e}{2m}E\tau\exp(-\mathrm{i}\omega t_0)\exp(-\mathrm{i}\omega_c\tau). \qquad \mathbf{5.53}$$

If the position the electron reaches at time $t_0+\tau$, in a plane perpendicular to the z-axis, is defined by the complex vector $r = x+\mathrm{i}y$, one finds, to the same approximation as above,

$$r = \int_0^\tau u\,d\tau \simeq \frac{\mathrm{i}}{2}\frac{eE}{m\omega_c}\tau\exp(-\mathrm{i}\omega t_0)\exp(-\mathrm{i}\omega_c\tau). \qquad \mathbf{5.54}$$

If one follows a particular electron which enters the field at an instant t_0, one sees that its distance $|r|$ from the axis increases linearly with its time of flight τ and its position vector rotates with the angular velocity ω_c. To observe the point at which the electrons cross a fixed plane perpendicular to the z-axis implies keeping τ = constant and watching the passage of electrons which entered the deflecting field at different times t_0. These points describe a circle, $|r|$ = constant, at an angular velocity ω.

Next we wish to examine the properties of the beam after it leaves the deflecting field and to describe these properties in terms commonly used in the discussion of wave propagation phenomena. In particular, we wish to describe the transverse components of the velocity of the electrons as a function of the position measured along the z-axis and the time measured at this position. These variables replace the previously used 'time of entry' t_0 and 'time of flight' τ. Denoting the velocity vector $u_x\mathbf{e}_x+u_y\mathbf{e}_y+u_0\mathbf{e}_z = \mathbf{u}$, where u_0 = constant, but u_x and u_y are functions of t and z, one has

$$\frac{d\mathbf{u}}{dt} = \frac{\partial\mathbf{u}}{\partial t}+(\mathbf{u}.\nabla)\mathbf{u}.$$

In the presence of the magnetic field, $\mathbf{B} = B\mathbf{e}_z$, the equations of motion yield for the transverse components

$$\frac{\partial u_x}{\partial t}+u_0\frac{\partial u_x}{\partial z} = \omega_c u_y \quad \text{and} \quad \frac{\partial u_y}{\partial t}+u_0\frac{\partial u_y}{\partial z} = -\omega_c u_x. \qquad 5.55$$

If we choose $z = 0$ at the exit of the deflecting field, one can write, according to equation **5.53**,

$$u_x(t, 0) = u_1 \cos \omega t,$$
$$u_y(t, 0) = -u_1 \sin \omega t. \qquad 5.53a$$

To find the solution of equations **5.55**, we use the trial solutions

$$u_x = u_1 \exp[-\mathrm{i}(\omega t-\beta z)],$$
$$u_y = a u_1. \qquad 5.56$$

Substituting the solutions **5.56** into equations **5.55**, one readily finds that

$$a = \pm \mathrm{i},$$
$$\beta = u_0^{-1}(\omega \pm \omega_c). \qquad 5.57$$

Thus it is seen that the conditions imposed by the magnetic field are compatible with the propagation of two waves with different velocities:

$$w_s = u_0 \frac{\omega}{\omega+\omega_c},$$
$$w_f = u_0 \frac{\omega}{\omega-\omega_c}. \qquad 5.58$$

In other words there are two cyclotron waves possible, one 'slow' and the other 'fast', as in the analogous case of space-charge waves. If the cyclotron waves are excited as described previously, only the fast wave will occur. This is shown by the initial conditions **5.53a**, which are satisfied only if one takes $a = -\mathrm{i}$, and hence the phase velocity w_f. Note that the kinetic energy carried by the beam, when excited in the manner described, is larger than that carried by the beam without excitation. The excess energy is that of orbital rotation. Thus the fast cyclotron wave is a 'positive energy carrier' in the sense explained above.

It is interesting to note a difference of practical significance between the space-charge waves, introduced previously, and the cyclotron waves. In the case of the space-charge waves, the separation between the phase velocities of the slow and fast waves is determined by the plasma frequency. This quantity is fixed by the space-charge density, and in practical situations it stays much below the signal frequencies of interest: $\omega_p \ll \omega$. Hence the two phase velocities, w_f and w_s, are close together. In the case of the cyclotron waves, the relevant quantity is the cyclotron frequency which is determined by the external field, and can be chosen to match the signal frequency ($\omega_c \simeq \omega$); hence $w_f \gg w_s$. This condition is exploited in a beam-type 'parametric' amplifier which will be discussed in section 5.7.

5.5 Amplification by coupling of modes of wave propagation (travelling-wave tubes)

It is well known that if two resonant systems of the same eigenfrequency are coupled together, the energy from one may be transferred to the other. If one of the resonant systems, say (1), is initially excited at a given energy level, its oscillation amplitude gradually decreases, while the initially unexcited system (2) starts oscillating with an increasing amplitude. The amplitude of the oscillations in (1) eventually vanishes, while that of (2) reaches a maximum, and the whole process repeats in the reverse direction. Note that if each resonator is loss-free, the total energy in the coupled system as a whole is conserved. If the amplitude of the oscillation of (1) is maintained by feeding in energy from an external source, the amplitude of the oscillations in (2) will increase monotically.

An analogous process takes place in two structures coupled along their entire lengths, if the two structures are loss-free and can carry a wave of the same frequency with the same phase velocity. In this case, the total energy propagating along the two structures is conserved, since the energy carried by one wave is drawn from the energy carried by the other. If the wave propagating in one of the structures is not allowed to decay, the amplitude of the wave propagating in the other structure will again grow monotonically.

It was seen in the preceding section that an electron can be considered as a structure capable of carrying space-charge waves accompanied by electromagnetic waves. In fact it was seen that there are two such waves, a 'fast' and a 'slow' one. The electromagnetic waves enable coupling to be established between the electron beam and any other structure which is capable of carrying electromagnetic waves, such as a transmission line or another electron beam.

Suppose a transmission line is designed to carry an electromagnetic wave of the

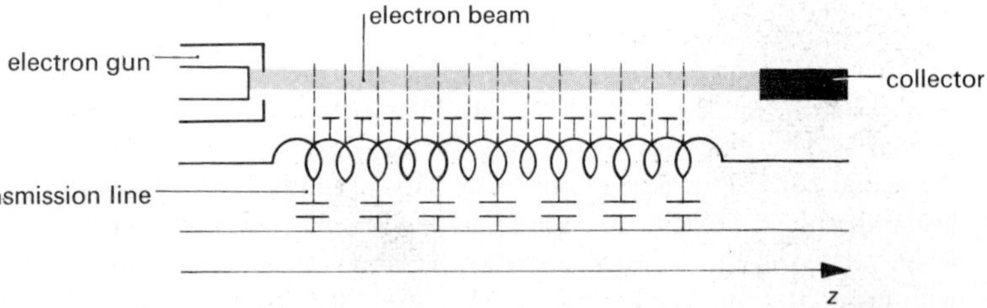

(a)

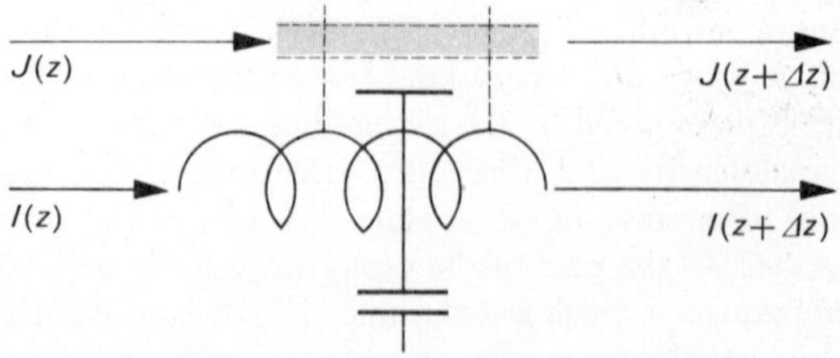

(b)

Figure 72

same frequency and phase velocity as that of the *slow* space-charge wave. If coupling is established between the electron beam and the transmission line, the slow space-charge wave will excite electromagnetic waves on the transmission line, and the latter will draw its energy from the former. Recall, however, that as energy is extracted from a slow wave, its amplitude is increasing rather than decaying. Consequently, as the slow space-charge wave is propagating along the beam with increasing amplitude, the electromagnetic wave travelling along the transmission line will have an increasing amplitude too. The energy gained by the electromagnetic wave along the transmission line is balanced by the loss of the *mean* kinetic energy of the electrons in the beam. The same argument shows that this mechanism cannot work for exciting growing waves on the transmission line by coupling them to the fast space-charge waves because, as energy is extracted from the fast wave, its amplitude decays.

Quantitative considerations show that the qualitative description of the process given above somewhat oversimplifies the picture. For simplicity's sake it will be assumed that the conditions are quasi-stationary, and that the transmission line is loss free. In this case, the line can be characterized by its series inductance per unit length, L, and its shunt capacitance per unit length, C (see Figure 72a). The beam flows parallel to the line in its immediate vicinity. The potential difference U across the line is a function of time and the coordinate measured along the line; it arises because the time-dependent current flowing in the line produces a voltage drop along the series inductance:

$$\frac{\partial U}{\partial z} = -L\frac{\partial I}{\partial t}. \qquad \textbf{5.59}$$

Another relationship between current and voltage is obtained if the divergence condition, $\nabla . (\mathbf{j}+\varepsilon_0\, \partial\mathbf{E}/\partial t) = 0$, is applied to a narrow slice of the system consisting of the transmission line and the beam between two surfaces a distance dz apart. (Figure 72b). Taking the volume integral for this slice

$$\int \nabla . \mathbf{j}\, dV = -\varepsilon_0 \frac{\partial}{\partial t}\int_N \nabla . \mathbf{E}\, dV. \qquad \textbf{5.60}$$

The left-hand side of this expression can be transformed, by means of Gauss's theorem, into a surface integral, which gives the difference between the current flowing in and that flowing out of the slice. Denoting the instantaneous beam current by J, one obtains $\int_V \nabla . \mathbf{j}\, dV = \partial/\partial z(I+J)\, dz$. Since $\nabla . \mathbf{E} = -\rho/\varepsilon_0$, the right-hand side of equation **5.60** represents the rate of change of the charge in the slice. In terms of the shunt capacitance of the line and the potential difference across it one can write $\varepsilon_0 \int_V \nabla . \mathbf{E}\, dV = CU\, dz$, and therefore equation **5.60** becomes

$$\frac{\partial I}{\partial z} = -C\frac{\partial U}{\partial t} - \frac{\partial J}{\partial z}. \qquad \textbf{5.61}$$

Note that in the absence of the beam ($J = 0$) equations **5.59** and **5.61** reduce to the conventional transmission-line equations, according to their solution, the transmission line can carry a voltage and a current wave, which depend on the time t and the coordinate z, as $\exp \mathrm{i}(\omega t - \beta_l z)$, with

$$\beta_l = \omega(LC)^{\frac{1}{2}}.$$

It is customary to specify the transmission line by its characteristic, or wave, impedance, defined as

$$Z_0 = \left(\frac{L}{C}\right)^{\frac{1}{2}},$$

so that one can write

$$\omega L = \beta_l Z_0 \qquad \text{and} \qquad \omega C = \beta_l Z_0^{-1}. \qquad \textbf{5.62}$$

To find a solution of the simultaneous differential equations **5.59** and **5.61** in the presence of the beam ($J \neq 0$) it will be assumed that the variable quantities are of the form $\exp(\mathrm{i}\omega t - \gamma z)$, where γ is a *complex* number. Hence one obtains from equations **5.59** and **5.61**

$$\gamma U = \mathrm{i}\beta_l Z_0 I, \qquad \textbf{5.59a}$$

$$\gamma I = \mathrm{i}\beta_l Z_0^{-1} U - \gamma J, \qquad \textbf{5.61a}$$

where the relations in **5.62** were used to eliminate L and C. By eliminating I from equations **5.59a** and **5.61a**, one finds the relation between the *line* voltage U and the *beam* current J:

$$U = -\mathrm{i}\frac{\beta_l \gamma}{\gamma^2 + \beta_l^2} Z_0 J. \qquad \textbf{5.63}$$

In the preceding section, equation **5.42** gave a relation between the space-charge current density j_1 and the field E_z for the one-dimensional problem. Since none of the variables depends on coordinates in the plane perpendicular to the z-axis, one can integrate **5.42** across the beam. Since $j_0 = \rho_0 u_0$,

$$J_1 = -\frac{\mathrm{i}\omega}{u_0} \frac{(e/m) J_0}{(\omega + \mathrm{i}\gamma u_0)^2} E_z, \qquad \textbf{5.42a}$$

where J_1 is the alternating component of the beam current, J_0 is its constant component (mean value) and β is replaced by $-\mathrm{i}\gamma$. If the beam passes in the immediate neighbourhood of the transmission line, the electric field component E_z, which occurs in equation **5.42a**, must be identical with the potential gradient $-\partial U/\partial z$ along the line, and the alternating beam current J_1, in equation **5.42a**, must be identical with the beam current J in equation **5.63**. Introducing

$$E_z = -\frac{\partial U}{\partial z} = \gamma U,$$

and expressing the mean velocity in terms of the initial accelerating voltage $\frac{1}{2}mu_0^2 = eU_0$, one finds from equation **5.42a**

$$J_1 = -\frac{i\omega}{u_0}\frac{\gamma J_0}{2U_0(\omega/u_0+i\gamma)^2}U. \qquad \textbf{5.64}$$

The condition required for consistency between equations **5.63** and **5.64** is $J = J_1$, and hence

$$\left(\gamma - i\frac{\omega}{u_0}\right)^2[\gamma^2-(i\beta_l)^2] = \frac{J_0 Z_0}{2U_0}\beta_l\frac{\omega}{u_0}\gamma^2. \qquad \textbf{5.65}$$

Let us choose the parameters of the transmission line, L, C, in such a manner that the phase velocity of the wave on the line matches the mean velocity of the electron beam,

i.e. $\quad \dfrac{\omega}{\beta_l} = u_0.$

In this case, equation **5.65** becomes

$$(\gamma - i\beta_l)^3(\gamma+i\beta_l) = 2K^3\beta_l^2\gamma^2, \qquad \textbf{5.65a}$$

where the notation,

$$K^3 = \frac{J_0 Z_0}{4U_0},$$

was introduced for convenience. K is a dimensionless quantity, and in practical circumstances $K \ll 1$. If $K = 0$, the roots of equation **5.65a** are readily found. There is one solution, $\gamma = -i\beta_l$, and three coincident solutions, $\gamma = i\beta_l$.

If $0 < K \ll 1$, it can be assumed that γ is in the vicinity of $\pm i\beta_l$. Suppose that $\gamma = \alpha + i\beta_l$, where $\alpha \ll \beta_l$ so that $\gamma + i\beta_l \simeq 2i\beta_l$ and $\gamma^2 \simeq -\beta_l^2$. In this case, equation **5.65a** yields

$$\gamma - i\beta_l = i^{\frac{1}{3}}\beta_l K = \beta_l K \exp\left[\tfrac{1}{3}(\tfrac{1}{2}\pi + 2n\pi)\right] \qquad (n = 0, 1, 2, \ldots).$$

Hence (see Figure 73) three different values are obtained for the propagation constant:

$$\gamma_1 = i\frac{\omega}{u_0}(1+\tfrac{1}{2}K) - \frac{\sqrt{3}}{2}K,$$

$$\gamma_2 = i\frac{\omega}{u_0}(1+\tfrac{1}{2}K) + \frac{\sqrt{3}}{2}K,$$

$$\gamma_3 = i\frac{\omega}{u_0}(1-K).$$

A fourth value is obtained in the neighbourhood of $\gamma \simeq -i\beta_l$:

$$\gamma_4 = -i\frac{\omega}{u_0}(1-\tfrac{1}{4}K^3).$$

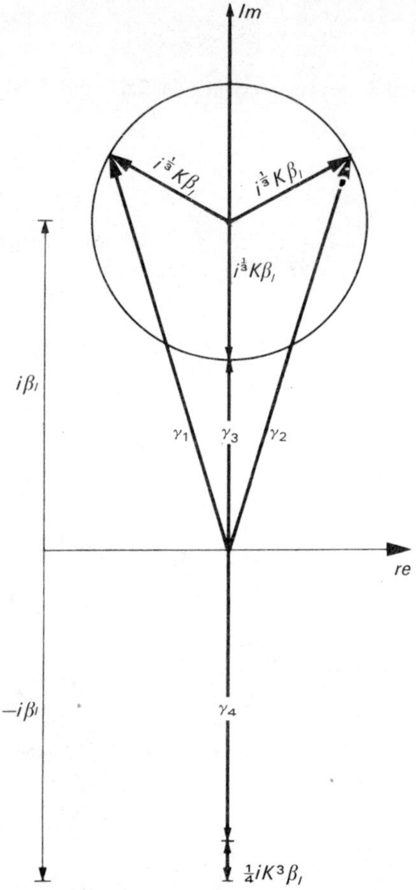

Figure 73

The solutions γ_r (r = 1, 2, 3, 4) substituted into the function exp $(\mathrm{i}\omega t-\gamma_r z)$ give four different types of travelling waves. With γ_3 and γ_4, one obtains two waves travelling in opposite directions, with different phase velocities. Both velocities are larger than the mean velocity of the electrons since the phase velocity is $\mathrm{i}\omega/\gamma$. With γ_1 and γ_2, one finds

$$\exp(\mathrm{i}\omega t-\gamma_{1,2}z) = \exp\left(\pm\frac{\sqrt{3}}{2}Kz\right)\exp\left[\mathrm{i}\omega\left(t-(1+\tfrac{1}{2}K)\frac{z}{u_0}\right)\right].$$

This shows that there is one wave with exponentially growing amplitude, and one with exponentially decaying amplitude; both waves have the same phase velocity

$$w = u_0(1+\tfrac{1}{2}K)^{-1} < u_0,$$

that is, both are slow waves.

Summarizing the results, one can say that the coupling between the electron beam and the transmission line, under the conditions set out above, gives rise to four travelling waves, one of which is a slow wave with growing amplitude. The picture obtained by qualitative arguments correctly predicted the excitation of the growing wave, although not that of the other waves.

The process described above is the basis of the mechanism of the amplification in '*travelling-wave tubes*'. In building such a tube, two conditions are of fundamental importance. One is contained in the condition $\omega/\beta_l = u_0$; in the absence of the beam the line must be able to carry electromagnetic waves with a phase velocity equal to the expected mean electron velocity, which is much below the velocity of light. The second condition, implicit in the whole argument, is that the electromagnetic wave on the line must have an electric field component parallel to the direction of propagation. There are a variety of ways in which these conditions can be satisfied. One example is a wave-guide loaded with regularly spaced apertures and excited in a TM-mode. This is the type mentioned in the discussion of linear accelerators. Figure 74 shows another scheme which is often used in travelling wave tubes. In this case the slow wave structure is a wire helix.

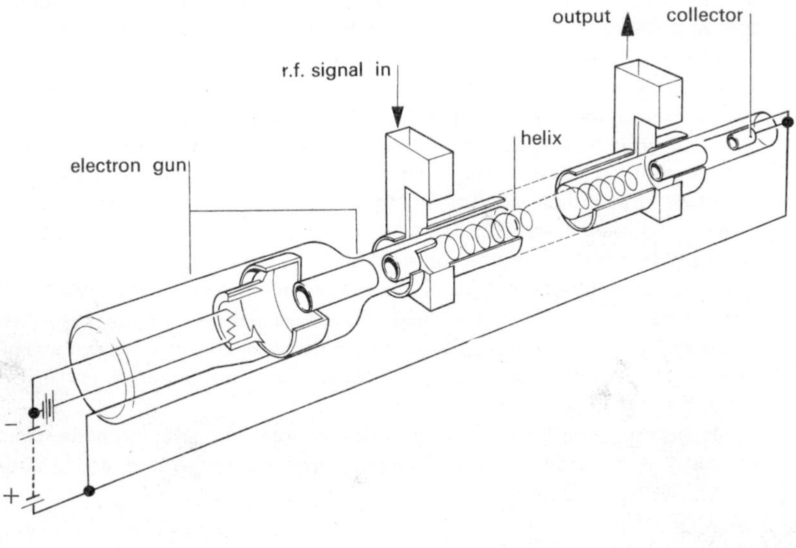

Figure 74

One may assume that a transverse plane electromagnetic wave is travelling along the helix, so the wave front follows a helical path and propagates with the velocity of light *in vacuo*. This picture leads to the intuitive conclusion that the electric field – between the turns of the helix, as it were – has an axial component. The phase velocity relevant to the tube's operation is the velocity of the wave measured along the z-axis, and this is a fraction of the velocity of light corresponding to the ratio of the pitch of the helix to its circumference. If a small input signal is applied, the incident electron beam acquires an initial velocity modulation and, by virtue of the processes analysed above, an increased amount of power can be obtained at the far end of the helix.

It has already been mentioned that the coupling of modes of propagation between an electron beam and a transmission line is one of several possibilities. Instead of a transmission line, for example, one can take another electron beam. If the mean velocity of the electrons is different in the two beams, one can arrange that the slow space-charge wave in one beam has the same phase velocity as the fast wave in the other. In this case, energy transfer will take place between the synchronous pair of waves, and both will grow exponentially in the direction of propagation along the beam axes. In a 'double-beam' amplifier, based on this principle, no transmission line or wave-guide structure is required. An initial velocity modulation is produced by an input cavity, and the increased output power can be coupled out by means of a second cavity placed further downstream.

5.6 The magnetron

The magnetron is the oldest of the microwave valves, but the greatest difficulty has been found in obtaining a satisfactory description of the mechanism of its operation. It will be seen that it is, in many ways, closely related to travelling wave tubes, but at the same time it has features which give its operation a distinctly different character. In particular, there is one fundamental difference which makes the efficiency of the magnetron superior to the valves discussed so far. Magnetrons are mainly used as oscillators, and operate at very high space-charge current densities. Under these conditions, approximations useful in the treatment of other valves are not permissible, and this leads to great mathematical difficulties, particularly in the description of the process by which the oscillations build up and reach their stationary state. In this section only two extreme situations will be considered, namely when the magnetron is not oscillating at all, and when it is in a stationary state of oscillation. To elucidate the most conspicuous features of the magnetron operation, we shall first concentrate our attention on processes involving individual electrons.

Magnetrons usually consist of a cylindrical cathode with an anode structure around it (see Figure 75). The whole system is immersed in a strong uniform

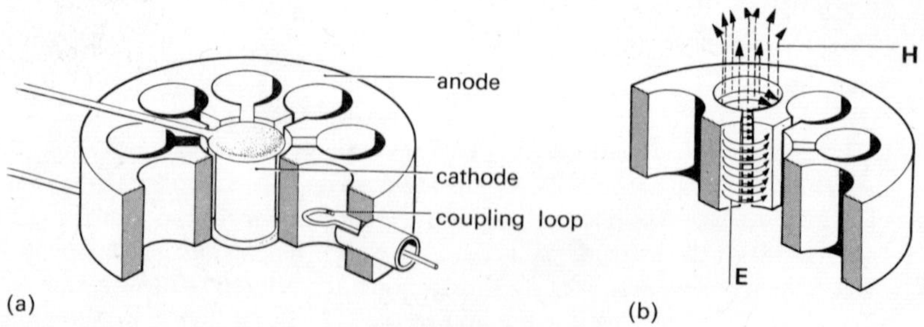

Figure 75

magnetic field, parallel to the axis of the electrode system. The anode structure, essentially a cylindrical block with regularly spaced radial recesses, is centred on the cathode axis. Its shape and size are chosen in such a manner that a resonant cavity is formed in which the electric field is located predominantly between the cathode and the anode block. The oscillating field can be considered as the superposition of two travelling waves, propagating in opposite directions around the cathode. The cylindrical structure imposes boundary conditions which define the possible wavelengths of the travelling waves, since at any instant the phase shift on a closed path encircling the cathode must be an integral multiple of 2π.

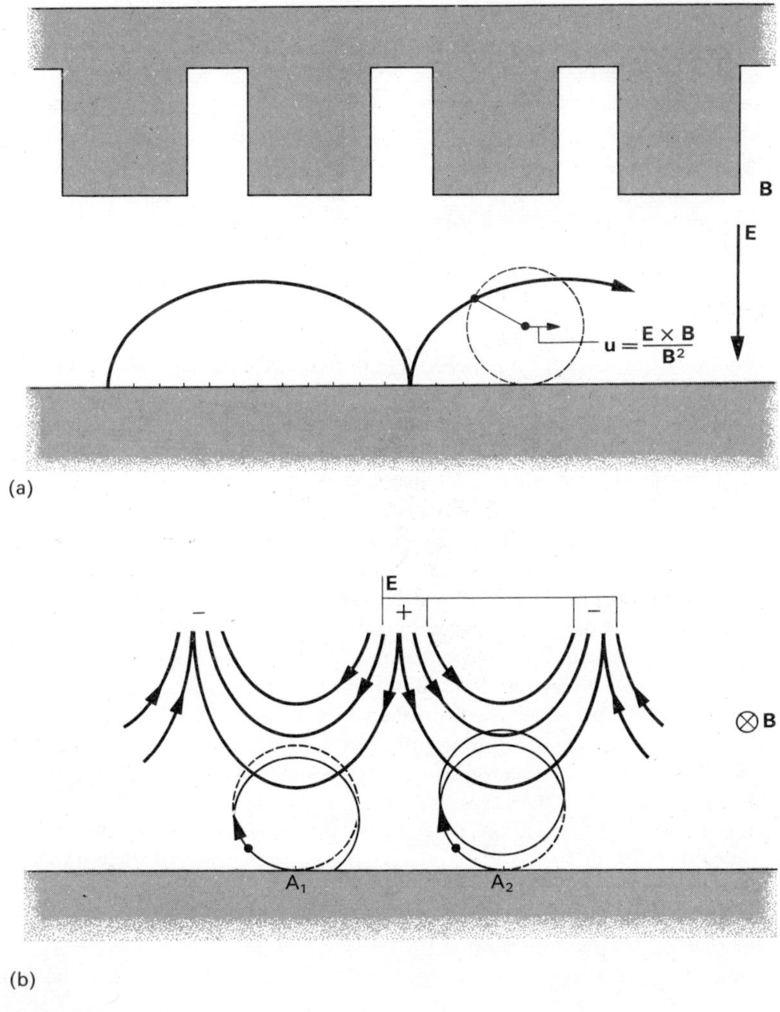

Figure 76

The essential features of the electron motion in the magnetron remain the same, but its description is generally simplified, if one imagines the magnetron unfolded into a linear structure (see Figure 76) with a uniform magnetic field perpendicular to the section illustrated in the drawing. If there is a constant potential difference U_0 between the cathode and the anode, the electric field is essentially uniform and perpendicular to the magnetic field. In this case the electron motion will take place as described at the end of section 1.1. If, for example, the electrons are emitted by the cathode with zero initial velocity (see Figure 76a), each will describe a cycloidal orbit generated by a point of a rolling circle of radius R. One defines R by the relation $R\omega_c = E/B$, where ω_c is the cyclotron frequency and E is the electric field intensity. Since the maximum distance the electron can move from the cathode is $2R$, the electrons cannot reach the anode if $2R < d$; in this case the anode current in the 'linear' magnetron is said to be 'cut off'. Note that the maximum kinetic energy the electrons can acquire under these circumstances is $2eER \simeq eU(2R/d)$, where d is the cathode–anode distance. Provided that the magnetic field is sufficiently high ($2R \ll d$) this kinetic energy is a small fraction of the energy the electrons would acquire if they could reach the anode.

It should be recalled that if the electron motion is described with reference to a coordinate system which is moving with a velocity $u = E/B$ in a direction perpendicular to the plane defined by the electric and magnetic fields, one finds that, in this coordinate system, the electric field vanishes and the electron trajectory is a circle. Let us now consider the electron motion in this moving coordinate system when the cavity is excited to high-frequency oscillations. In the cavity's own frame the oscillating field can be considered as the superposition of two waves travelling in opposite directions. Because of the periodic structure of the cavity these waves have phase velocities far below the velocity of light and consequently, when the waves are observed from the moving frame, the phase velocities appear different. By suitable choice of the structure of the cavity it can be arranged that the waves are travelling, relative to the cavity, with velocities $\pm(E/B)$, and therefore one of the two waves appears in the moving coordinate system as a stationary field. While the perturbation due to the travelling wave averages out, the field which appears stationary in the moving frame will have a cumulative effect on the electron motion, and the orbits, which are circles in the absence of the high-frequency excitation of the cavity, will change significantly.

The distribution of the electric field which appears stationary in the moving frame is shown schematically in Figure 76b. Let us first consider the motion of an electron which leaves the cathode near the point A_1 and which moves clockwise on a circular orbit in the absence of the high-frequency excitation of the cavity. Under the defined conditions such an electron is acted upon by an apparently stationary electric field, which is predominantly parallel to the cathode, and which makes the electron trajectory drift in the direction of $\mathbf{E} \times \mathbf{B}$, that is, *towards the cathode*. Thus the electron re-enters the cathode and ceases to participate in any further process. Furthermore, A_1 represents a point of unstable phase in the sense in which this phrase was used in the study of resonance accelerators. Electrons emitted by points on either side of A_1 will drift away laterally, because there are electric field components perpendicular to the cathode. In particular, electrons emitted by points to the left of A_1 tend to drift further to the left, into a region where the field distribution is increasingly unfavourable for their survival. An electron emitted near A_2 is subject to a field which makes the electron orbit drift *towards the anode*, while its *mean* kinetic energy stays essentially constant. A_2 is a point of phase stability: electrons emitted by points in the neighbourhood of A_2 tend to drift nearer to the plane perpendicular to the cathode at A_2, because of the direction of the effective electric field. Thus there are two processes running in parallel: there is a partial selection and a partial phase focusing. The resulting effect is that only those electrons which drift towards the anode, and have a constant mean kinetic energy, can survive. These electrons transfer energy to the electromagnetic field in a way we have not encountered before. Looking at the process from the point of view of an observer fixed in the laboratory, one can see that the electrons move from a region of high potential energy (cathode region) to a region of low potential energy (anode region), without acquiring the corresponding kinetic energy. The potential energy difference is *directly* transferred to the high-frequency field. The source of this energy is clearly the d.c. supply, which maintains the anode potential with respect to the cathode. With this model of the magnetron operation one can readily estimate the efficiency of the energy conversion process.

Observed in the magnetron's (rest) frame, the kinetic energy of the electrons on any trochoidal loop of the orbit varies between 0 and $\frac{1}{2}m(2E/B)^2$. If the d.c. anode voltage is U_0, and it is assumed that the electron reaches the anode with maximum kinetic energy, the energy transferred to the high-frequency field is $eU_0 - 2m(E/B)^2$, and hence the efficiency is

$$\eta = 1 - 2\frac{m}{e}\frac{E}{B^2 d},$$

where the voltage U_0 has been expressed in terms of the field intensity E, and cathode–anode distance d: $U_0 = Ed$. The 'cut-off' values of the fields, E_c and B_c, are defined as the values which just prevent the electrons from reaching the anode in the absence of the high-frequency field. Thus

$$d = 2R = \frac{2mE_c}{eB_c^2}, \qquad \textbf{5.66}$$

and hence

$$\eta = 1 - \frac{U/B^2}{U_c/B_c^2}. \qquad \textbf{5.67}$$

This equation suggests that if the magnetron operates at values U and B, such that the ratio U/B^2 is much smaller than the corresponding ratio of the cut-off values, the efficiency of the magnetron approaches 100 per cent. This result, in spite of the simplicity of the arguments behind it, describes the actual state of affairs with remarkable accuracy; magnetrons have been operated with higher than 90 per cent efficiency. The much lower efficiency limit of other valves is due to the fact that the initial potential energy eU_0 is first converted into kinetic energy, and then most of this is dissipated as heat in the electrode which collects the electrons after their interaction with the high-frequency field.

Considering the cylindrical magnetron, as it is built in practice, but still neglecting space-charge effects, the analysis is complicated only by the fact that the overall motion is a rotating one. What was a drift parallel to the cathode in the previous scheme becomes a rotation about the valve axis, and the motion perpendicular to the cathode becomes a radial motion. Thus the electron trajectories are generated by a point of a circle rolling on the cathode surface in the plane perpendicular to the magnetic field. To obtain the cut-off conditions one must recall the angular momentum theorem as applicable in the presence of a uniform magnetic field. According to equation **1.26** an electron starting at a radial position r_0 with zero initial velocity, and moving in a uniform magnetic field has, at a radial position r, an azimuthal velocity

$$\dot{\theta} = \omega_L\left(1-\frac{r_0^2}{r^2}\right),$$

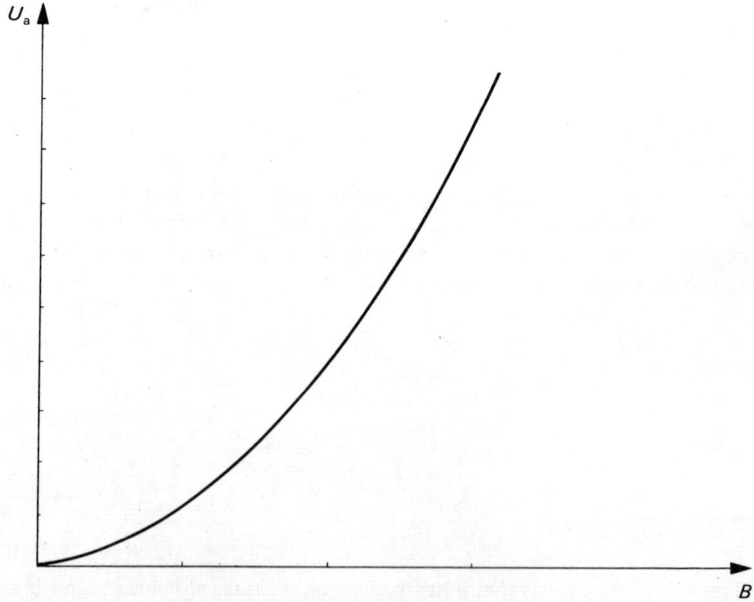

Figure 77

where $\omega_L = \frac{1}{2}(eB/m)$ is the Larmor frequency. Under static conditions (that is, in the absence of high-frequency excitation) the anode current is cut off if the electrons can just reach the anode with grazing incidence. In this case all the kinetic energy arises from the azimuthal motion and therefore energy conservation requires that

$$eU_a = \tfrac{1}{2}m\dot{\theta}^2 R_a^2, \tag{5.68}$$

where U_a is the potential difference between anode and cathode and R_a is the radius of the anode. If the cathode radius is R_c, with the aid of equation **1.26** one obtains from equation **5.68**

$$eU_a = \tfrac{1}{2}m\omega_L R_a^2\left(1-\frac{R_c^2}{R_a^2}\right)^2,$$

and hence
$$U_a = \frac{1}{8}\frac{e}{m}R_a^2\left(1-\frac{R_c^2}{R_a^2}\right)^2 B^2. \tag{5.69}$$

Plotting U_a against B for given geometric parameters R_a and R_c, one obtains a curve which is usually referred to as the 'cut-off parabola' of the static magnetron (see Figure 77). Note that if the cylindrical cathode and anode are unfolded, that is, $R_c, R_a \to \infty$, while keeping $R_a - R_c = d =$ constant, equation **5.69** reduces to

$$U_a = \frac{eB^2}{8m}d^2\left(1+\frac{R_a}{R_c}\right)^2 \to \frac{eB^2}{2m}d^2,$$

which is the cut-off condition for the linear magnetron (see equation **5.66**).

If the cathode can supply a high electron current density, the motion of the electrons will be affected by their mutual interaction. It is reasonable to expect that, because of 'collisions' between them, electrons cannot complete full cycloidal orbits and there will be a stationary state in which the electrons drift in concentric circles about the axis of the valve without a further rotary motion superimposed. Since the static field is conservative, electrons emitted by the cathode with zero initial velocity will have an angular velocity Ω satisfying the relation

$$\Omega = \omega_L\left(1-\frac{R_c^2}{r^2}\right) \tag{5.70}$$

at any orbit radius r. If the orbits are concentric circles of radius r about the valve axis, energy conservation requires

$$e\,U(r) = \tfrac{1}{2}m(r\Omega)^2,$$

where $U(r)$ is the potential at a radial position r; $U(R_c) = 0$. Because of the cylindrical symmetry of the system, the potential is a function of r only, and is obtained by substituting Ω from equation **5.70**:

$$U(r) = \frac{m}{2e}\omega_L^2 r^2\left(1-\frac{R_c^2}{r^2}\right)^2 = \frac{eB^2}{8m}r^2\left(1-\frac{R^2}{r^2}\right)^2. \tag{5.71}$$

This equation means that the potential distribution *at every point within the space-charge region* satisfies a condition which is identical with the *cut-off condition.* If the anode potential and the magnetic field have values satisfying the cut-off condition, the space charge will fill the whole ring-shaped region between cathode and anode. Let such a pair of values be denoted by U_{a0} and B_0. If the magnetron is strongly cut off, $U(R_a) = U_a < U_{a0}$ and $B > B_0$, the space charge will occupy a cylindrical ring between the radii $r = R_c$ and $r = R < R_a$. Within this region, that is, for $R_c \leqslant r \leqslant R$, the space-charge distribution is given by equation **5.70**, and for $R \leqslant r \leqslant R_a$, it is given by the equation valid in free space:

$$U_0(r) = \frac{U_a - U(R)}{\ln(R_a/R)} \ln r + \frac{U(R)\ln R_a - U_a \ln R}{\ln(R_a/R)}. \qquad \textbf{5.72}$$

R is defined by the condition that the potential and the field intensity on the boundary between the space-charge region and free space must be continuous:

$$U(R) = U_0(R) \qquad \text{and} \qquad \left(\frac{dU}{dr}\right)_R = \left(\frac{dU_0}{dr}\right)_R.$$

This condition is readily visualized in the case of the plane magnetron for which, say, $r - R_c = s$, $R - R_c = S$, $R_a - R_c = d$ while R_c, R, $R_a \to \infty$. In this case,

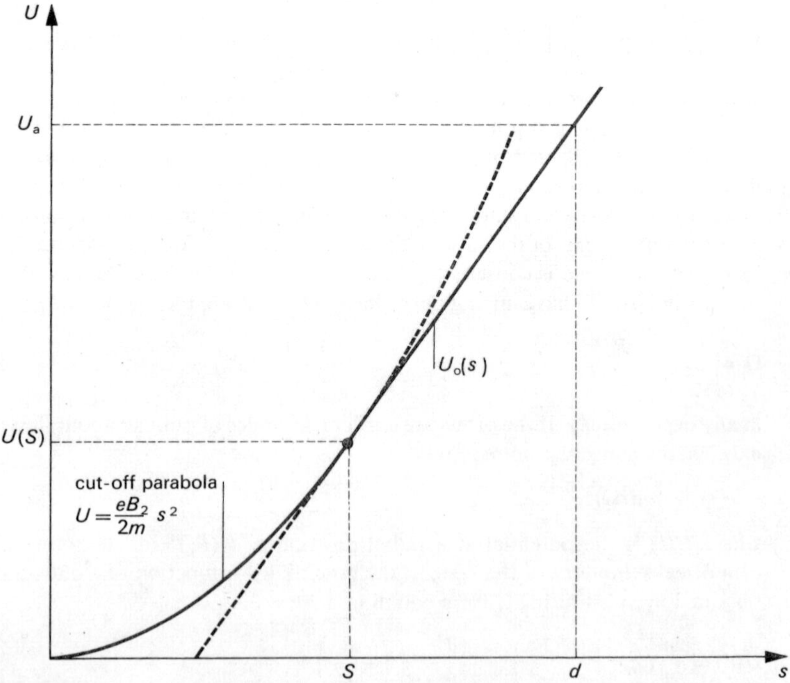

Figure 78

equation **5.71** reduces to

$$U(s) = \frac{eB^2}{2m} S^2 \qquad \textbf{5.71a}$$

and equation **5.72** reduces to

$$U_0(s) = U(S) + \frac{U_a - U(S)}{d - S}. \qquad \textbf{5.72a}$$

Matching the boundary conditions in this case means that the straight line, representing equation **5.72a** in a plot of U against s (see Figure 78), must be a tangent to the parabola representing equation **5.71a**. If required, the space-charge density distribution can be determined from the Poisson equation, $\nabla^2 U(r) = -\rho/\varepsilon_0$, by substituting expression **5.71** for $U(r)$. Remember that in cylindrical polar coordinates

$$\nabla^2 U(r) = \frac{1}{r}\frac{d}{dr}\left(r\frac{dU}{dr}\right).$$

It must be emphasized that the solution of the space-charge problem presented above describes the simplest possible case. From a more general treatment other more complicated solutions are also obtained. The discussion of these goes beyond the scope of our considerations.

Using the results obtained above, one can construct a crude picture of the events in an oscillating magnetron, and of the conditions of oscillation. It has already been mentioned that the magnetron anode is usually a cylindrical block, with N radial recesses at regular intervals. Thus the structure is invariant with respect to a rotation of the anode block about the cathode axis by an angle $2\pi/N$; or in other words, the boundary conditions in the interaction region are identical in 'cells' of the shape of cylindrical sectors of azimuthal aperture $2\pi/N$. The whole interaction region acts as a cavity resonator, and can oscillate in different modes, characterized by different phase shifts between corresponding points of the neighbouring cells. The field, however, must be a single-valued function of the azimuth angle, and hence the phase shift between points at an azimuth angle $2\pi/N$ apart is $\theta_n = n(2\pi/N)$; n is an integer, and is called the 'order of the mode' of oscillation. The field distribution of a few modes of a magnetron cavity is schematically shown in Figure 79. The standing wave pattern of a mode can be looked upon as the superposition of two waves, travelling in opposite directions, and can be described by functions of the type

$$f(r) \exp[i(\omega t \pm n\theta)].$$

Hence the fields are rotating about the axis of the valve with an angular velocity

$$\frac{d\theta}{dt} = \pm\frac{\omega}{n},$$

where ω is the frequency of oscillation in the cavity.

The wave, which rotates in synchronism with the space charge about the

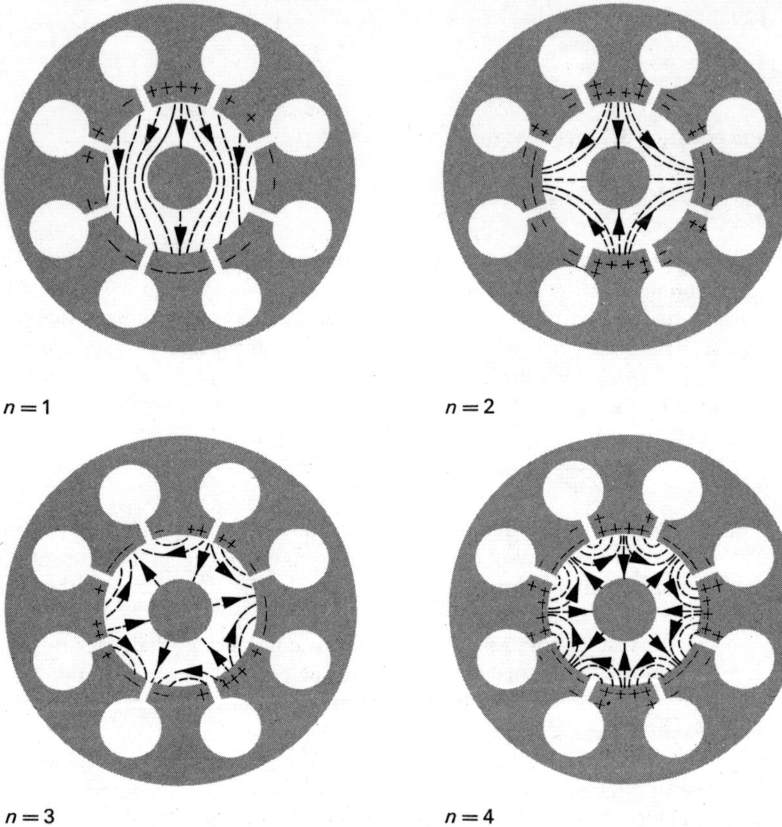

Figure 79

cathode, will have a cumulative effect. As explained before, electrons in a region about a particular phase of the wave will be extracted from the space-charge layer and will move towards the anode. The condition of synchronism, $d\theta/dt = \Omega$, indicates that electrons drift towards the anode, provided

$$\frac{\omega}{n} \geqslant \omega_L\left(1-\frac{R_c^2}{R^2}\right). \qquad 5.73$$

If the equality applies, electrons are extracted from the boundary of the space-charge layer, where the electron orbit radius is R; in the case of inequality, the electrons come from deeper layers where the orbit radius is $r < R$. In practice, the resonant frequency of the magnetron cavity, ω, and the mode of operation n are fixed. To satisfy the condition **5.73**, ω_L must be set by adjusting the magnetic field and the thickness of the space-charge layer, that is, R, must be set by adjusting the anode voltage U_a. The inequality **5.73** suggests that the conditions of oscillation are not critically dependent on the magnetic field and anode voltage. The

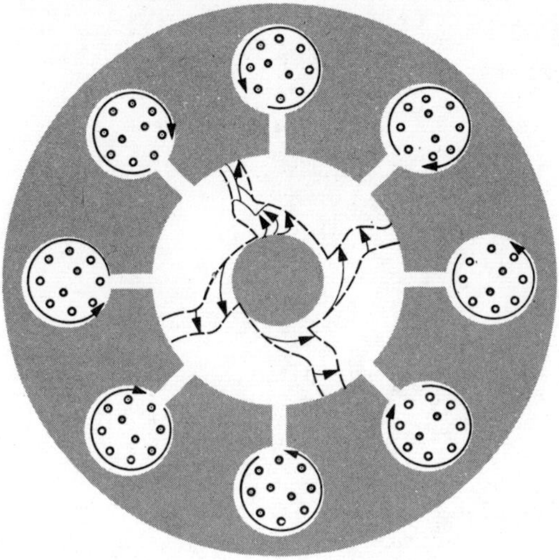

Figure 80

instantaneous shape of the space-charge cloud in the oscillating magnetron looks something like that shown in Figure 80, and this pattern rotates in synchronism with the travelling electromagnetic wave. It should be remembered that, in spite of an appearance reminiscent of a rotating wheel, the energy transferred to the electromagnetic field does not come from the kinetic energy of the rotation, but from the loss of potential energy of the electrons which reach the anode.

5.7 Parametric amplifiers

In all the devices discussed so far, amplification is obtained by virtue of the interaction between a slow space-charge wave and an electromagnetic wave propagating in synchronism. The question arises whether it is possible to use the fast space-charge wave as a means of achieving amplification. In fact this is possible, but the mechanism of the process must be fundamentally different from that discussed so far.

It has been emphasized repeatedly that in the slow-wave devices the amplification arises by the conversion of the kinetic or potential energy of the electrons into electromagnetic energy. In each case the ultimate source of the energy is a *constant* potential difference. To use a mechanical analogy, one may think of a pendulum clock in which the movement of the pendulum is maintained by a weight which loses its potential energy as it drops lower and lower. This method of driving a pendulum is not the only possible one. For example, an entirely different 'mechanism' is involved in the process by which a child works a swing. The child, standing on a swing, bends his knees rhythmically and enjoys 'pumping up' and then maintaining the elongations of the swing. He finds out very quickly that the rhythm of the movement has to be matched to the swing on which he is standing.

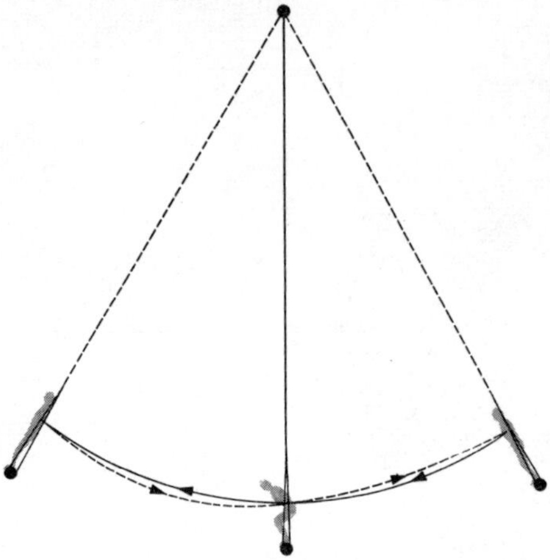

Figure 81

What happens is that by bending his knees, he shifts the centre of gravity of the swing up and down (see Figure 81), and thus the effective length of the pendulum is varied periodically. If the pendulum length is long on its downward swing and short on its upward swing, it will reach ever-increasing heights. The energy increment of the swing is equal to the work done when lifting the centre of gravity of the swing, which takes place *twice* during one complete period of the pendulum motion. Thus the energy necessary to amplify the oscillations is not drawn from a constant source, but from one which operates *periodically* and varies a characteristic parameter of the oscillator. For this reason, the process is called 'parametric' amplification of oscillations.

The mechanism outlined above is suitable for boosting the amplitude of any harmonic oscillator. Consider, for example, a resonant circuit which consists of a fixed inductance L and a parallel-plate capacitor, whose capacitance C can be varied by a small amount ΔC by changing the separation of its plates (see Figure 82). Suppose that $\Delta C \ll C$, so that the frequency of the resonant circuit can be taken to be constant and equal to the unperturbed resonance frequency $\omega_0 = (LC)^{-\frac{1}{2}}$. If the condenser plates are pulled apart when the voltage across the plates is a maximum, work must be done against the force attracting the plates. In return, the capacitance is decreased, and therefore the instantaneous charge on the capacitor results in an increased potential difference between the plates, so that the oscillation continues with an increased amplitude. At the instant when there is no field between the plates, the original separation between them can be restored without performing work. If this process continues cyclically, the amplitude of the oscillation is gradually increased.

It should be noted that the frequency at which the capacitance is varied is twice the resonant frequency of the unperturbed circuit and, in order to obtain the

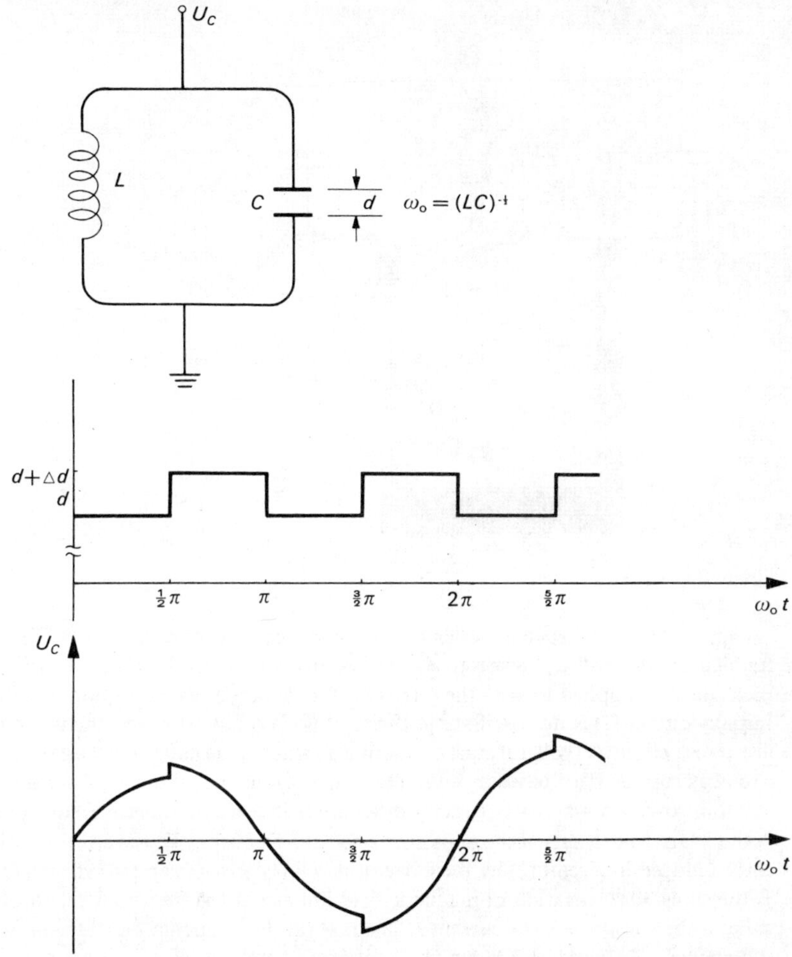

Figure 82

maximum gain, the phase of the potential difference on the condenser and the phase of the variation of the condenser must be related in the way described above. If the phase relation is reversed, the energy stored in the resonant circuit will gradually decrease. In other words, the direction of the energy flow between the resonant circuit and the source (which controls the capacitance as well as the rate of the flow) depends on the phase relation. The restriction on the relative phase can be removed if one has *two* loosely coupled circuits with different resonance frequencies, say ω_1 and ω_2, and the external agency varies the coupling between the circuits at a frequency $\Omega = \omega_1 + \omega_2$.

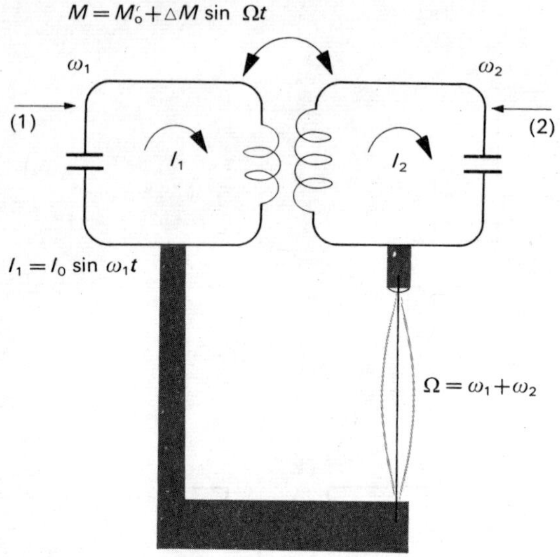

Figure 83

Figure 83 shows such a scheme: two loss-free resonant circuits, tuned to frequencies ω_1 and ω_2 respectively, are coupled inductively, and an external mechanism is applied to vary the relative distance of the two coils. Suppose that initially circuit (2) is not oscillating. If circuit (1) is oscillating, and the two coils are at rest, circuit (2) will not start oscillating, in spite of the coupling, because each circuit is supposed to have an infinitely sharp resonance, and therefore cannot maintain oscillations at frequencies other than its own resonance frequencies. If the relative positions of the coils are varied at a frequency Ω, the electromotive force induced in circuit (2) by the current in coil (1) arises from two effects. One is the sinusoidal variation of magnetic field linkage at the frequency Ω, and the other is the variation of the current I_1 in coil (1) at the frequency ω_1. If the mutual inductance is $M = M_0 + \Delta M \sin \Omega t$, the induced e.m.f., $M(dI_1/dt)$, contains components of the frequencies ω_1, $\Omega+\omega_1 = 2\omega_1+\omega_2$ and $\Omega-\omega_1 = \omega_2$. The first two components are ineffective, but the last one occurs at the resonance frequency of circuit (2), and will excite oscillations in it. The process in circuit (1) is similar, because the varying mutual inductance, and the current I_2 flowing in circuit (2) at the frequency ω_2, produce in circuit (1) an e.m.f. which consists of components of frequencies ω_2, $\Omega+\omega_2 = 2\omega_1+\omega_2$ and $\Omega-\omega_2 = \omega_1$. The phase relations are adjusted automatically, by virtue of Lenz's law, in such a manner that the currents in both circuits oppose the change in the relative position of the coils. In other words, work must be done to maintain the variation of the mutual inductance. In return, the amplitude of the oscillation in *each* circuit will increase steadily, irrespective of the starting phase of the variation of the coupling, provided that the frequency condition $\Omega = \omega_1+\omega_2$ is satisfied.

The quantitative description of coupled resonant systems with time-dependent coupling can best be given in terms of the analysis of the 'normal modes' of vibration of such systems. The development of this theory is, however, outside the scope of this book, and the reader should consult specialist literature (see Further Reading list) for a more extensive treatment.

It must be emphasized that the process described is independent of the details of the nature of the resonant system, which can be electrical, mechanical, or any other kind, always provided that it can be described mathematically as an harmonic oscillator. Furthermore, the mechanism by which the oscillators are coupled can be chosen arbitrarily. The agency which drives the variable coupling is usually called the '*pump*' and Ω is the pump frequency. If the initial oscillation of frequency ω_1 is the '*signal*', the oscillation produced at the frequency ω_2 is a waste-product, and it is usually called the '*idler*'. In the rest of this section, we shall be concerned with a particular example of parametric amplification, in which the resonant system consists of electrons which form a helical shaped beam in a uniform magnetic field.

First of all, it must be realized that electron beams of this kind, which carry fast cyclotron waves, are suitable for the transmission of signals between a generator and a load. It was shown in section 5.4 that cyclotron waves are excited in superposed crossed electric and magnetic fields (see Figure 71), by varying the former at a frequency ω, near the cyclotron frequency ω_c. Consider now a second pair of condenser plates (see Figure 84), which form a part of a passive resonant circuit,

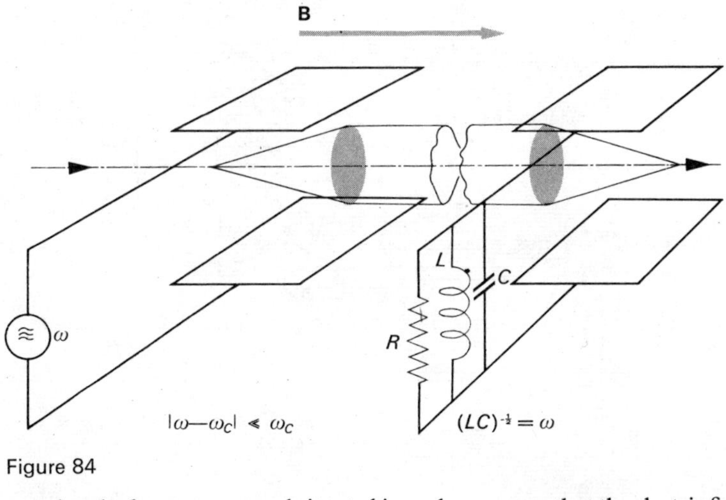

Figure 84

tuned to the frequency ω, and situated in such a manner that the electric field they can support is perpendicular to the magnetic field. If the beam carrying the cyclotron wave passes between these plates, the electrons will induce currents in the external circuit by virtue of the component of their motion which is perpendicular to the plates. Although the individual electrons are orbiting at an angular velocity equal to the cyclotron frequency, the frequency of electron motion as it appears in a fixed plane perpendicular to the beam axis is ω. Therefore the induced current will excite the resonant circuit and, as in the analogous situations discussed previously, the field produced between the condenser plates will oppose the transverse motion of the electrons. Thus the electrons will gradually transfer energy to the resonant circuit at the expense of their rotational energy. If the time spent in the first electric field, which excites the cyclotron waves, is equal to that spent in the second field, where energy is extracted, all the rotational energy is recovered by the output resonant circuit.

In the process described above, there is neither loss nor gain in energy. In order to achieve amplification, the energy stored in the beam in the form of the rotational energy must be increased in the region between the exciting field and the extractor field. This is achieved by increasing the radius of the orbital motion. As shown by equation **5.54**, the radius of the electron orbit is proportional to the amplitude of the field which excites the cyclotron waves, that is, to the amplitude of the input signal. In order to obtain output signal amplitudes proportional to the input amplitudes, it is necessary that in the region where the rotational energy is boosted, the growth of the orbital radius should be proportional to its initial value. This condition is achieved in the 'Adler tube' by letting the orbiting electrons pass through a radio-frequency field which varies at a frequency equal to twice the cyclotron frequency and which has the spatial distribution of a quadrupole field centred on the beam axis (Figure 85).

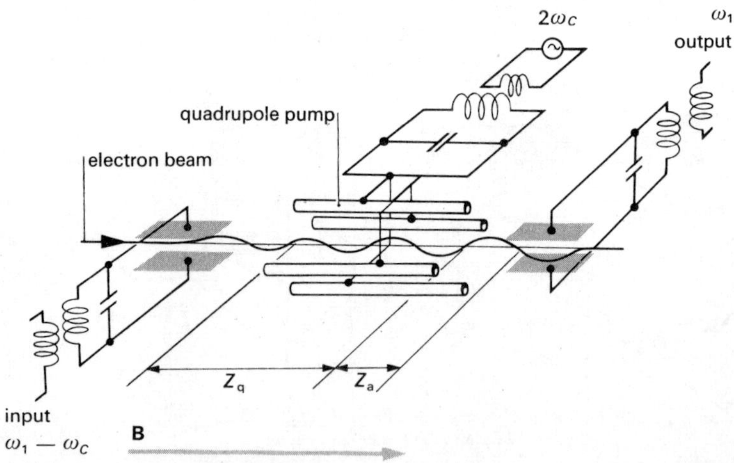

Figure 85

Expressed in the terminology of cyclotron waves, the task is to increase the amplitude of the *fast* waves, and therefore the electromagnetic field must propagate along the beam axis in synchronism with the fast cyclotron wave. It should be recalled that, if the excitation frequency is near the cyclotron frequency, the phase velocity of the fast wave approaches infinity (see equations **5.58**). Thus the electromagnetic wave must also have infinite phase velocity. This is the case if the electromagnetic field is independent of the coordinate measured along the beam axis. Taking a Cartesian coordinate system with its z-axis parallel to the beam axis, and thus parallel to the magnetic field, the required field distribution (see Figure 86) can be derived from a time-dependent potential of the form

$$U(x, y, t) = \frac{1}{2}\frac{E_0}{a}(y^2 - x^2)\sin 2\omega_c t,$$

yielding the field components

$$E_x(t) = \frac{x}{a}E(t) \quad \text{and} \quad E_y(t) = -\frac{y}{a}E(t), \qquad \textbf{5.74a}$$

where

$$E(t) = E_0 \sin 2\omega_c t = -\tfrac{1}{2}\mathrm{i}E_0\{\exp 2\mathrm{i}\omega_c t - \exp(-2\mathrm{i}\omega_c t)\}. \qquad \textbf{5.74b}$$

Let us investigate the motion of the electrons in the quadrupole region. If the magnetic field is $\mathbf{B} = B\mathbf{e}_z$, the equations of motion are

$$\ddot{x} = \frac{e}{m}\frac{x}{a}E(t) + \omega_c \dot{y},$$
$$\ddot{y} = -\frac{e}{m}\frac{y}{a}E(t) - \omega_c \dot{x}. \qquad \textbf{5.75}$$

Consider the (x, y) plane as a complex plane, that is, the position of the electron shall be defined by the complex quantity

$$\mathbf{r} = x + \mathrm{i}y.$$

Thus the two equations **5.75** can be combined to give

$$\mathbf{r} + \mathrm{i}\omega_c \mathbf{r} = \frac{e}{m}\frac{E(t)}{a}\mathbf{r}^*, \qquad \textbf{5.76}$$

where $\mathbf{r}^* = x - \mathrm{i}y$. In the *absence of excitation* of the quadrupole $(E(t) = 0)$ equation **5.76** reduces to $\mathbf{r} + \mathrm{i}\omega_c \mathbf{r} = 0$, the solution of which is

$$\mathbf{r} = r_0 \exp[-\mathrm{i}(\omega_c t + \phi)], \qquad \textbf{5.77}$$

where r_0 and ϕ are real quantities determined by the initial conditions. If the electron beam is excited as described in section 5.4, and enters the quadrupole at a time t_0, the initial conditions are

$$x_0 + \mathrm{i}y_0 = r_0 \exp(-\mathrm{i}\omega t_0),$$

i.e. $\quad \phi = (\omega - \omega_c)t_0,$

where ω is the frequency of excitation. Equation **5.77** shows the expected result that in the absence of excitation on the quadrupole, the electrons continue to rotate with a constant radius, and with an angular velocity equal to the cyclotron frequency.

One can anticipate that in the presence of the alternating quadrupole field the two components described by the two terms in equation **5.74b** have different effects on the electron motion. It is plausible that the component rotating with the angular frequency $-2\omega_c$ will have a cumulative effect on the electrons orbiting at an angular velocity $-\omega_c$ because of the two-fold mirror symmetry of the field. The component rotating at the angular velocity $2\omega_c$ will produce a

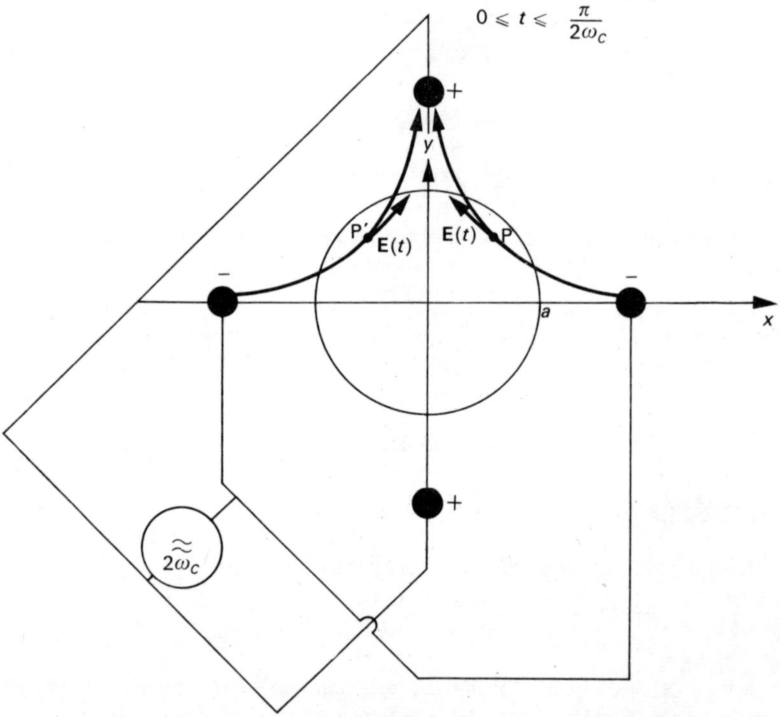

Within the circle of radius a the potential distribution is

$$U(x, y; t) \simeq \frac{E_0}{2a}(y^2 - x^2)\sin 2\omega_C t$$

Figure 86

periodic perturbation only. With reference to Figure 86, one can say that if an electron enters at the position P when the field has a maximum intensity, it will be accelerated as in a cyclotron, and an electron entering, at the same time, at P′ will undergo the reverse process. Therefore the term $\exp(2i\omega_c t)$ in equation **5.74b** will be disregarded. Thus the motion of an electron which enters the quadrupole field at an instant t_0 is described the equation of motion **5.76**, with

$t = t_0+\tau$

and substituting what remains of equation **5.74b**, one finds

$$\dot{r}+\mathrm{i}\omega_c r = \frac{\mathrm{i}}{2}\omega_c \frac{E_0}{aB} r^* \exp[-2\mathrm{i}\omega_c(t+\tau)]. \qquad \textbf{5.78}$$

The considerations below are restricted to the case when the quadrupole is excited with such a small oscillation amplitude that

$$\frac{E_0/B}{2a\omega_c} \ll 1$$

and we look for a solution of the equation of motion **5.78** which is distinguished from the solution **5.77** by a slowly varying amplitude only. Therefore we take

$$r = r_0 \exp(k\tau)\exp[-\mathrm{i}\omega_c(t_0+\tau)]\exp[-\mathrm{i}(\omega-\omega_c)t_0], \qquad \textbf{5.79}$$

where $k \ll \omega_c$. Substitution of equation **5.79** into **5.78** gives an equation for k, and hence one finds, under the conditions stated,

$$k = -k_0 \exp 2\mathrm{i}(\omega-\omega_c)t_0,$$

$$\text{with} \quad k_0 = \frac{1}{2}\frac{E_0}{aB}. \qquad \textbf{5.80}$$

This equation, together with equations **5.79**, indicates that

$$r = \pm r_0 \exp(k_0\tau) \exp[-\mathrm{i}\omega_c(t_0+\tau)] \qquad \text{if } (\omega-\omega_c)t_0 = n\pi,$$

$$r = \pm r_0 \exp(-k_0\tau) \exp[-\mathrm{i}\omega_c(t_0+\tau)] \qquad \text{if } (\omega-\omega_c)t_0 = (n+\tfrac{1}{2})\pi.$$

In other words, the electron is rotating in the quadrupole field with the cyclotron frequency; the radius of the orbit is increasing if the first phase condition applies, but the radius decreases exponentially if the second phase condition is valid. The meaning of these phase conditions is made clear in Figure 86. To obtain the optimum condition for the increase of the orbit radius, the electron must enter at the point P at a time when the field has the maximum intensity, that is, $\omega t_0 = \frac{1}{4}\pi+n_1\pi$ and $2\omega_c t_0 = \frac{1}{2}\pi+2n_2\pi$, and hence $(\omega-\omega_c)t_0 = (n_1-n_2)\pi$. This is the first condition above. Similarly, the second condition corresponds to entry at P′, in which case the electrons suffer the maximum retardation in their orbital rotation.

The variation of the orbit radius of an electron entering under general conditions can be described as the linear superposition of the special solutions, with coefficients which are determined by the initial conditions:

$$r = \{r_1 \exp(k_0\tau)+\mathrm{i}r_2 \exp(-k_0\tau)\}\exp\{-\mathrm{i}\omega_c(t_0+\tau)\}, \qquad \textbf{5.81}$$

where r_1 and r_2 are real, and the initial conditions require that

$$\boldsymbol{r}(0) \equiv r_0 \exp(-\mathrm{i}\omega t_0) = (r_1 + \mathrm{i}r_2)\exp(-\mathrm{i}\omega_c t_0). \qquad \textbf{5.82}$$

Equation **5.81** shows that an electron entering the quadrupole field at an instant t_0 moves with an angular velocity ω_c, and, in general, its motion is the superposition of two components in time quadrature. One of the components is growing and the other is decaying exponentially at the same rate.

From equation **5.82** and its complex conjugate, one can readily express r_1 and $\mathrm{i}r_2$ in terms of r_0 and exponential functions of $\mathrm{i}(\omega-\omega_c)t_0$. Substituting these values into equation **5.81**, one obtains

$$\boldsymbol{r} = r_0\{\cosh k_0\tau \exp[-\mathrm{i}(\omega t_0 + \omega_c\tau)] + \sinh k_0\tau \exp[-\mathrm{i}(2\omega_c - \omega)t_0 - \mathrm{i}\omega_c\tau]\}. \qquad \textbf{5.83}$$

Here one must recall that from the point of view of the amplification of the signal which produces the initial orbital radius r_0, the relevant question is how much energy the beam can transfer to the extractor field. This was shown to depend on the motion of the electrons at the entrance plane to the output 'coupler'. All the electrons which arrive at this plane have spent the same period $\tau = Z_q/u_0$ in the quadrupole field, Z_q being the length of the quadrupole and u_0 the electrons' drift velocity parallel to the beam axis. On leaving the quadrupole, the electrons drift for a period Z_d/u_0 before reaching the output coupler (Z_d = length of the drift space), but during this time their orbit radius stays constant. Therefore the position of the electrons, in the entrance plane of the extractor field, is given by an expression which depends on the radius r_0 and the time t_0 characterizing the entry into the quadrupole field:

$$\frac{\boldsymbol{r}}{r_0} = \exp\left(\frac{-\mathrm{i}\omega_c Z_d}{u_0}\right)\left[\cosh\left(\frac{k_0 Z_q}{u_0}\right)\exp\left\{-\mathrm{i}\left(\omega t_0 + \frac{\omega_c Z_q}{u_0}\right)\right\} + \right.$$
$$\left. + \sinh\left(\frac{k_0 Z_q}{u_0}\right)\exp\left\{-\mathrm{i}(2\omega_c - \omega)t_0 - \mathrm{i}\frac{\omega_c Z_q}{u_0}\right\}\right]. \qquad \textbf{5.84}$$

Thus one finds that the electron beam sweeps the entrance plane of the output coupler in a circle, and the motion of the point of intersection between the beam and this plane consists of two components:

one has a frequency $\omega_1 = \omega$
and the other $\omega_2 = 2\omega_c - \omega$.

Observe that the sum of the two frequencies is $\omega_1 + \omega_2 = \Omega$, which is the relation between the frequencies of the 'signal', 'idler' and 'pump' in a parametric amplifier.

Suppose the plates of the output coupler are perpendicular to the x-axis of our coordinate system, and they form a part of a resonant circuit tuned to the signal frequency ω. The current I induced in the external circuit will be proportional to $x = \mathcal{R}\boldsymbol{r}$:

$$I \propto r_0\left[\cosh\left(\frac{k_0 Z_q}{u_0}\right)\cos(\omega t_0 + \phi) + \sinh\left(\frac{k_0 Z_q}{u_0}\right)\cos\{(2\omega_c - \omega)t_0 + \phi\}\right], \qquad \textbf{5.85}$$

where $\phi = \frac{\omega_c}{u_0}(Z_q + Z_d).$

In the special case when $\omega = \omega_c$, equation **5.85** reduces to

$$I \propto r_0 \exp\left(\frac{k_0 Z_q}{u_0}\right) \cos(\omega t_0 + \phi),$$

and this current excites the output coupler at its resonant frequency. If all the energy of the orbital motion of the electrons is extracted, the amplifier has a gain of $\exp(k_0 Z_q/u_0)$. In general, however, the output signal is excited only by the current represented by the first term in expression **5.85**, and the gain is given by

$$\cosh\left(\frac{k_0 Z_q}{u_0}\right) = \cosh\left(\frac{1}{2}\frac{E_0}{B}\frac{Z_q}{a}\frac{1}{u_0}\right),$$

where the parameter k_0 is written explicitly to show the role of the different physical parameters in the amplification.

The second term in expression **5.85** gives the induced current which produces the idle oscillations. Their amplitude is proportional to $\sinh(k_0 Z_q/u_0)$, but the output coupler will support such oscillations only if the frequency $2\omega_c - \omega$ is sufficiently near to its resonant frequency ω. In this case, one can observe the beats between the amplified signal and the idler. Otherwise the idler remains ineffective.

The question arises as to whether the parametric method of amplification has any *advantage* over more conventional methods. The decisive argument in favour of parametric amplifiers is that their inherent *noise* level is lower than that of the conventional types. We shall look at this problem very briefly in the next section.

5.8 Concluding remarks on electronic amplifiers and oscillators

All electronic amplifiers fall into the two categories discussed in the preceding sections, in the sense that their operation can be interpreted in terms of one of two fundamental processes. One can call amplifiers 'conventional', if the gain arises by the transfer of the kinetic or potential energy of electrons to the alternating electromagnetic field, *and* the source of this energy is a constant potential difference. The other category is that of parametric amplifiers, in which the source of the energy which the electrons can transfer to an alternating field is itself time-dependent. All the amplifiers can be turned into self-excited oscillators by feeding back to the input a fraction of the amplified output in an appropriate phase, provided that the over-all gain around the closed 'feedback loop' is sufficient to cover the losses of the external circuitry. In the final analysis, the output power of amplifiers and oscillators can be identified with the radiation emitted by the electrons, and the distinction between the two categories corresponds to two fundamentally different mechanisms by which electrons can radiate.

5.8.1 It was found convenient to refer to the properties of space-charge waves in the conventional amplifiers operating at high frequencies, and this approach was particularly successful in the interpretation of travelling wave type devices. It was seen that the relevant process was the interaction between a 'slow' space-charge wave and an electromagnetic wave of matching phase velocity. Since the energy transferred to the electromagnetic wave is usually taken from the kinetic energy of the electrons, one is inclined to look at the process as akin to the radiation of decelerating electrons, for example, bremsstrahlung. This analogy, however, cannot be correct. In order to maintain the growth of the electromagnetic wave of constant phase velocity in a travelling wave type tube (including the magnetron) the electrons in the beam must be trapped in the neighbourhood of a particular phase of the electromagnetic wave. This implies that the mean kinetic energy of the electrons travelling in bunches must stay constant.

The process in which electrons of essentially constant kinetic energy radiate is known as Cherenkov radiation. This phenomenon is usually discussed in the context of the emission of light by charged particles moving in an optically dispersive medium, but there is no reason to restrict the considerations to the visible range of the spectrum of electromagnetic radiation. In complete generality, one can say that the condition for the production of Cherenkov radiation is that electrons should travel in a medium with a higher velocity than the phase velocity of the radiation in that medium.

A development of the analogy between the processes in a travelling wave amplifier and Cherenkov radiation would go beyond the scope of this book. Suffice it to say that the analogy hinges on the recognition that the slow wave structures required in travelling wave type amplifiers, such as helical lines, periodically loaded wave-guides, etc., can and should be considered as dispersive media, simply because, as in optical media, the phase velocity of the radiation is a function of its frequency. It must be mentioned that there are significant differences between the dispersive properties of an infinite uniform medium, which is usually considered in the discussion of Cherenkov radiation as an optical phenomenon, and the dispersive properties of, say, a periodically loaded wave-guide. The first one supports a continuous spectrum of modes of vibration, the latter has a discrete spectrum of modes. As a result the Cherenkov radiation, under conventional circumstances, has a continuous spectrum, but the radiation produced by the same mechanism in a slow wave structure used in travelling wave type valves, has a discrete spectrum.

Once the mechanism of Cherenkov radiation is considered responsible for the gain produced in electronic amplifiers, it becomes reasonable to make a deliberate attempt to use this process for the production of electromagnetic radiation in the spectral range of practical interest, such as microwaves, or the range between these and infra-red radiation. Although several schemes of this kind were shown to be feasible, they have not achieved great practical importance, mainly because of the advent of 'quantum-electronic' devices.

5.8.2 The operation of parametric amplifiers cannot be understood by seeking an analogy with Cherenkov radiation, or with bremsstrahlung for that matter. Parametric amplification requires a system which is capable of performing harmonic oscillations at two frequencies, say ω_1 and ω_2, and in order to increase the amplitude of oscillation at the frequency ω_1, energy must be supplied to the system in the form of oscillations at a frequency Ω which satisfies the condition

$$\Omega = \omega_1 + \omega_2. \qquad \textbf{5.86}$$

As a by-product of the process, the energy available at the frequency ω_2 will also increase. Putting it in different terms, one can say that the system under consideration has the property of being able to *absorb* energy from an oscillatory source at the frequency Ω, and to store or to re-emit it as oscillatory energy at the frequencies ω_1 and ω_2, provided that the three frequencies satisfy equation **5.86**.

The 'sum rule' (equation **5.86**) is identical with the 'Ritz combination principle' for atomic line spectra. In the light of quantum theory, this relationship is seen to be an energy relation. If a system has three stationary energy states with energies

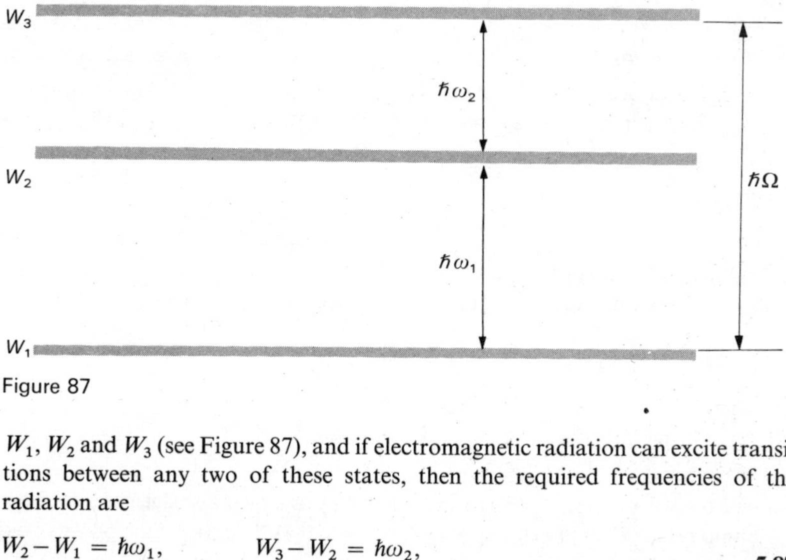

Figure 87

W_1, W_2 and W_3 (see Figure 87), and if electromagnetic radiation can excite transitions between any two of these states, then the required frequencies of the radiation are

$$W_2 - W_1 = \hbar\omega_1, \qquad W_3 - W_2 = \hbar\omega_2,$$
$$W_3 - W_1 = \hbar\Omega = \hbar(\omega_1 + \omega_2), \qquad \textbf{5.87}$$

where h is Planck's constant and $\hbar = h/2\pi$. The last of the equations **5.87** contains equation **5.86**.

The analogy between a quantized system with three stationary states and a parametric amplifier is a fundamental one, and it is not mere coincidence that the energy relation applies to parametric amplifiers. Harmonic oscillators, as any other physical system, are always subject to the rules of quantum mechanics, irrespective of whether they have atomic dimensions or they are large on such a scale. The difference is that in the former case, their dominant features can be described only in quantum-mechanical terms, whereas in the latter case, classical (non-quantum-mechanical) description gives adequate results. This is the essence of the 'correspondence principle'. It happens, only rarely, that calling upon quantum-mechanical arguments helps in elucidating the analysis of macroscopic systems. Certain properties of parametric amplifiers present a case when this does happen.

Let us consider, as an example, the power relation between the signal and the idler in a parametric amplifier. If the parametric amplifier is considered as a quantized system with three stationary states, the energy stored in the system will be constant if the population of each state is constant. With reference to the population of level (2), this implies that the *net* rate of transition between levels (3) and (2), and the net rate of transition between (2) and (1) totals zero.

$$\frac{\partial N_{32}}{\partial t}+\frac{\partial N_{21}}{\partial t} = 0.$$

This equation may be written as

$$\frac{\hbar\omega_2}{\omega_2}\frac{\partial N_{32}}{\partial t}+\frac{\hbar\omega_1}{\omega_1}\frac{\partial N_{21}}{\partial t} = 0. \qquad \textbf{5.88}$$

Now $\hbar\omega_2(\partial N_{32}/\partial t) = -P_2$ is the energy emitted in unit time in the form of radiation of frequency ω_2, and $\hbar\omega_1(\partial N_{21}/\partial t) = -P_1$ has a similar meaning. Thus equation **5.88** relates the power radiated at the respective frequencies:

$$\frac{P_1}{\omega_1}+\frac{P_2}{\omega_2} = 0. \qquad \textbf{5.89}$$

The same argument leads to similar relations with respect to any pair of ratios P_j/ω_j. With reference to level (1), for example, one finds

$$\frac{P_p}{\Omega}+\frac{P_1}{\omega_1} = 0$$

or $$-P_1 = \frac{\omega_1}{\Omega}P_p,$$

which gives the signal power available (hence the negative sign) in return for the pump power P_p supplied to the system. Note that the ratio of the powers is equal to the ratio of the corresponding frequencies.

Equation **5.89** is known as the Manley–Rowe relation, and was originally derived as an energy transformation relation in non-linear reactive networks. The derivation by classical arguments is much more laborious.

Looking at parametric amplifiers as quantized systems is a first step in building a bridge which leads to the domain of 'quantum-electronic' amplifiers and oscillators, that is, masers and lasers. The description of these devices requires quantum-mechanical considerations, because the processes involved in their operations are transitions between stationary states of genuine quantized systems, such as atoms, molecules or solids. A classical (non-quantum-mechanical) approach to the description of these devices is totally inadequate, because the electrons responsible for the maser or laser action cannot be considered free to any permissible degree of approximation. Since we set out to study problems which can be interpreted in terms of the behaviour of free electrons, the treatment of quantum-electronic devices is outside the scope of this book.

5.8.3 There is one feature which makes the performance of parametric amplifiers superior to that of conventional amplifiers. This becomes apparent if one studies the inherent 'noise' of amplifiers. The difference in the noise properties is a direct consequence of the fundamentally different mechanism of operation. Without attempting to analyse the noise of amplifiers in quantitative terms, one can see from qualitative arguments that the contribution of the electron beam to the noise observed at the output is smaller in parametric amplifiers than in conventional ones.

In all our considerations, it was assumed that the initial conditions under which electrons are emitted by a source are completely under our control, and the initial parameters are identical for all the electrons which form a beam. This assumption, however, is not strictly valid. Electron emission from any source is a random process, and the effect of this manifests itself in two ways. The number of electrons emitted by any source in unit time shows random fluctuations, and the initial velocity of the electrons has no sharply defined value, either in magnitude or in direction. As a result, the current density measured in a beam shows random fluctuations as well, and this is what one means by saying that an electron beam is inherently 'noisy'. To express this in the terminology of space-charge (or cyclotron) waves, one can say that as a result of the random fluctuations of the emission process, and the spread in the initial electron velocities, an electron beam always carries an infinitely wide spectrum of space-charge and cyclotron waves. The amplitudes of these waves are the same over the whole spectrum, their phases are distributed at random, and slow and fast waves are equally present. Suppose such a noisy electron beam interacts with a resonant cavity. The waves whose frequencies coincide with the resonance frequency will excite the cavity, and in this process the slow-wave amplitude will grow and the fast-wave amplitude will be attenuated, as explained in section 5.4. Similarly, if the interaction takes place between the beam and a structure capable of supporting travelling waves of a given phase velocity, the slow wave of this phase velocity will grow and the fast wave of the same velocity will decay, as energy is transferred from space-charge wave to the electromagnetic field. In other words, the interaction with the electromagnetic field supported by the external circuit leads to an increase of the noise carried in the shape of slow waves, and to a decrease of the noise carried by fast waves.

One should recall here that in a conventional amplifier the input signal gives rise to both slow and fast space-charge waves of equal amplitude, but the amplification is the result of the growth of the slow waves. Therefore the initial noise of the beam and the signal are necessarily amplified together, that is, the signal-to-noise ratio at the output will not be more favourable than that at the input. In contrast to this situation, the input coupler of a parametric amplifier, as explained above, acts as a sink for the power carried by the fast waves. Thus, while it excites cyclotron waves of a definite phase, representing a signal, it attenuates the noise. Since the amplification takes place by virtue of the interaction between the quadrupole field and the fast cyclotron waves, and the input coupler effectively strips the beam of the noise beforehand, the signal-to-noise ratio at the output is significantly higher than the ratio of the input signal was to the initial beam noise. In other words, the limit set to the minimum detectable signal by the inherent noise is significantly lower in the case of parametric amplifiers than in the case of conventional ones.

Further reading

K. R. Spangenberg, *Vacuum Tubes*, McGraw-Hill, 1948.

The mechanism of the operation of low-frequency thermionic valves is discussed in an easily readable style. The role of transit-time effects is considered. The mechanism of klystron and reflex klystron operation are treated mainly in terms of the kinematic description of velocity modulation and the resulting charge-density modulation. The behaviour of thermionic valves as linear network elements is related to the mechanism of their operation.

W. W. Harman, *Fundamentals of Electron Motion,* McGraw-Hill, 1953.

The concept of space-charge waves is introduced and used as a basis for the analysis of the operation of klystrons, travelling-wave tubes and magnetrons.

A. H. W. Beck, *Thermionic Valves*, Cambridge University Press, 1953.

A monograph which gives a thorough theoretical account of the behaviour of thermionic high-vacuum devices, with strong emphasis on the phenomena relevant at high (microwave) frequencies.

J. C. Slater, *Microwave Electronics* (see references on p. 178).

Chapter 9 deals with the general principles of electronic devices; Chapters 10, 12 and 13 are devoted to microwave amplifiers and oscillators. The results of the analysis of electron motion in radio-frequency electromagnetic fields are 'translated' into the language of network analysis.

W. H. Louisell, *Coupled Mode and Parametric Electronics*, Wiley, 1960.

The theory of coupled modes of oscillations and of coupled modes of wave propagation is developed at an advanced level. This provides the formalism by means of which all kinds of travelling-wave tubes, parametric amplifiers and oscillators are analysed following a unified point of view.

Appendix

A.1 Simple field distributions

A.1.1 *Potential distribution near a plane of mirror symmetry*

Let the $y = 0$ plane of a Cartesian coordinate system be a plane of mirror symmetry (see Figure 88). It is assumed that the electrodes are infinitely long, and

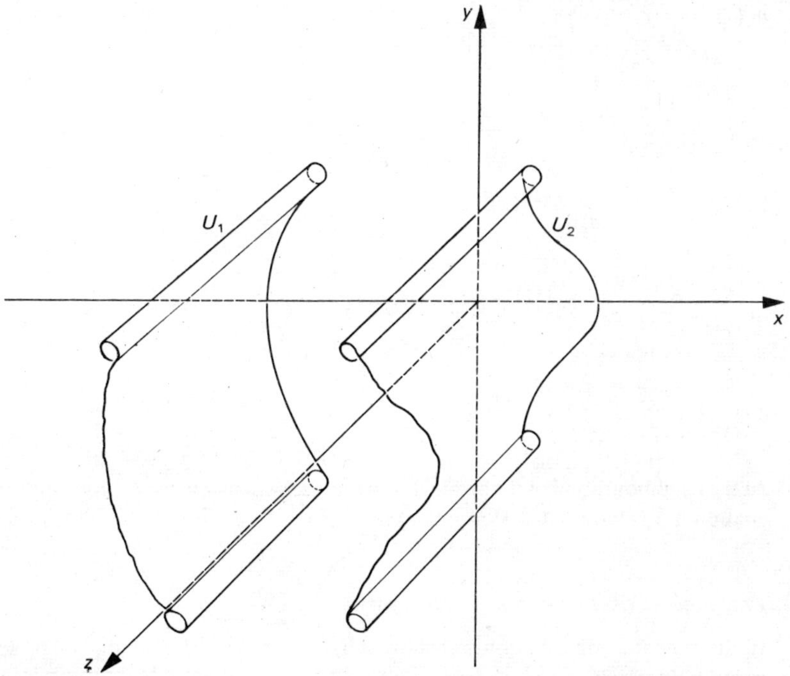

Figure 88

hence the potential distribution is independent of z, furthermore there is no space charge present. In this case $\Phi(x, -y) = \Phi(x, y)$ and therefore

$$\Phi(x, y) = \sum_{n=0}^{\infty} a_n(x) y^{2n}, \qquad \textbf{A.1}$$

subject to the condition $\nabla^2\Phi = 0$ (Laplace equation).

Hence $a_n''(x)y^{2n}+2n(2n-1)a_n(x)y^{2(n-1)} = 0,$ **A.2**

where $a'' = \partial^2 a/\partial x^2$.

Rearranging equation **A.2** according to powers of y, the equation holds valid for any arbitrary value of y if the coefficient of each term vanishes. This condition gives the recurrence relation

$$a_{n-1}''(x) = -2n(2n-1)a_n(x) \qquad (n = 1, 2, \ldots). \tag{A.3}$$

From equation **A.1** $a_0(x) = \Phi(x, 0) \equiv U(x)$

is the potential at a point x on the symmetry plane, and relation **A.3** yields

$$a_1(x) = -\frac{a_0''(x)}{2\times 1} = -\frac{U''(x)}{2\times 1},$$

$$a_2(x) = \frac{U^{(4)}(x)}{4\times 3\times 2\times 1},$$

.

.

.

$$a_n(x) = (-1)^n \frac{1}{(2n)!} U^{(2n)}(x),$$

where $U^{(2n)} = \dfrac{\partial^{2n}U}{\partial x^{2n}}.$

Thus one obtains

$$\Phi(x, y) = \sum_{n=0}^{\infty} (-1)^n \frac{U^{(2n)}(x)}{(2n)!} y^{2n}. \tag{A.4}$$

In the neighbourhood of the symmetry plane, the expansion gives a good approximation if it is terminated after a few terms.

A.1.2 *Potential distribution of rotationally symmetric fields*

If the potential distribution is rotationally symmetrical about the z-axis of a coordinate system, the choice of cylindrical polar coordinates is appropriate, and the potential becomes independent of the azimuth angle θ:

$$\Phi(\rho, z) = \sum_{n=0}^{\infty} a_n(z)\rho^{2n},$$

subject to the condition

$$\nabla^2\Phi = \frac{1}{\rho}\frac{\partial}{\partial\rho}\left(\rho\frac{\partial\Phi}{\partial\rho}\right)+\frac{\partial^2\Phi}{\partial z^2} = 0.$$

By the same arguments as before, one finds

$$\Phi(\rho, z) = \sum_{n=0}^{\infty} (-1)^n \frac{U^{(2n)}(z)}{(n!)^2} (\tfrac{1}{2}\rho)^{2n}, \qquad \text{A.5}$$

$U^{(2n)}(z) = \frac{\partial^{2n} U}{\partial z^{2n}}$

where $U(z)$ is the potential on the symmetry axis.

A.1.3 *Vector potential of rotationally symmetric fields*

The vector potential of a rotationally symmetrical magnetic field can be determined by taking account of the condition that such fields are generated by rotationally symmetrical current distributions. Therefore the current density, as well as the vector potential, has azimuthal components only. In most cases of practical interest, one is concerned with regions about the symmetry axis, where there are no currents (and no magnetic materials) present. In this case the magnetic field can be derived from a scalar potential $\phi_m(\rho, z)$:

$$\mathbf{B}(\rho, z) = -\nabla\phi_m(\rho, z) = -\left[\frac{\partial \phi_m}{\partial \rho}\mathbf{e}_\rho + \frac{\partial \phi_m}{\partial z}\mathbf{e}_z\right].$$

Due to the rotational symmetry, $\phi_m(\rho, z)$ can be expanded in a power series of the form given by equation **A.5**, and therefore

$$B_z(\rho, z) = \sum_{n=0}^{\infty} (-1)^n \frac{B_0^{(2n)}(z)}{(n!)^2} (\tfrac{1}{2}\rho)^2, \qquad \text{A.6}$$

because $\quad B_0(z) \equiv B_z(0, z) = -\left(\frac{\partial \phi_m}{\partial z}\right)_{\rho=0}$

Similarly $\quad B_\rho(\rho, z) = \sum_{n=0}^{\infty} (-1)^n \frac{B_0^{(2n-1)}(z)}{n!(n-1)!} (\tfrac{1}{2}\rho)^{2n-1}. \qquad$ **A.7**

In a region where there are no currents, or no magnetic materials, the vector potential can be expressed in terms of the flux density on the axis, $B_0(z)$. The only non-vanishing component of the vector potential, A_θ, is independent of the azimuth angle, and therefore one obtains from Stokes's theorem:

$$\int_0^{2\pi} A_\theta \rho \, d\theta = \int_0^{\rho}\int_0^{2\pi} (\nabla \times \mathbf{A})_z r \, dr \, d\theta = 2\pi \int_0^{\rho} B_z r \, dr$$

and hence $\quad A_\theta = \frac{1}{\rho}\int_0^{\rho} B_z r \, dr.$

Substituting B_z from equation **A.6**, the integration yields

$$A_\theta = \sum_{n=0}^{\infty} (-1)^n \frac{B_0^{(2n)}(z)}{n!(n+1)!} (\tfrac{1}{2}\rho)^{2n+1}. \qquad \text{A.8}$$

A.2 Some properties of the matrix relating object and image space in geometrical electron optics ('transfer matrix')

First of all, it must be noted that a general linear homogeneous second-order differential equation,

$$\frac{d^2\rho}{dz^2}+a(z)\frac{d\rho}{dz}+b(z)\rho = 0$$

can always be brought to the form

$$\frac{d}{dz}\left[A(z)\frac{d\rho}{dz}\right]+B(z)\rho = 0 \qquad \textbf{A.9}$$

by multiplying the previous equation by $A(z) \equiv \exp(\int a\, dz)$, and denoting $b(z)A(z) = B(z)$.
Let us denote the column vector with $\rho'(z) = d\rho/dz$

$$\begin{bmatrix}\rho(z)\\ A(z)\rho'(z)\end{bmatrix} = \mathbf{R}(z),$$

and show that the transformation matrix $\mathrm{N}(z)$ in the equation

$$\mathbf{R}(z) = \mathrm{N}\,\mathbf{R}(0) \qquad \textbf{A.10}$$

has the property: $\det \mathrm{N} = 1$. From equation **A.10**, it follows that if N^{-1} is the inverse of N

then $\qquad \mathbf{R}(0) = \mathrm{N}^{-1}\mathbf{R}(z)$

and $\qquad \frac{d}{dz}\mathbf{R}(z) = \left(\frac{d}{dz}\mathrm{N}\right)\mathrm{N}^{-1}\mathbf{R}(z) = \frac{d}{dz}(\log \mathrm{N})\mathbf{R}(z). \qquad$ **A.11**

Note that if the operator L is defined as $\mathrm{L} = \log \mathrm{N}$, then

$$\mathrm{N} = \exp \mathrm{L} = \lim_{n\to\infty}\left(\mathrm{I}+\frac{1}{n}\mathrm{L}\right)^n,$$

where I is the unit matrix.
On account of the differential equation **A.9** we can also write

$$\frac{d}{dz}\mathbf{R}(z) = \begin{bmatrix}0 & \dfrac{1}{A(z)}\\ -B(z) & 0\end{bmatrix}\mathbf{R}(z).$$

Comparison with equation **A.11** gives

$$\mathrm{L} = \log \mathrm{N} = \begin{bmatrix}0 & \displaystyle\int\frac{dz}{A(z)}\\ -\displaystyle\int B(z)\,dz & 0^{+}\end{bmatrix}.$$

Note that

$$\mathrm{tr}\,\mathrm{L} = 0.$$

Therefore $\det N = \det(\exp L) = \det\left[\lim_{n\to\infty}\left(I+\frac{1}{n}L\right)^n\right]$

$$= \lim_{n\to\infty}\left[\det\left(I+\frac{1}{n}L\right)\right]^n$$

$$= \lim_{n\to\infty}\left[I+\frac{1}{n}\,\mathrm{tr}\,L+\frac{1}{n^2}\det L\right]^n$$

$$= \exp(\mathrm{tr}\,L)$$

and hence $\det N = 1.$

A.3 The Lagrange invariant of geometrical electron optics

In order to derive the Lagrange invariant, some definitions and simple results of matrix algebra must be recalled.

(a) If one defines

$$\mathbf{r}_i = \begin{bmatrix} u_i \\ v_i \end{bmatrix} \qquad \tilde{\mathbf{r}}_i = [u_i \quad v_i]$$

and A is a quadratic matrix, the transpose of which is denoted by $\tilde{A}$, then $A\mathbf{r}_i$ is a column vector and $\widetilde{A\mathbf{r}_i} = \tilde{\mathbf{r}}_i\tilde{A}$ is a row vector.

(b) If $\mathbf{r}_1$ and $\mathbf{r}_2$ are two vectors, of components (u_1, v_1) and (u_2, v_2) respectively, their scalar product is, by definition,

$$\mathbf{r}_1 . \mathbf{r}_2 = \tilde{\mathbf{r}}_1 \mathbf{r}_2.$$

(c) Given the matrix

$$S = \begin{bmatrix} 0 & -1 \\ 1 & 0 \end{bmatrix},$$

the modulus of a vector product is

$$|\mathbf{r}_1 \times \mathbf{r}_2| = \widetilde{S\mathbf{r}_1}\,\mathbf{r}_2 = \tilde{\mathbf{r}}_1 \tilde{S}\mathbf{r}_2 = \tilde{\mathbf{r}}_1(-S)\mathbf{r}_2 = -\tilde{\mathbf{r}}_1 S\mathbf{r}_2$$

and $\quad |\mathbf{r}_2 \times \mathbf{r}_1| = -|\mathbf{r}_1 \times \mathbf{r}_2|.$

(d) If A is a quadratic matrix, then direct calculation shows that

$$\tilde{A}SA = (\det A)S.$$

Let us consider now two vectors corresponding to two trajectories in the same field. If one defines

$$\mathbf{R}_i(z) = \begin{bmatrix} \rho_i(z) \\ V^{\frac{1}{2}}(z)\rho_i'(z) \end{bmatrix} \qquad (i = 1, 2) \qquad \mathbf{A.12}$$

then, according to equation **2.32**, these vectors satisfy the relation

$$\mathbf{R}_i(z) = \mathrm{N}\mathbf{R}_i(0)$$

where det N = 1.

For the vector product $|\mathbf{R}_2(z)\times\mathbf{R}_1(z)|$, one finds

$$|\mathbf{R}_2(z)\times\mathbf{R}_1(z)| = \tilde{\mathbf{R}}_1(z)\mathrm{S}\mathbf{R}_2(z) = \widetilde{\mathrm{N}\mathbf{R}_1}(0)\mathrm{SN}\mathbf{R}_2(0) =$$

$$\tilde{\mathbf{R}}_1(z)\tilde{\mathrm{N}}\mathrm{SN}\mathbf{R}_2(0) = \tilde{\mathbf{R}}_1(0)\mathrm{S}\mathbf{R}_2(0) = |\mathbf{R}_2(0)\times\mathbf{R}_1(0)|.$$

Since the vectors $\mathbf{R}_1(0)$ and $\mathbf{R}_2(0)$ are defined by the initial conditions, the result obtained above means that

$$|\mathbf{R}_2(z)\times\mathbf{R}_1(z)| = \text{constant.}$$

By substituting equation **A.12**, one finds explicitly

$$|\mathbf{R}_2(z)\times\mathbf{R}_1(z)| = V^{\frac{1}{2}}[\rho_1'(z)\rho_2(z)-\rho_1(z)\rho_2'(z)] = \text{constant.} \qquad \textbf{2.40a}$$

A.4 The effect of finite angular momentum on image formation

In section 2.3 the imaging properties of rotationally symmetric fields were discussed under the assumption that the canonical momentum component $\pi_\theta \equiv C = 0$. Let us consider the effect of lifting this restriction.

Equation **2.28** shows that removing the restriction $C = 0$ changes the angular velocity of the electron. An observer at rest in the meridian plane, itself rotating at the rate $\theta' = -(e/8mV)^{\frac{1}{2}}B_0$, will see the electron rotating about the z-axis at a rate

$$\theta' = C(2emV)^{-\frac{1}{2}}\rho^{-2}. \qquad \textbf{A.13}$$

If one considers the plane perpendicular to the z-axis as a complex plane, one may describe the position of the electron relative to the meridian plane, by a complex vector $\rho\exp(\mathrm{i}\theta)$, where $\rho = \rho(z)$, and $\theta = \theta(C, z)$ is a function of z, as well as the parameter C. By necessity, $\theta(0, z) = \text{constant}$, that is, θ' vanishes if $C = 0$, since, in this case, the electron stays in the meridian plane. Introducing the notation

$$\boldsymbol{w} = \rho\exp(\mathrm{i}\theta),$$

one expects this function to satisfy an equation of the type **2.31**:

$$\frac{d}{dz}(V^{\frac{1}{2}}\boldsymbol{w}') = -P(z)\boldsymbol{w}, \qquad \textbf{A.14}$$

since, for C = 0, the differential equation for $\boldsymbol{w}$ must reduce to equation **2.31**. The validity of equation **A.14** can be proved by carrying out the differentiations. Since $\boldsymbol{w}$ is a complex function, the resulting equation will contain real and imaginary terms. The equation for the imaginary terms, after substituting equation **A.13** for θ', gives an identity. The equation for the real terms leads to the following:

$$\frac{d}{dz}(V^{\frac{1}{2}}\rho') = -\left[P(z)-(2emV)^{-\frac{1}{2}}\left(\frac{C}{\rho^2}\right)^2\right]\rho,$$

in agreement with equation **2.31**. It follows from equation **A.14** that the complex function $\boldsymbol{w}$ satisfies the relation

$$\begin{bmatrix} \boldsymbol{w}(z) \\ V^{\frac{1}{2}}(z)\boldsymbol{w}'(z) \end{bmatrix} = \mathrm{N} \begin{bmatrix} \boldsymbol{w}(0) \\ V_0^{\frac{1}{2}}\boldsymbol{w}'(0) \end{bmatrix}$$

with the same matrix N as in equation **2.31**.

The trajectories of electrons which originate at the same point of the plane $z = -z_1$ in a field-free region of potential V_1 and pass through a rotationally symmetrical field of potential $V(z)$, $0 \leqslant z \leqslant Z$, into a field-free region of potential V_2, satisfy the equation

$$\mathbf{U}(z_2) = \overline{\mathrm{N}}\mathbf{U}(-z_1), \qquad \textbf{A.15}$$

where $\overline{\mathrm{N}}$ denotes the matrix defined by equation **2.33**, and

$$\mathbf{U}(z) = \begin{bmatrix} \boldsymbol{w}(z) \\ V^{\frac{1}{2}}(z)\boldsymbol{w}'(z) \end{bmatrix}.$$

Equation **A.15** is valid for any arbitrary value of the constant C.

Since $\boldsymbol{w}(z)$ is complex, $\mathbf{U}(z)$ has a real and an imaginary component:

$$\mathbf{U}(z) = \mathbf{U}_1(z)+\mathrm{i}\mathbf{U}_2(z),$$

and one finds by substituting $\boldsymbol{w}(z) = \rho(z)\exp[\mathrm{i}\theta(z)]$ that

$$\mathbf{U}_1(z) = \begin{bmatrix} \rho\cos\theta \\ V^{\frac{1}{2}}(\rho'\cos\theta-\rho\theta'\sin\theta) \end{bmatrix},$$

$$\mathbf{U}_2(z) = \begin{bmatrix} \rho\sin\theta \\ V^{\frac{1}{2}}(\rho'\sin\theta+\rho\theta'\cos\theta) \end{bmatrix},$$

and equation **A.15** must be valid for both vectors, $\mathbf{U}_1$ and $\mathbf{U}_2$, separately.

Without loss of the generality, one can choose the meridian plane, with reference to which the electron trajectories are described, to contain the source point, that is, $\theta(-z_1) = 0$. Hence

$$\mathbf{U}_1(-z_1) = \begin{bmatrix} \rho(-z_1) \\ V_1^{\frac{1}{2}}\rho'(-z_1) \end{bmatrix},$$

$$\mathbf{U}_2(-z_1) = \begin{bmatrix} 0 \\ V_1^{\frac{1}{2}}\rho(-z_1)\,\theta'(-z_1) \end{bmatrix}.$$

Thus, if the elements of the matrix $\overline{\mathrm{N}}$ are denoted by N_{jk}, the relation

$$\mathbf{U}_2(z_2) = \overline{\mathrm{N}}\mathbf{U}_2(-z_1)$$

gives

$$\rho(z_2)\sin\theta'(z_2) = N_{12}V_1^{\frac{1}{2}}\rho(-z_1)\,\theta'(-z_1).$$

If the planes $z = -z_1$ and $z = z_2$ are conjugate, then $N_{12} = 0$, and therefore $\sin\theta(z_2) = 0$ (since $\rho(z_2) \neq 0$). This means that if the source point is on the line of intersection of the object plane and the meridian plane, the image will also be located on the intersection of the image plane and the meridian plane, although the trajectory does not lie in the meridian plane. Under such circumstances the relation $\mathbf{U}_1(z_2) = \overline{\mathbf{N}}\,\mathbf{U}_1(-z_1)$ reduces to

$$\begin{bmatrix} \rho(z_2) \\ V_2^{\frac{1}{2}}\,\rho'(z_2) \end{bmatrix} = \overline{\mathbf{N}} \begin{bmatrix} \rho(-z_1) \\ V_1^{\frac{1}{2}}\,\rho'(-z_1) \end{bmatrix},$$

which is identical with the relation found in the case $C = 0$. Hence we conclude that all the imaging laws established in section 2.3, are valid, independently of the value of the constant C.

A.5 Hamiltonian equations of trajectories

For the study of electron trajectories adjacent to the equilibrium orbit in static electromagnetic fields, the differential equations of the trajectories will be developed in the Hamiltonian form. Although this can be done in even greater generality, the considerations below are specialized in two respects. Firstly, the treatment is non-relativistic, and secondly, the fields under consideration are restricted by imposing the conditions that they are symmetrical with respect to the plane $z = 0$, and rotationally symmetrical about the z-axis of a coordinate system.

The symmetry conditions suggest that the trajectories shall be described in terms of cylindrical polar coordinates, r, θ, z. In such a coordinate system, the electric field can be derived from a scalar potential which is independent of the azimuth angle θ, and has the symmetry properties $\Phi(r, z) = \Phi(r, -z)$. The magnetic field can be derived from a vector potential, which has only an azimuthal component, A_θ, independent of the azimuthal angle, and satisfies the symmetry properties $A_\theta(r, z) = A_\theta(r, -z)$. Thus the Lagrangian takes the form

$$L = \tfrac{1}{2}m(\dot{r}^2 + r^2\dot{\theta}^2 + \dot{z}^2) + er\dot{\theta}A_\theta - e\Phi.$$

To investigate the trajectories with reference to the equilibrium orbit, which is a circle of radius R in the $z = 0$ plane, a new orthogonal coordinate system will be introduced, so that the equilibrium orbit will play the part of one of the coordinate axes. The position of an electron will be defined by the radial distance from the equilibrium orbit, $\xi_1 = r - R$, its normal distance from the plane of the equilibrium orbit, $\xi_2 = z$, and the length of arc $s = R\theta$, measured along the equilibrium orbit from an arbitrary origin (see Figure 89). In terms of these new coordinates, the Lagrangian reads

$$L = \tfrac{1}{2}m[\dot{\xi}_1^2 + \dot{\xi}_2^2 + (1 + R^{-1}\xi_1)^2\dot{s}^2] + e\dot{s}A_s - e\Phi,$$

where the notation

$$A_s \equiv (1 + R^{-1}\xi_1)A_\theta$$

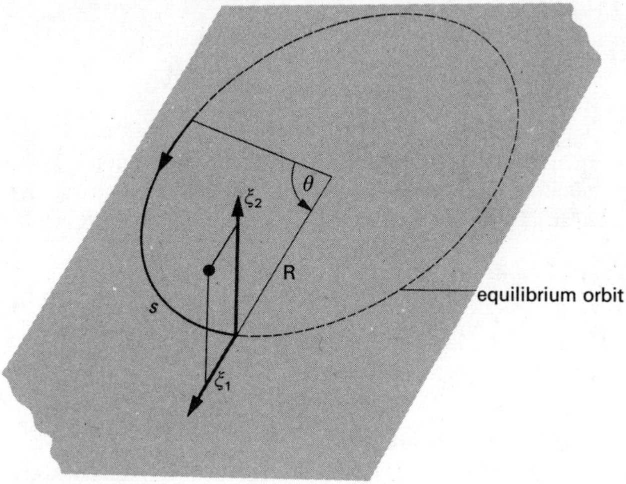

Figure 89

was introduced, although A_s is *not* the vector potential component tangential to the equilibrium orbit. Φ and $A\theta$ are now written in terms of the new coordinates ξ_1, ξ_2, s. These new coordinates will be considered as generalized coordinates, and the canonical momenta are determined from the Lagrangian in the usual manner, yielding

$$\pi_1 = \frac{\partial L}{\partial \dot{\xi}_1} = m\dot{\xi}_1 \equiv mv_1,$$

$$\pi_2 = \frac{\partial L}{\partial \dot{\xi}_2} = m\dot{\xi}_2 \equiv mv_2,$$

$$\pi_s = \frac{\partial L}{\partial \dot{s}} = (1+R^{-1}\xi_1)^2 m\dot{s} + eA_s; \qquad m\dot{s} = v_s.$$

Note that v_s *is* the velocity component tangential to the equilibrium orbit, but π_s is *not* the tangential component of the momentum.

The Hamiltonian is now formed as described in section 1.2:

$$H = \boldsymbol{\pi}.\mathbf{v} - L$$

$$= \frac{1}{2m}[\pi_1^2 + \pi_2^2 + (1+R^{-1}\xi_1)^{-2}(\pi_s - eA_s)^2] + e\Phi. \qquad \textbf{A.16}$$

The canonical equations of *motion* are formed with the aid of the Hamiltonian **A.16**:

$$\frac{d\xi_j}{dt} = \frac{\partial H}{\partial \pi_j} \qquad \frac{d\pi_j}{dt} = -\frac{\partial H}{\partial \xi_j} \quad (j = 1, 2)$$

$$\frac{ds}{dt} = \frac{\partial H}{\partial \pi_s} \qquad \frac{d\pi_s}{dt} = -\frac{\partial H}{ds} = 0$$

The differential equations of the *trajectories* can be obtained from the equations of motion by eliminating the time as independent variable and introducing the length of arc s instead. This may be done by using the equation $ds/dt = \partial H/\partial \pi_s$. One finds

$$\frac{d\xi_j}{ds} = \frac{d\xi_j/dt}{ds/dt} = \frac{\partial H/\partial \pi_j}{\partial H/\partial \pi_s} = \frac{\partial H}{\partial \pi_j}\frac{\partial \pi_s}{\partial H} = -\frac{\partial \pi_s}{\partial \pi_j} \qquad \textbf{A.17a}$$

and similarly $$\frac{d\pi_j}{ds} = \frac{\partial \pi_s}{\partial \xi_j}, \qquad \textbf{A.17b}$$

where π_s is expressed from the Hamiltonian **A.16**.

At this point, it must be emphasized that H does not depend explicitly on time, that is, H is a constant of motion, $H = W =$ constant. Furthermore, H does not depend on s either, and hence π_s is a constant of motion too.

Looking at equations **A.17a** and **b**, one observes that they are of the form of the canonical equations. One can take a new Hamiltonian $-\pi_s \equiv G$, which can be found explicitly from equation **A.16**.

$$G = -[2m(W-e\Phi)-\pi_1^2-\pi_2^2]^{\frac{1}{2}}(1+R^{-1}\xi_1)-eA_s. \qquad \textbf{A.18}$$

In terms of this new Hamiltonian, the differential equations of the trajectories take the standard 'canonical' form

$$\frac{d\xi_j}{ds} = \frac{\partial G}{\partial \pi} \qquad \frac{d\pi_j}{ds} = -\frac{\partial G}{\partial \xi_j} \quad (j = 1, 2). \qquad \textbf{A.19}$$

This is equation **2.56**, which formed the starting point for the derivation of the paraxial equation **2.58**. The argument concerning the derivatives of G, with respect to the coordinates and canonical momenta, can be directly tested by carrying out the necessary operations on G, given by **A.18**, and taking into account the symmetry properties of the fields. In particular, one must recognize that both Φ and A_θ are even functions of $\xi_2 = z$. Hence their partial derivatives with respect to z are odd functions, and therefore they vanish on the symmetry plane, $z = 0$. Furthermore,

$$\frac{\partial A_s}{\partial \xi_1} = \frac{\partial}{\partial \xi_1}(1+R^{-1}\xi_1)A_\theta = R^{-1}\frac{\partial}{\partial r}(rA_\theta)$$

and hence $$\left(\frac{\partial A_s}{\partial \xi_1}\right)_{\substack{\xi_1=0\\ \xi_2=0}} = \left(r^{-1}\frac{\partial}{\partial r}(rA_\theta)\right)_{\substack{r=R\\ z=0}} = B(R).$$

Similarly $$\left(\frac{-\partial \Phi}{\partial \xi_1}\right)_{\substack{\xi_1=0\\ \xi_2=0}} = -\left(\frac{\partial \Phi}{\partial r}\right)_{\substack{r=R\\ z=0}} = E(R)$$

To calculate the constant k_j^2, defined as

$$k_j^2 = \left(\frac{\partial^2 G}{\partial \xi_j^2}\right)_0 \left(\frac{\partial^2 G}{\partial \pi_j^2}\right)_0, \qquad \textbf{A.20}$$

the radii of curvature R_e and R_m, defined by equation **2.55**, are introduced and the radius of the equilibrium orbit, R, using equation **2.54**. Finally, it must be recalled that the linear momentum of the electron on its equilibrium orbit is given by $p = m\dot{s} = (\pi_s - eA_s)_{\xi_1 = 0}$.

The argument followed above leads to relativistically correct results, if one starts with the relativistic form of the Hamiltonian. Following the instructions presented in section 1.2, one finds

$$H = c[\pi_1^2 + \pi_2^2 + (1 + R^{-1}\xi_1)^{-2}(\pi_s - eA_s)^2 + m_0 c^2]^{\frac{1}{2}} + e\Phi,$$

in place of equation **A.16**. Consequently, the Hamiltonian G will be different from that given by equation **A.18**. When the constant k is calculated explicitly from its definition **A.20**, one obtains

$$k_2^2 = -R^{-2}\left\{RR_e^{-1}\left[1 + \left(\frac{\partial E/\partial r}{E/R}\right)_R\right] + RR_m^{-1}\left(\frac{\partial B/\partial r}{B/R}\right)_R\right\},$$

$$k_1^2 = R^{-2}\left[1 + (RR_e^{-1})^2\left(1 - \frac{v^2}{c^2}\right)\right] - k_2^2.$$

Equations **2.59a** and **b** give k_j^2 in the limit when $v \ll c$.

A.6 Floquet's theorem

Mathieu's equation (**3.16**) is a special case of linear differential equations, in which the coefficient of the dependent variable w is a one-valued periodic function of the independent variable u (with period π):

$$\frac{d^2 w}{du^2} + (a + 2q \cos 2u)w = 0. \qquad \textbf{A.21}$$

This equation has two independent solutions, $w_1(u)$ and $w_2(u)$, so that any particular solution can be constructed by the linear superposition,

$$w(u) = c_1 w_1(u) + c_2 w_2(u), \qquad \textbf{A.22}$$

where c_1 and c_2 are constants.

The differential equation **A.21** is invariant with respect to shifting the value of u by π, and therefore if $w_i(u)$ is a solution, then $w_i(u+\pi)$ is a solution also. This does *not* mean that $w_i(u+\pi) = w_i(u)$, but it does require that

$$\begin{bmatrix} w_1(u+\pi) \\ w_2(u+\pi) \end{bmatrix} = \begin{bmatrix} m_{11} & m_{12} \\ m_{21} & m_{22} \end{bmatrix} \begin{bmatrix} w_1(u) \\ w_2(u) \end{bmatrix}, \qquad \textbf{A.23}$$

where the matrix elements m_{jk} are constants. Consequently, equation **A.23** implies that

$$w(u+\pi) = (c_1 m_{11} + c_2 m_{21})w_1(u) + (c_1 m_{12} + c_2 m_{22})w_2(u). \qquad \textbf{A.24}$$

Equation **A.23** indicates that if the functions $w_i(u)$ are given, the matrix elements m_{jk} are determined. The constants c_1 and c_2, however, can be chosen arbitrarily because $w(u)$ defined by equation **A.23** remains a particular solution of equation **A.21**. Thus if we choose

$$\begin{aligned} c_1 m_{11} + c_2 m_{21} &= \lambda c_1 \\ c_1 m_{12} + c_2 m_{22} &= \lambda c_2, \end{aligned}$$

the constant λ is subject to the condition

$$\begin{vmatrix} m_{11} - \lambda & m_{21} \\ m_{12} & m_{22} - \lambda \end{vmatrix} = 0$$

and λ is given by the roots of this equation which can be real, imaginary or complex. By extracting λ, equation **A.24** takes the form

$$w(u+\pi) = \lambda[c_1 w_1(u) + c_2 w_2(u)] = \lambda\, w(u). \qquad \textbf{A.24}$$

The constant λ may be written in the form $\lambda = \exp \pi\mu$ and substituting

$$w(u) = \exp(\mu u)\, f(u), \qquad \textbf{A.26a}$$

into equation **A.20** one finds

$$\exp \mu(u+\pi)\, f(u+\pi) = \exp(\mu u + \pi\mu)\, f(u)$$

and hence $\quad f(u+\pi) = f(u).$

Thus the function $f(u)$ turns out to be a periodic function with period π.

Equation **A.26a** gives a particular solution of Matthieu's equation as the product of an exponential and a periodic function. Since the differential equation **A.21** is invariant with respect to the reflection $u \rightarrow -u$, the function

$$w(-u) = \exp(-\mu u)\, f(-u) \qquad \textbf{A.26b}$$

is a valid solution too. Furthermore, solutions **A.26a** and **b** are linearly independent of one another, and therefore the general solution is a linear combination of these:

$$A_1 \exp(\mu u)\, f(u) + A_2 \exp(-\mu u)\, f(-u).$$

This is the result quoted in section 3.2 as 'Floquet's theorem'.

A.7 Useful relativistic relationships

One of the most often used relativistic equations relates the total energy of a particle, W, to its momentum $\mathbf{p}$, both observed in the same inertial frame:

$$W^2 = (pc)^2 + (m_0 c^2)^2,$$

where m_0 is the rest mass of the particle: $E = m_0 c^2$ is the rest energy. The kinetic energy W_k of the particle is the difference between the total energy and the rest energy: $W_k = W - E$. Measuring all quantities which have the dimension of energy in units of the rest energy, the equation above gives the simple relationship

$$\left(\frac{W}{m_0 c^2}\right)^2 - \left(\frac{p}{m_0 c}\right)^2 = 1.$$

In terms of the velocity of the particle, v, measured in the inertial frame, the momentum and the total energy are given by

$$p = m_0 c\beta\gamma \qquad \text{and} \qquad W = m_0 c^2 \gamma,$$

$$\text{where} \qquad \beta = \frac{v}{c} \leqslant 1 \qquad \gamma = \left(1 - \frac{v^2}{c^2}\right)^{-\frac{1}{2}} \geqslant 1.$$

Also $W = mc^2$ with $m = m_0 \gamma$, where the latter expression describes the relativistic change of mass with velocity. From these relations

$$\beta = \frac{p}{W} \qquad \text{and} \qquad \gamma = \frac{m}{m_0} = \frac{W}{m_0 c^2}.$$

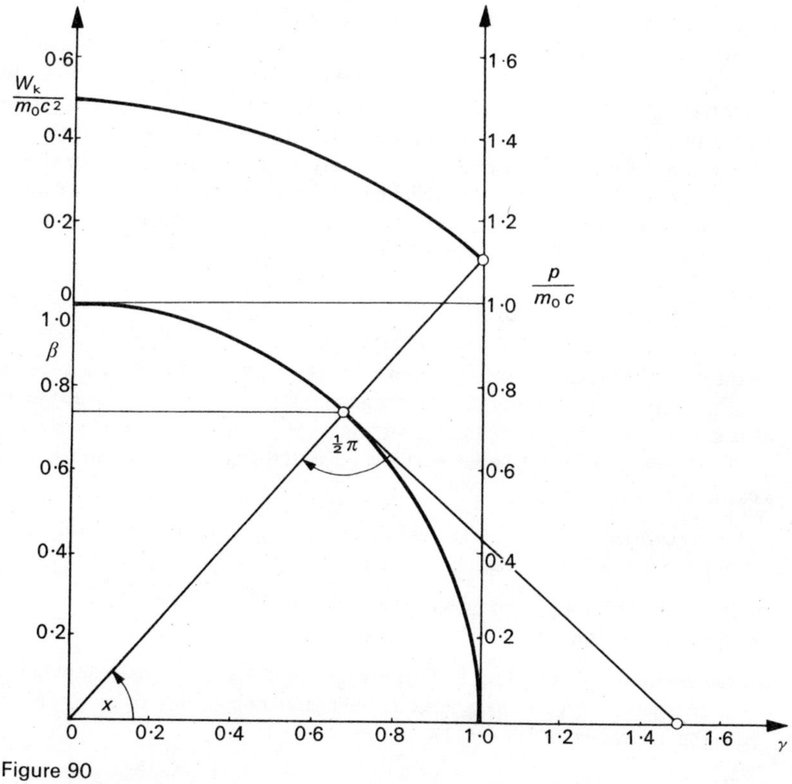

Figure 90

It is good to remember that in numerical calculations trigonometric tables can be used because defining

$$\beta = \sin x \qquad \left(0 \leqslant x \leqslant \frac{\pi}{2}\right),$$

one finds $\quad \gamma = \sec x \qquad \beta\gamma = \tan x.$

Hence the kinetic energy, in units of the rest energy,

$$\frac{W_k}{m_0 c^2} = \frac{W}{m_0 c^2} - 1 = \sec x - 1$$

and finally $\quad \dfrac{p}{m_0 c} = \tan x.$

All these relations are summarized by means of a nomogram shown in Figure 90.

In collision problems, whether the process is elastic or not, it is advantageous to express the energy and the momentum of the colliding partners in a reference frame in which the centre of mass is at rest. If energy and momentum are given in the laboratory frame which is an inertial frame, one must carry out a Lorentz transformation to another inertial frame, defined by the condition that in the new reference frame the total momentum is zero.

The total energy of the particles, as observed in the centre-of-mass system, can be determined without an explicit knowledge of the velocity of the centre of mass, if one recalls that the modulus of the momentum four-vector, as the scalar product of any two four-vectors, is invariant under Lorentz transformation. If the total energy and the momentum of a system in any inertial frame are denoted by W and $\mathbf{p}$ respectively, the four-vector is defined by its components $[p_1, p_2, p_3, (i/c)W]$; the first three (space-like) components being the three orthogonal momentum components, and the fourth (time-like) one is W/c. The quantity

$$\left(\frac{W}{c}\right)^2 - p^2 = (M_0 c)^2 = \text{invariant.} \qquad \textbf{A.27}$$

In the case of one particle the invariant is the rest energy of the particle expressed in terms of the particle's uniquely determined rest mass. In the case of a system of particles, the invariant is written in an analogous form, in terms of a mass M_0, which may be considered as the total rest mass of the system, concentrated in the centre of mass.

Let us consider now *two* particles of rest mass m_a and m_b respectively. In the laboratory coordinate system, let m_a be in motion, m_b at rest, so that:

m_a has a momentum $\mathbf{p}_a$, and hence an energy $E_a^2 = (p_a c)^2 + (m_a c^2)^2$,
m_b has a momentum 0, and hence an energy $E_b = m_b c^2$.

In the centre-of-mass system both particles are moving and have some energy E'_a and E'_b respectively but, by definition, their total momentum is zero, $\mathbf{p}'_a + \mathbf{p}'_b = 0$. Thus the invariance relation **A.27**, gives

$$c^{-2}(E'_a + E'_b)^2 = c^{-2}(E_a + E_b)^2 - p_a^2$$

and hence $(E'_a - E'_b)^2 = E_a^2 - (p_a c)^2 + E_b^2 + 2E_a E_b$

$$= (m_a c^2)^2 + (m_b c^2)^2 + 2E_a m_b c^2.$$

By introducing the rest energies, $m_a c^2 = E_{a0}$, and $m_b c^2 = E_{a0}$, one gets

$$W'^2 = (E'_a + E'_b)^2 = E_{a0}^2 + 2E_a E_{b0}. \qquad \textbf{A.28}$$

One can express E_a in terms of the kinetic energy W_{ka} of the particle of rest energy $E_{a0} : E_a = W_{ka} + E_{a0}$. Substituting in equation **A.28**,

$$W'^2 = (E_{a0} + E_{b0})^2 + 2W_{ka} E_{b0}. \qquad \textbf{A.29}$$

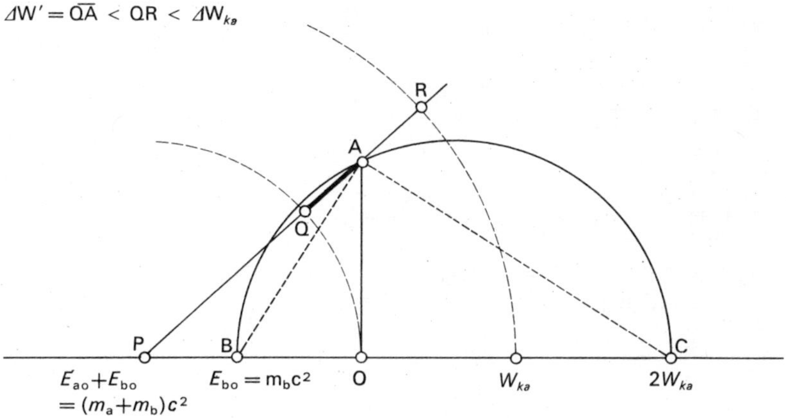

Figure 91

This is the equation quoted in section 4.1, in the footnote. Figure 91 shows a simple diagram from which the energy

$$\Delta W' = W' - (E_{a0} + E_{b0})$$

available for a reaction in the centre-of-mass system can be directly seen.

Since the triangle ABC is a rectangular one:

$$\overline{\mathrm{OA}} = (2W_{ka} E_{b0})^{\frac{1}{2}}$$

and $\overline{\mathrm{PA}} = [(E_{a0} + E_{b0})^2 + 2W_{ka} E_{b0} = W'.$

Finally $\Delta W' = \overline{\mathrm{QA}} < \overline{\mathrm{QR}} = W_{ka}.$

Thus one can see that the energy available in the centre-of-mass system is always smaller than the kinetic energy of the projectile observed in the laboratory frame. Note that to obtain an excess energy $\Delta W'$ in the centre-of-mass frame comparable with the rest energy of the colliding particles, the kinetic energy of the projectile in the laboratory frame must be considerably higher than $\Delta W'$.

A two-body collision process is fully characterized by the motion of the particles in the centre-of-mass system after the collision. On the other hand, the experimenter is at rest in the laboratory system and observes a fixed target hit by a projectile. In order to relate theoretical predictions expressed with reference to the centre-of-mass frame and the results of experiments, the relative velocity V of the two frames of reference must be known. To determine this quantity, we postulate that the centre of mass as observed in the laboratory system behaves as a particle of rest mass $M_0 = W'/c^2$. The total momentum in the laboratory system is p_a which, in terms of the rest mass of the mass centre, is

$$p_a = M_0 c\gamma_V \beta_V,$$

where $\beta_V = \dfrac{V}{c}$, $\gamma_V = \left(1-\dfrac{V^2}{c^2}\right)^{-\frac{1}{2}}$

The total energy in the laboratory system, again in terms of the rest mass M_0 of the mass centre, is

$$W \equiv E_a+E_{b_0} = M_0 c^2\gamma_V \qquad \textbf{A.30}$$

and hence $\beta_V = \dfrac{p_a c}{W} = p_a c(E_a+m_b c^2)^{-1}.$ **A.31**

Since $M_0 c^2 = W'$ is the total energy in the centre-of-mass system, one finds from equation **A.30** directly

$$\gamma_V = \frac{W}{W'}. \qquad \textbf{A.32}$$

A.8 The Hamiltonian of a charged particle in a travelling electromagnetic wave

The relativistic Lagrangian of a particle is

$$L = -m_0 c^2\left(1-\frac{v^2}{c^2}\right)^{\frac{1}{2}}+e\mathbf{v}.\mathbf{A},$$

where $\mathbf{v} = \dot{x}_1\mathbf{e}_1+\dot{x}_2\mathbf{e}_2+\dot{z}\mathbf{e}_z.$

If the position of the electron (see equation **4.4**) is given by

$$X_1 = x_1 = 0,\ X_2 = x_2 = 0,\ X_3 = Z = z-v_0 t,$$

the Lagrangian takes the form

$$L = -m_0 c^2\left[1-\left(\frac{\dot{Z}+v_0}{c}\right)^2\right]^{\frac{1}{2}}+e(\dot{Z}+v_0)\,A(Z), \qquad \textbf{A.33}$$

where $A(Z) = \dfrac{E_0}{\omega}\cos\dfrac{\omega Z}{v_0}.$ **[4.6]**

From the Lagrangian **A.33** one obtains the canonical momentum components, by definition

$$P_1 = \frac{\partial L}{\partial \dot{X}_1} = 0 \qquad P_2 = \frac{\partial L}{\partial \dot{X}_2} = 0,$$

$$P_Z = \frac{\partial L}{\partial \dot{Z}} = \gamma m_0(\dot{Z}+v_0)+eA, \qquad \textbf{A.34}$$

with $\gamma = \left(1-\frac{\dot{z}^2}{c^2}\right)^{-\frac{1}{2}} = \left[1-\left(\frac{\dot{Z}+v_0}{c}\right)^2\right]^{-\frac{1}{2}}.$

Note that $\gamma m_0(\dot{Z}+v_0) = p$ is the linear momentum of a particle moving along the z-axis with velocity $v = \dot{Z}+v_0$ and therefore

$$P_Z = p+eA. \qquad \textbf{A.35}$$

From equations **A.33** and **A.34**, one finds for the Hamiltonian:

$$H = P_Z\dot{Z}-L = \gamma m_0 c^2-\gamma m_0 v_0(v_0+\dot{Z})-ev_0 A.$$

The first term of this expression is the total energy of a particle moving with velocity v, that is, in terms of its linear momentum $\mathbf{p} = \mathbf{e}_z p$,

$$\gamma m_0 c^2 = [(m_0 c^2)^2+p^2c^2]^{\frac{1}{2}} = [(m_0 c^2)^2+(P_Z-eA)^2c^2]^{\frac{1}{2}}.$$

The other two terms give $P_Z v_0$, and therefore

$$H = [(m_0 c^2)^2+c^2(P_Z-eA)^2]^{\frac{1}{2}}-P_Z v_0. \qquad \textbf{A.36a}$$

This is the Hamiltonian of the system in terms of the canonically conjugate quantities Z and P_Z, defined by equation **A.34**. We observe, however, that if P_Z is substituted from equation **A.35**, yielding

$$H = [(m_0 c^2)^2+c^2p^2]^{\frac{1}{2}}-pv_0-ev_0 A(Z), \qquad \textbf{A.36b}$$

one obtains an alternative form of the Hamiltonian, where the canonically conjugate pair of quantities are Z and the linear momentum p. Equation **A.36b** is identical with equation **4.5**.

The validity of both forms of the Hamiltonian is readily seen if one compares the Hamiltonian equations of motion. From equation **A.36a**,

$$\frac{dZ}{dt} = \frac{\partial H}{\partial P_Z} \quad \text{and} \quad \frac{dP_Z}{dt} = -\frac{\partial H}{\partial Z}, \qquad \textbf{A.37a}$$

and from equation **A.36b**

$$\frac{dZ}{dt} = \frac{\partial H}{\partial p} \quad \text{and} \quad \frac{dp}{dt} = -\frac{\partial H}{\partial Z}. \qquad \textbf{A.37b}$$

From the first one of either set of equations, one finds

$$\frac{dZ}{dt} = \frac{dz}{dt}-v_0,$$

which is in agreement with the choice of the moving coordinate system, defined by equation **4.4**. Since from equation **A.35**,

$$\frac{dP_Z}{dt} = \frac{dp}{dt} + e\dot{Z}\frac{\partial A}{\partial Z},$$

the second one of either set of equations gives, with equation **4.6**,

$$\frac{dp}{dt} = ev_0\frac{\partial A}{\partial Z} = eE_0 \sin\frac{\omega Z}{v_0},$$

or, substituting equation **4.4**,

$$\frac{dp}{dt} = eE_0 \sin \omega\left(t - \frac{z}{v_0}\right).$$

This is Newton's equation of motion for a particle of charge e, in the field given by equation **4.2**.

Index

Index

Aberration 49
Accelerators
cyclic 153–75, 177
linear 139–53, 176
resonance 138
Action 51
Adiabatic
invariants 26
theorem 30, 152
Adler tube 236
Admittance
beam 185, 202
diagram 204
Amplifiers 179, 241–4
travelling wave 216–22
triode 192–5
see also Klystron, Parametric amplifier
Angular momentum
conservation of 24
effect on image formation 252

Betatron 160–64
oscillation 166, 175
Boundary conditions 113
Brillouin flow 99
Bunching 132, 150
parameter 135, 196, 201
Busch's theorem 26

Canonical equations, *see* Hamilton's equation of motion
Canonical momentum 19, 23, 24, 51, 143, 255, 263
Cardinal elements
of electric lenses 58, 60, 63
of magnetic lenses 47
of quadrupole lenses 68
of rotationally symmetric fields 55
of sector fields 81
see also Focal length, Foci, Principal planes
Cavity resonator 126
modes of excitation of 128
Centre of mass coordinate system 138, 260
Cherenkov radiation 242
Child–Langmuir law 99, 191, 192
Conservation laws 17–26
see also Angular momentum, Energy
Coulomb gauge 112
Coupling 216–22
Current density 111, 182, 189, 191
modulation 135, 194, 196, 201
wave 207, 210
Cyclotron 154–6
frequency 16, 154, 155, 213, 214, 215
waves 212–15, 236

Deflection by alternating fields 117–31
Dipole particles
motion of, in static fields 99–107
Diode 185–8
Dispersion 84–90
Doppler shift 209
Drift
space 118
tube 141

Electron lenses
electric 59–64
magnetic 44–9
quadrupole 68–70
rotationally symmetric 49–58
sector field 79–89
weak 58
Electron motion in static fields
crossed 16
magnetic 14, 36–41

Electron motion in static fields-*contd*
quadrupole 68-70
sector 79-84
Energy
conservation 22, 114, 183
conversion 197
limit of accelerators 175-8
loss by radiation 175
transfer 161, 181, 225
transport by space-charge waves 211
Equilibrium orbit 71, 170
motion relative to 75, 77
radius of 72
stability conditions of 78
Equivalent network 184, 198, 203
Euler-Lagrange differential equations 53

Fast waves 209, 210, 216
Feedback loop 199, 241
Field index 158
Fields
alternating 109, 111-14, 126, 128, 142
static
alternating gradient 170
constant gradient 65
plane 22, 247-8
quadrupole 65, 67
rotationally symmetric 23, 248-9
sector 79-84
six-pole 104
Floquet's theorem 122, 257
Focal length 48, 55
Focal plane 47
Foci, positions of 47, 55
Focusing
general conditions of 90-99
ring 41-4, 70
strong 169-75

Gauge transformation 112
Group velocity 209

Hamiltonian equations
of motion 21
of trajectories 77, 254-7
Hamiltonians 20, 21, 23, 143
of particles in a travelling wave 144, 146, 262
Hamilton's principle 52
Harmonic frequency 136
generation 198
Helmholtz-Lagrange relation 56

Idler 235, 241
Image formation
by quadrupole fields 68-70
by rotationally symmetric fields 49-58
by sector fields 80-82, 84-9
by uniform magnetic fields 44-9

Kinetic power flow 211
Klystron 196-8, 206, 211
see also Reflex klystron

Lagrange invariant 56, 251
Lagrangian 18, 126
equation of motion 18, 126
Larmor frequency 26, 46, 51, 227
Lorentz condition 112
gauge 112

Magnetic bottle 31-3
Magnetic lenses
see Electron lenses
Magneto-gyric ratio 100
Magnetron 222-31
cut-off 225
energy transfer in 225
oscillation
conditions of 229
modes of 229
space charge in 228
Magnification 46, 56
Manley-Rowe relations 244
Mass spectrometer
radio frequency 119-25
static 87-90
Matthieu equations 121-4
Matrix, *see* Transfer matrix
Maxwell's equations 111
Meridian plane 24
Microtron 158-60

Noise 245

Orbit stability 78
in accelerators 157, 163, 167

Oscillator 194, 199, 229
see also Reflex oscillator

Parametric amplifier 231–41
frequency-sum rule in 233, 243
gain of 241
idler in 235, 241
pump in 235
Paraxial ray equation 53, 77
Phase
focusing 131–8, 150, 165, 195
integral 30
oscillation 148, 152, 166
plane 27, 144
space 28, 148
stability 142, 152
velocity 144, 148, 151, 209, 237
Plasma frequency 208
Polarization of neutral spin particles 105
Potential, electromagnetic 112
see also Scalar potential, Vector potential
Poynting vector 115, 183
Principal planes 48, 55
see also Cardinal elements
Principle of least action 35, 51
Propagation constant 209
Pump, *see* Parametric amplifier

Quadrupole fields
alternating 121
static 65–6

Reflex klystron 199–206
electronic tuning of 205
equivalent network of 203
modes of oscillation 204
Reflex oscillator 202, 204
Relativistic
energy 72, 147, 148, 156, 175, 258
mass 14
momentum 14, 175, 258
Resolving power 70
of static sector fields 84–90
of alternating quadrupole fields 125
Ring focusing 41–4, 70

Scalar potential distribution
of plane fields 22, 247
of quadrupole fields 65
of rotationally symmetric fields 23, 248
Six-pole magnets 104
Slow waves 209, 210, 217, 242
Space-charge flow 96–9, 188–95
Space-charge waves 206–11
coupling of 216–22
fast 209, 210
slow 209, 210, 242
travelling 211
Stability
diagrams 128, 172
of orbits 78, 157, 163, 167
of oscillation 75, 78, 125, 172
of phase 142, 151
Stern–Gerlach experiment 101, 103
Strong focusing 169–75
Synchro-cyclotron 156–8
Synchronotron 164–9
oscillation 166

Transfer matrix
general properties of 91–6, 250
of electric lenses 58, 60, 62
of periodic structures 91–6, 171
of quadrupole fields 68
of rotationally symmetric fields 54
of sector fields 79, 86, 88
Transit time effects 110, 118, 185–8
Trapping
in magnetic fields 31–3
in travelling waves 146
Transmission 105
line 216–17
Travelling wave amplifier, *see* Amplifiers
Triode, *see* Amplifiers

Unit cell 91, 167, 171

Vector potential 23, 25, 39, 67, 71, 103, 112, 127, 130, 143
Velocity modulation 131–6, 195
Velocity selection 83
by r.f. gating 118–20
Virtual cathode 191

Wave propagation, *see* Space charge waves, Transmission lines
Wien filter 83